QUANTUM MECHANICS OF
MOLECULAR RATE
PROCESSES

QUANTUM MECHANICS OF
Molecular Rate Processes

RAPHAEL D. LEVINE

OXFORD
AT THE CLARENDON PRESS
1969

Oxford University Press, Ely House, London W. 1

GLASGOW NEW YORK TORONTO MELBOURNE WELLINGTON
CAPE TOWN SALISBURY IBADAN NAIROBI LUSAKA ADDIS ABABA
BOMBAY CALCUTTA MADRAS KARACHI LAHORE DACCA
KUALA LUMPUR SINGAPORE HONG KONG TOKYO

© OXFORD UNIVERSITY PRESS 1969

PRINTED IN GREAT BRITAIN

Preface

THE aim of this book is to survey the quantum-mechanical theory of rate processes and its application to molecular problems. The basic (and formal) theory is now sufficiently developed, and further progress in this field can only come from a close cooperation between the theoretician and the experimentalist. The increasing concern of both with the basic theory and with increasing experimental sophistication has prompted this monograph.

The discussion assumes familiarity with the contents of a conventional introductory undergraduate course in quantum mechanics, as is given in most universities. In order to formulate the theory in a compact (and elegant) way it is convenient to follow the Dirac and von Neumann formulations of quantum mechanics. The necessary concepts are introduced in the first part of the book, in terms of a discussion of physical attributes and their average values as measured in an ensemble of identical, non-interacting systems. The concepts of a state and its representation by an amplitude and the spectral representation of operators are discussed as tools designed for the evaluation of average values of attributes. A second chapter is devoted to the equations of motion of the average values and the Green's operator.

The second part of the book is devoted to the quantal collision theory. The boundary condition, that before the collision the molecules are far apart and non-interacting, is used to rewrite the Schrödinger (differential) equation as the Lippmann–Schwinger (integral) equation in Chapter 2.1. In Chapter 2.2 the Lippmann–Schwinger equation for structureless particles (potential scattering theory) is discussed and solved for the case of separable interactions. The concepts of scattering amplitude and cross-section are introduced and evaluated exactly for separable interactions and, in general, to the lowest order in the interaction (the Born approximation). The method of partial-wave analysis is described and applied to scattering by a central potential.

Chapter 2.3 concludes the introductory discussion of collision theory with a derivation of the equations for relative motion during a collision of molecules with internal structure. Application to rotational excitation and the Born approximation are also included in the chapter. At this point, the reader interested in further applications can proceed directly to Chapters 3.4 and 3.5 in Part 3.

The expectation values of observables that refer to the non-interacting molecules (i.e. the internal energy) may change as a result of a collision. In Chapter 3.4 we derive a general expression for the rate of this change (the generalized Ehrenfest theorem) and prove the optical theorem as a direct consequence of the conservation of probability. As applications, we derive expressions for collision rates and cross-sections, consider collisions where the initial state is not a pure state, and discuss the relaxation equation.

Chapter 2.5 presents the operator formulation of collision theory and introduces the concept of the S matrix and the techniques used in the theory of scattering by two and by several potentials. The discussion is extended to cover reactive collisions in Chapter 2.6. In particular, some attention is given to a re-derivation of the main results, working explicitly in the coordinate representation.

The formulation of collision theory from a time-dependent point of view is dealt with in Chapter 2.7. The technique of time ordering is used to reintroduce the wave operator that transforms the wavefunction for the system in the absence of interaction to the actual wavefunction. The use of symmetry in collisions, in particular, time-reversal invariance and its consequences in the form of the reciprocity and the microscopic reversibility theorems, is discussed in Chapter 2.8.

The application and extension of the theory are considered in Part 3. Chapter 3.1 introduces the partitioning technique, which can be used to focus attention on those variables that are of interest in a given collision. It is used in Chapter 3.2, which describes, in terms of 'equivalent potentials', various aspects of molecular encounters. The adiabatic approximation is introduced, and its breakdown due to internal excitation is discussed. The separation of the reaction amplitude into a 'direct' part and a contribution due to 'compound' processes is used in discussing the theory of direct reactions. In Chapter 3.3 the partitioning technique is applied to the operator formulation of collision theory.

Various models in collision theory, which are designed to replace an exact solution for the dynamics of the collision, are described in Chapters 3.4 and 3.5.

Chapter 3.6 is concerned with the theory of unimolecular reactions.

The evolution of an activated state from $t = 0$ onwards is discussed and the separation of the total wavefunction into an activated state portion and a portion that represents the products is considered. To treat the excitation process we introduce the Markoffian approximation, which enables us to distinguish between a microscopic and a macroscopic time-scale.

The general problem of rate processes of ensembles can also be cast in terms of a study of time-correlation functions. Chapter 3.7 consists of a formal discussion of time-correlation functions and their application in the study of the linear response of a system to an external perturbation. A further development is the introduction of the Liouville operator, which enables us to employ the partitioning technique in the solution of the quantum-mechanical equations of motion.

In the text, equations are numbered consecutively for each chapter and appendix. Reference to equations in a different part is made by prefixing the part number. Thus (2.4.10) is in Part 2, Chapter 2.4, and is the tenth equation in that chapter.

The references quoted are to those papers that have influenced the discussion in the text, to review articles, and to papers that supplement the text. The main purpose in giving references, apart from acknowledging the origin of the ideas, is to enable the reader to read further. In view of the enormous literature available I have certainly missed some key references, for which I apologize. I wish to thank the authors quoted for permission to use their figures.

Selected portions of the subject were first given as a graduate course in the Mathematical Institute, Oxford. I am indebted to Professor C. A. Coulson, F.R.S., who suggested that the course be given and who encouraged me to compile my lecture notes as a monograph, for his continuous advice and support. Subsequently, similar graduate courses were given in the University of Wisconsin, The Hebrew University, and The Ohio State University.

I would like to thank Professor J. O. Hirschfelder, director of the Theoretical Chemistry Institute of the University of Wisconsin, for the kind hospitality and support extended to me while the monograph was being written. During that period I also had the benefit of innumerable discussions with Professor R. B. Bernstein, who has also read most of the text. For stimulating my interest in molecular rate processes, I am indebted to my two teachers, Professor G. Stein, who introduced me to experimental chemical kinetics and Professor G. G. Hall, who taught me that physical ideas can be cast in a quantitative form.

Several other colleagues have commented on various parts of the text and I am indebted to them all. The responsibility for any errors or inaccuracies remains mine.

R. D. L.

Department of Physical Chemistry
Hebrew University
Jerusalem
April 1968

Contents

PART 1
THE FORMULATION OF QUANTUM MECHANICS

Introduction   1

1.1. STATES AND OBSERVABLES   2

1.2. THE EQUATIONS OF MOTION   13
    1.2.1. The Green's operator for the Schrödinger equation   15
    APPENDIX 1.A. The two-body problem and the addition of angular momentum   21

PART 2
COLLISION THEORY

Introduction   25

2.1. STATIONARY COLLISION THEORY—THE LIPPMANN–SCHWINGER EQUATION   27

2.2. THE LIPPMANN–SCHWINGER EQUATION FOR 'STRUCTURELESS' MOLECULES   30
    2.2.1. The cross-section   33
    2.2.2. Separable interactions   39
    2.2.3. Partial wave analysis   43
    2.2.4. The Born approximation   50

2.3. INTERNAL EXCITATION IN COLLISIONS   54
    2.3.1. The theory of rotational excitation   60
    2.3.2. The Born approximation   64

2.4. THE RATE OF CHANGE OF OBSERVABLES   70
    2.4.1. Collision rates and cross-sections   73
    2.4.2. Collision rates in ensembles   80
    2.4.3. The relaxation equation   83

CONTENTS

2.5. FORMAL COLLISION THEORY ... 87
    2.5.1. The S matrix ... 90
    2.5.2. Scattering by two potentials ... 95
    2.5.3. The density of states ... 101
    2.5.4. Multiple-scattering theory ... 106

2.6. REACTIVE COLLISIONS ... 110
    2.6.1. Reaction rates ... 113
    2.6.2. Operator formulation of the theory of reactive collisions ... 117
    2.6.3. The yield function—absolute rate theory ... 120
    2.6.4. The theory of reactive collisions in the coordinate representation ... 123

2.7. TIME-DEPENDENT COLLISION THEORY ... 129
    2.7.1. The time evolution ... 132
    2.7.2. The wave operator ... 134
    2.7.3. The change in observables ... 136
    2.7.4. Formal theory of reactive collisions—the Jauch resolution ... 137

2.8. SYMMETRY ... 142
    2.8.1. Time reversal ... 142
    2.8.2. Reciprocity and microscopic reversibility ... 145
    2.8.3. Constants of motion ... 148
    2.8.4. Permutation symmetry in collisions ... 150
    APPENDIX 2.A. Normalization of the solution of the L.S. equation ... 154
    APPENDIX 2.B. On the convolution theorem and linear systems ... 155
    APPENDIX 2.C. Intertwinning and the S operator for reactive collisions ... 157
    APPENDIX 2.D. On the long-time behaviour and adiabatic switching in collisions ... 158
    APPENDIX 2.E. The transition amplitude density method ... 159

PART 3
MOLECULAR RATE PROCESSES

Introduction ... 161

3.1. THE PARTITIONING TECHNIQUE ... 162

3.2. MOLECULAR ENCOUNTERS ... 168
    3.2.1. The adiabatic approximation ... 169
    3.2.2. Internal excitation in collisions ... 178
    3.2.3. Theories of direct reactions ... 192

3.3. OPERATOR PARTITIONING THEORY	201
3.3.1. Variational principles	205
3.4. MODELS IN COLLISION THEORY	215
3.4.1. The opacity function—the optical model	219
3.4.2. The impact parameter method	231
3.4.3. The adiabatic theory of reactive collisions	239
3.5. STATISTICAL THEORIES	246
3.5.1. The statistical approximation	252
3.5.2. The optical potential	259
3.5.3. Statistical theory for overlapping resonances	265
3.6. UNIMOLECULAR REACTIONS	268
3.6.1. Unimolecular breakdown	269
3.6.2. Collision theory in ensembles	283
3.6.3. Excitation processes	288
3.7. THE TIME-CORRELATION METHOD	293
3.7.1. Time-correlation functions	294
3.7.2. Linear response theory	300
3.7.3. The Liouville operator	306
REFERENCES	314
AUTHOR INDEX	327
SUBJECT INDEX	331

PART 1
The formulation of quantum mechanics

Introduction

THE quantitative formulation of a physical theory can be considered as having two parts. First, a mathematical framework, which can be formulated without any reference to a possible application and second, a set of rules of correspondence which identifies the mathematical constructs with physical concepts. In our use of quantum mechanics we shall follow the mathematical structure pioneered by Dirac (1958) and von Neumann (1955), and the rules of correspondence as discussed by these authors.

In the following chapter we present a brief summary of these ideas. We shall outline the rules of correspondence while introducing the mathematical notation. Further discussion can be found in most books on quantum mechanics. In particular, the following references contain additional material: Golden (1968), Gottfried (1966), Furry (1966), Kaempffer (1965), Roman (1965), Feynman and Hibbs (1965), and Messiah (1961). Further discussion of the mathematical framework can be found in Akhiezer and Glazman (1961), Halmos (1958), and Stone (1932).

The equations of motion of quantum mechanics are discussed in section 1.2.0, and Appendix 1.A considers the formulation of the problem of the relative motion of two bodies.

1.1. States and observables

WE follow the view that quantum mechanics can only predict the probabilities of various outcomes when measurements are made on a very large number of systems. For simplicity we shall generally refer to a *system* by which we mean a member of the above ensemble. If, when observations are made, we obtain the same, definite, results we shall say that the system is in a particular dynamical state (a *pure state* or simply, for abbreviation, a *state*) and associate with it a *state vector*, denoted $|n\rangle$. Here n is the set of numbers (i.e. the results of the observations) necessary to specify completely the state. For example, the internal state of a hydrogen atom is specified by the set consisting of the principal quantum number, the orbital quantum number (l), the azimuthal quantum number (m_l) and the spin component (m_s). In other situations some of these numbers can vary in a continuous fashion. Thus the state of a free structureless particle can be denoted $|\mathbf{p}\rangle$, where $\mathbf{p}$ is the momentum of the particle.

The state vector is an abstract entity. To make the connection with measurements we need an algorithm for the computation of probabilities. We begin with the following conceptual experiment. Let A be some physical attribute and $\{\alpha\}$ be the (for the moment—discrete) set of values (the *spectrum*) that the attribute A can assume. We now perform measurements of the attribute A on individual systems of the ensemble, recording 1, whenever the particular value α is observed in a measurement and 0 otherwise. The average value of our results, which is necessarily between 0 and 1, is regarded as the probability of observing the value α for the attribute A. In the ensemble whose members are all in the same dynamical state n, the average value of our score will be denoted as

$$|F_n(\alpha)|^2 = |\langle \alpha | n \rangle|^2. \tag{1.1}$$

$\langle \alpha | n \rangle$ is the *scalar product* and is thus regarded as a (possibly complex) number. In the context of the probability interpretation, the scalar product is referred to as a *probability amplitude* or simply *amplitude*. $\langle \alpha | n \rangle$ is thus the amplitude§ for observing the value α for the attribute A

---

§ Strictly speaking one should further qualify this statement. If, using eqn (1.2), we write $|F_n(\alpha)|^2 = \langle n|\alpha\rangle\langle \alpha|n\rangle$, then $|F_n(\alpha)|^2$ is the number that represents the average scored, irrespective of the nature of the attribute A. However, if the number α does not completely specify a dynamical state of the system, then $\langle \alpha|n \rangle$ need not be a number (cf. eqn (1.41)).

when the system is in the state n. Conversely, $\langle n|\alpha\rangle$ is the amplitude for observing the value n (for the attribute N) when the system is in the state $|\alpha\rangle$.

We assume that the scalar product satisfies

$$\langle \alpha|n\rangle = \langle n|\alpha\rangle^*, \tag{1.2}$$

$$\langle n|n\rangle \geqslant 0,$$

and
$$\langle \alpha|(c|n\rangle) = c\langle \alpha|n\rangle, \tag{1.3}$$

where c is a (complex) number. The scalar product $\langle n|n\rangle$ is referred to as the *norm* of the state $|n\rangle$. Using eqn (1.3) we can ensure, by a suitable selection of constants that $\langle \alpha|\alpha\rangle = 1$. Provided that

$$\langle \alpha|\alpha'\rangle = \delta_{\alpha,\alpha'}, \tag{1.4}$$

where $\delta_{\alpha,\alpha'}$ is the Kronecker delta, the set of states $|\alpha\rangle$ is referred to as an *orthonormal* set. Equation (1.4) has the obvious interpretation that if we know that the system is in the dynamical state $|\alpha'\rangle$ and we perform a measurement of A as outlined above, then we shall consistently score 1 if $\alpha = \alpha'$ and 0 otherwise.

A basic principle is the principle of 'superposition'. Mathematically this is the statement that the space of state vectors is a *linear space*, so that if $|n\rangle$ and $|m\rangle$ are state vectors, so is $|l\rangle$,

$$|l\rangle = c|n\rangle + d|m\rangle, \tag{1.5}$$

where c and d are arbitrary complex numbers. The amplitude $\langle \alpha|l\rangle$ is given by the requirement that the operation of performing the scalar product be linear, so that

$$\langle \alpha|l\rangle = c\langle \alpha|n\rangle + d\langle \alpha|m\rangle. \tag{1.6}$$

It is possible to regard $\langle \alpha|$ as a state vector in a *dual* vector space (Dirac 1958, Halmos 1958). With each state vector $|l\rangle$ we can then associate, in a one-to-one fashion, a state vector $\langle l|$ by the correspondence

$$\langle l| = c^*\langle n| + d^*\langle m|. \tag{1.7}$$

In view of the notation of the scalar product (bracket) one often refers to $|l\rangle$ as a *ket* and to $\langle l|$ as a *bra*.

When the spectrum is continuous (i.e. position) one can only define a probability density. Thus, for a free particle $|\langle \mathbf{R}|\mathbf{p}\rangle|^2 d\mathbf{R}$ is the probability of finding the particle (whose momentum is $\mathbf{p}$) at the volume element $d\mathbf{R}$, around the position $\mathbf{R}$. Here $|\langle \mathbf{R}|\mathbf{p}\rangle|^2$ is the probability density and $\langle \mathbf{R}|\mathbf{p}\rangle$ is the (probability density) amplitude.

In the usual wavefunction notation we write

$$\psi_\mathbf{p}(\mathbf{R}) = \langle \mathbf{R}|\mathbf{p}\rangle \tag{1.8}$$

and refer to $\psi_\mathbf{p}(\mathbf{R})$ as the *wavefunction* of the particle. Strictly speaking, $\psi_\mathbf{p}(\mathbf{R})$ is the wavefunction 'in the coordinate representation', since it is the amplitude to observe the particle at a given position. Other representations are possible. The wavefunction for a free particle in the 'angular momentum representation' is discussed below (eqn (1.A.29)). In general we shall use the notation

$$\psi_n(\alpha) = \langle \alpha | n \rangle. \tag{1.9}$$

The concept of a representation can be considered an analogue to that of a coordinate system in a three-dimensional vector space. It is possible to formulate the theory of vectors without the explicit introduction of a coordinate system, but for the purpose of computation one has to introduce a coordinate system, which can, however, be selected to suit a particular problem.

The previous conceptual experiment can be considered as a measurement of an attribute of the system, namely the attribute that the value α will be found for the attribute A. We shall denote the average value of this attribute by $\langle P_\alpha \rangle \equiv |F_n(\alpha)|^2$. Using eqn (1.2),

$$|F_n(\alpha)|^2 = \langle n | \alpha \rangle \langle \alpha | n \rangle = \langle n | P_\alpha | n \rangle, \tag{1.10}$$

where we have introduced the definition

$$P_\alpha = |\alpha\rangle\langle\alpha| \tag{1.11}$$

so that
$$P_\alpha |n\rangle = |\alpha\rangle\langle\alpha|n\rangle. \tag{1.12}$$

Thus when P_α operates on a state vector ($|n\rangle$) it transforms it to another state vector (in this case, $|\alpha\rangle$). Moreover, from the linearity of the scalar product, we have, in the notation of eqn (1.5),

$$P_\alpha |l\rangle = c|\alpha\rangle\langle\alpha|n\rangle + d|\alpha\rangle\langle\alpha|m\rangle = cP_\alpha|n\rangle + dP_\alpha|m\rangle. \tag{1.13}$$

These two properties (eqns (1.12) and (1.13)) imply that P_α is a *linear operator*.

A further property of P_α is that for any $|n\rangle$

$$P_\alpha(P_\alpha|n\rangle) = |\alpha\rangle\langle\alpha|\alpha\rangle\langle\alpha|n\rangle = P_\alpha|n\rangle \tag{1.14}$$

or, in an operator notation,
$$P_\alpha^2 = P_\alpha. \tag{1.15}$$

Equation (1.15) is simply the statement that eqn (1.14) holds for an arbitrary state vector $|n\rangle$. Moreover, from (1.4),

$$P_\alpha P_{\alpha'} = P_\alpha \delta_{\alpha,\alpha'}. \tag{1.16}$$

We shall use the notation $\langle P_\alpha \rangle = \langle n | P_\alpha | n \rangle$ for the average (or 'expectation') value of the attribute P_α as found by measurement. It follows immediately from the Schwarz inequality

$$|\langle \alpha | n \rangle|^2 \leq \langle \alpha | \alpha \rangle \langle n | n \rangle \qquad (1.17)$$

that
$$0 \leq \langle P_\alpha \rangle \leq 1; \qquad (1.18)$$

the upper limit is realized when $|\alpha\rangle = |n\rangle$ and the lower when $\langle \alpha | n \rangle = 0$. In the latter case the states $|\alpha\rangle$ and $|n\rangle$ are *orthogonal*.

If A is a physical attribute its measurement must yield some value, and so

$$\begin{aligned}
\sum_\alpha \langle P_\alpha \rangle &= 1 \\
&= \sum_\alpha \langle n | \alpha \rangle \langle \alpha | n \rangle \\
&= \langle n | \sum_\alpha | \alpha \rangle \langle \alpha | n \rangle \\
&= \langle n | I | n \rangle. \qquad (1.19)
\end{aligned}$$

Since $\langle n | n \rangle = 1$, I is the *identity operator*,

$$I|n\rangle = |n\rangle \qquad (1.20)$$

for any $|n\rangle$, and
$$I = \sum_\alpha |\alpha\rangle\langle\alpha| = \sum_\alpha P_\alpha. \qquad (1.21)$$

Equation (1.21) is known as the *resolution of the identity*. Clearly the discussion relies on the fact that the set of numbers α (the spectrum) is the totality of the possible values of the attribute A.

The expectation value of A, denoted $\langle A \rangle$ as measured in the ensemble, can now be written as an average,

$$\langle A \rangle = \sum_\alpha \alpha |F_n(\alpha)|^2 = \sum_\alpha \alpha \langle P_\alpha \rangle, \qquad (1.22)$$

where the summation is over all the possible values of α. Proceeding as in eqn (1.19),
$$\begin{aligned}
\langle A \rangle &= \sum_\alpha \alpha \langle n | \alpha \rangle \langle \alpha | n \rangle \\
&= \langle n | \left\{ \sum_\alpha \alpha |\alpha\rangle\langle\alpha| \right\} |n\rangle \qquad (1.23) \\
&= \langle n | A_{\text{op}} | n \rangle.
\end{aligned}$$

Equation (1.23) defines the (linear) operator A_{op},

$$A_{\text{op}} = \sum_\alpha \alpha |\alpha\rangle\langle\alpha| = \sum_\alpha \alpha P_\alpha, \qquad (1.24)$$

such that, when every system in the ensemble is in the same dynamical state $|n\rangle$,
$$\langle A \rangle = \langle n | A_{\text{op}} | n \rangle. \qquad (1.25)$$

The resolution (1.24) is known as the *spectral resolution*. (For a rigorous discussion see Akhiezer and Glazman (1961) or other texts in functional analysis dealing with the spectral theorem.)

Equations (1.24) and (1.4) imply the relation
$$A_{op}|\alpha\rangle = \alpha|\alpha\rangle, \tag{1.26}$$
known as an eigenvalue relation. $|\alpha\rangle$ is then referred to as an *eigenstate* and α as the *eigenvalue* of the operator A. The set of eigenvalues is thus equivalent to the spectrum. The *trace* of A_{op} is defined by
$$\operatorname{Tr}\{A_{op}\} = \sum_\alpha \langle\alpha|A_{op}|\alpha\rangle = \sum_\alpha \alpha, \tag{1.27}$$
provided the sum is finite.

In a similar fashion, if we measure some function of A, f(A), then
$$\langle \mathrm{f}(A) \rangle = \sum_\alpha \mathrm{f}(\alpha)\langle P_\alpha \rangle, \tag{1.28}$$
so that in a formal fashion we can define
$$\mathrm{f}(A_{op}) = \sum_\alpha \mathrm{f}(\alpha) P_\alpha \tag{1.29}$$
such that
$$\langle \mathrm{f}(A) \rangle = \langle n|\mathrm{f}(A_{op})|n\rangle. \tag{1.30}$$
In particular, if c is some number,
$$\delta(c-A_{op}) = \sum_\alpha \delta(c-\alpha) P_\alpha. \tag{1.31}$$

The derivation of eqn (1.24) was restricted to the case when α assumes discrete values only, but a similar argument can be made in the general case. Consider, for example, the attribute 'position'. Since a system must be 'somewhere' we require that
$$\int d\mathbf{R}\, |\psi_n(\mathbf{R})|^2 = 1$$
$$= \int d\mathbf{R}\, \langle n|\mathbf{R}\rangle\langle\mathbf{R}|n\rangle, \tag{1.32}$$
or, by analogy to eqn (1.21),
$$I = \int |\mathbf{R}\rangle\, d\mathbf{R}\, \langle\mathbf{R}|. \tag{1.33}$$
In writing eqn (1.33) we have adopted the normalization
$$\langle \mathbf{R}'|\mathbf{R}\rangle = \delta(\mathbf{R}'-\mathbf{R}), \tag{1.34}$$
so that $I|\mathbf{R}\rangle = |\mathbf{R}\rangle$.

Equation (1.21), or (1.33), is also referred to as a *completeness relation* since an arbitrary state vector $|n\rangle$ can then be written
$$|n\rangle = \sum_\alpha |\alpha\rangle\langle\alpha|n\rangle \tag{1.35a}$$
or
$$|n\rangle = \int |\mathbf{R}\rangle\, d\mathbf{R}\, \langle\mathbf{R}|n\rangle. \tag{1.35b}$$
The linear operator
$$P(\mathbf{R}) = |\mathbf{R}\rangle\langle\mathbf{R}|, \tag{1.36a}$$
$$P(\mathbf{R})|n\rangle = |\mathbf{R}\rangle\langle\mathbf{R}|n\rangle$$

can be regarded as the analogue of the operator P_α for the discrete spectrum. The relation analogous to eqn (1.16) is

$$P(\mathbf{R})P(\mathbf{R}') = P(\mathbf{R})\delta(\mathbf{R}-\mathbf{R}'). \tag{1.36b}$$

The transformation from one representation to another can also be expressed in terms of eqns (1.35). Thus, for example,

$$\begin{aligned}\psi_n(\mathbf{p}) &= \langle \mathbf{p}|n\rangle \\ &= \int \langle \mathbf{p}|\mathbf{R}\rangle \, d\mathbf{R} \, \langle \mathbf{R}|n\rangle \\ &= \int \langle \mathbf{p}|\mathbf{R}\rangle \, d\mathbf{R} \, \psi_n(\mathbf{R}).\end{aligned} \tag{1.37}$$

In terms of amplitudes, eqn (1.37) expresses the amplitude to observe the momentum $\mathbf{p}$ when the system is in the state $|n\rangle$, as the integral over all space of the amplitude to observe the system at the position $\mathbf{R}$ times the amplitude ($\langle \mathbf{p}|\mathbf{R}\rangle$) to observe the momentum $\mathbf{p}$ when the system is at the position $\mathbf{R}$.

An extension of the notation is necessary when the measurement of the attribute A does not completely determine the dynamical state of the system. If A and B are two independent attributes of the system and together are sufficient to specify a dynamical state, we shall denote the state as $|\alpha\rangle|\beta\rangle$ or, more simply, $|\alpha, \beta\rangle$. For example, A can be the momentum and B the internal state of an isolated molecule. Then $|\mathbf{p}, n\rangle$ will denote the state in which the molecule possesses the (translational) momentum $\mathbf{p}$ and is in the internal state $|n\rangle$.

The state vector $|\alpha, \beta\rangle$ is regarded as a state-vector in a *product space*. The identity operator in this space can be written

$$I = \sum_{\alpha,\beta} |\alpha, \beta\rangle\langle\alpha, \beta|, \tag{1.38}$$

where

$$|\alpha, \beta\rangle\langle\alpha, \beta| = (|\alpha\rangle\langle\alpha|)(|\beta\rangle\langle\beta|),$$

while

$$\begin{aligned}P_\alpha &= \sum_\beta |\alpha, \beta\rangle\langle\alpha, \beta| \\ &= |\alpha\rangle\Big\{\sum_\beta |\beta\rangle\langle\beta|\Big\}\langle\alpha| \\ &= |\alpha\rangle I_\beta \langle\alpha|,\end{aligned} \tag{1.39}$$

where I_β is the identity operator in the subspace in which the attribute B completely determines the dynamical state. (For example, the momentum completely determines the state of translational motion.) If $|m\rangle$ is some dynamical state then

$$|m\rangle = \sum_{\alpha,\beta} |\alpha, \beta\rangle\langle\alpha, \beta|m\rangle \tag{1.40}$$

so that
$$\langle\alpha|m\rangle = \sum_\beta |\beta\rangle\langle\alpha,\beta|m\rangle$$
is not a number, but
$$\langle m|(|\alpha\rangle\langle\alpha|)|m\rangle = \langle m|(|\alpha\rangle I_\beta\langle\alpha|)|m\rangle$$
$$= \sum_\beta |\langle m|\alpha,\beta\rangle|^2 \qquad (1.41)$$
is the average value obtained in the experiment in which P_α is measured.

It follows from the construction of the states $|\alpha,\beta\rangle$ that
$$A_{\text{op}}|\alpha,\beta\rangle = \alpha|\alpha,\beta\rangle \qquad (1.42\,\text{a})$$
and
$$B_{\text{op}}|\alpha,\beta\rangle = \beta|\alpha,\beta\rangle \qquad (1.42\,\text{b})$$
where
$$A_{\text{op}} = \sum_\alpha \alpha|\alpha\rangle\langle\alpha|,$$
and similarly for B_{op}.

Multiplying eqn (1.42a) by B_{op} and eqn (1.42b) by A_{op} we obtain, for any state $|\alpha,\beta\rangle$,
$$(A_{\text{op}} B_{\text{op}} - B_{\text{op}} A_{\text{op}})|\alpha,\beta\rangle = 0, \qquad (1.43)$$
or
$$A_{\text{op}} B_{\text{op}} - B_{\text{op}} A_{\text{op}} \equiv [A_{\text{op}}, B_{\text{op}}] = 0_{\text{op}} \qquad (1.44)$$
where 0_{op} is the null operator.

Equation (1.44) defines the *commutator* of two operators§ and our discussion implies that the necessary condition for two operators to be simultaneously measurable is that their commutator equals the null operator. We then say that the two operators 'commute with one another'. In denoting the states as $|\alpha,\beta\rangle$ we have assumed eqns (1.42) and hence assumed that the operators A_{op} and B_{op} commute.

In the general discussion that follows we shall continue to assume (unless otherwise noted) that the attribute(s) A are sufficient to specify completely the state of the system. If A stands for a group of attributes then it is necessary (and sufficient) that the separate attributes commute with one another.

The completeness relations can also be used to provide different representations of operators. For example,
$$B_{\text{op}} = \sum_\alpha |\alpha\rangle\langle\alpha|B_{\text{op}}\left\{\sum_{\alpha'} |\alpha'\rangle\langle\alpha'|\right\} = \sum_{\alpha,\alpha'} |\alpha\rangle\langle\alpha|B_{\text{op}}|\alpha'\rangle\langle\alpha'|, \qquad (1.45)$$
where B_{op} is some operator that corresponds to the attribute B. The elements $\langle\alpha|B_{\text{op}}|\alpha'\rangle$ are sometimes denoted $B(\alpha,\alpha')$ or $B_{\alpha,\alpha'}$. The latter notation is particularly useful when the index α is discrete, so that the set of elements $B_{\alpha,\alpha'}$ can be regarded as elements of a matrix **B**.

§ In general, the commutator of two linear operators is a linear operator which need not be the null operator.

This matrix is referred to as the matrix representation of the operator. Given the matrix elements one can always represent the operator as in eqn (1.45). For this reason one sometimes uses the term the 'B matrix' rather than the 'B operator'. If the spectrum is continuous we have, for example,
$$B_{\text{op}} = \iint |\mathbf{R}'\rangle \, \mathrm{d}\mathbf{R}' \, \langle \mathbf{R}'|B_{\text{op}}|\mathbf{R}\rangle \, \mathrm{d}\mathbf{R} \, \langle \mathbf{R}|. \tag{1.46}$$
An operator that satisfies
$$\langle \mathbf{R}'|B_{\text{op}}|\mathbf{R}\rangle \equiv B(\mathbf{R}',\mathbf{R}) = B(\mathbf{R})\delta(\mathbf{R}'-\mathbf{R}) \tag{1.47}$$
is known as a *local* operator. It is easy to see that, for example, the identity operator is a local operator.

Taking a representation of the state vector $B_{\text{op}}|n\rangle$ we have
$$\begin{aligned}
\langle \alpha|B_{\text{op}}|n\rangle &= \langle \alpha|B_{\text{op}}\Big\{\sum_{\alpha'}|\alpha'\rangle\langle\alpha'|n\rangle\Big\} \\
&= \sum_{\alpha'} \langle\alpha|B_{\text{op}}|\alpha'\rangle\langle\alpha'|n\rangle \\
&= \sum_{\alpha'} B_{\alpha,\alpha'}\psi_n(\alpha')
\end{aligned} \tag{1.48}$$
and, in a similar fashion,
$$\begin{aligned}
\langle n|B_{\text{op}}|n\rangle &= \sum_{\alpha,\alpha'} \langle n|\alpha\rangle\langle\alpha|B_{\text{op}}|\alpha'\rangle\langle\alpha'|n\rangle \\
&= \sum_{\alpha,\alpha'} \psi_n^*(\alpha) B_{\alpha,\alpha'}\psi_n(\alpha').
\end{aligned} \tag{1.49}$$

Thus $B(\alpha,\alpha')$ is sometimes referred to as 'the B operator in the α-representation'. For example, in the coordinate representation
$$\langle \mathbf{R}|B_{\text{op}}|n\rangle = \int B(\mathbf{R},\mathbf{R}')\psi_n(\mathbf{R}') \, \mathrm{d}\mathbf{R}' \tag{1.50a}$$
and if B_{op} is a local operator
$$= B(\mathbf{R})\psi_n(\mathbf{R}), \tag{1.50b}$$
so that
$$\langle n|B_{\text{op}}|n\rangle = \int \psi_n^*(\mathbf{R})B(\mathbf{R})\psi_n(\mathbf{R}) \, \mathrm{d}\mathbf{R}. \tag{1.51}$$

Comparing eqns (1.51) and (1.49) we see that in the discrete spectrum case one can define a property analogous to eqn (1.47), by the property that the matrix representation of the operator B is diagonal, i.e.
$$B_{\alpha,\alpha'} = B_\alpha \delta_{\alpha,\alpha'}.$$

In eqns (1.49) or (1.51) we have expressed the expectation value of B in terms of operations on numbers (namely on amplitudes) which can thus be performed, given the relevant amplitudes.

As an example we consider the kinetic energy operator K (from now on we drop the subscript 'op'). The kinetic energy of a free particle of

momentum $\mathbf{p}$ and mass m is $p^2/2m$. Thus, using the normalization
$$\langle \mathbf{p}'|\mathbf{p}\rangle = (2\pi\hbar)^3 \delta(\mathbf{p}-\mathbf{p}') \tag{1.52}$$
and eqn (1.24),
$$K = \int (p^2/2m)|\mathbf{p}\rangle(2\pi\hbar)^{-3}\,\mathrm{d}\mathbf{p}\,\langle\mathbf{p}| \tag{1.53}$$
or
$$K|\mathbf{p}\rangle = (p^2/2m)|\mathbf{p}\rangle. \tag{1.54}$$
In the coordinate representation
$$\psi_\mathbf{p}(\mathbf{R}) = \langle\mathbf{R}|\mathbf{p}\rangle = \exp(i\mathbf{p}\cdot\mathbf{R}/\hbar) \tag{1.55}$$
so that (cf. eqn (1.34))
$$\begin{aligned}\langle\mathbf{R}|\mathbf{R}'\rangle &= (2\pi\hbar)^{-3}\int \langle\mathbf{R}|\mathbf{p}\rangle\,\mathrm{d}\mathbf{p}\,\langle\mathbf{p}|\mathbf{R}'\rangle \\ &= \int \exp[i(\mathbf{R}-\mathbf{R}')\cdot\mathbf{p}/\hbar](2\pi\hbar)^{-3}\,\mathrm{d}\mathbf{p} \\ &= \delta(\mathbf{R}-\mathbf{R}').\end{aligned} \tag{1.56}$$
Taking the coordinate representation of eqn (1.53),
$$\begin{aligned}\langle\mathbf{R}|K|\mathbf{R}'\rangle &= \int (p^2/2m)\exp[i(\mathbf{R}-\mathbf{R}')\cdot\mathbf{p}/\hbar](2\pi\hbar)^{-3}\,\mathrm{d}\mathbf{p} \\ &= -(\hbar^2/2m)\frac{\mathrm{d}^2}{\mathrm{d}\mathbf{R}^2}\int \exp[i(\mathbf{R}-\mathbf{R}')\cdot\mathbf{p}/\hbar](2\pi\hbar)^{-3}\,\mathrm{d}\mathbf{p} \\ &= -(\hbar^2/2m)\frac{\mathrm{d}^2}{\mathrm{d}\mathbf{R}^2}\delta(\mathbf{R}-\mathbf{R}'),\end{aligned} \tag{1.57}$$
where we used eqn (1.56) in obtaining the last line. Thus
$$\begin{aligned}\langle n|K|n\rangle &= \iint \mathrm{d}\mathbf{R}\mathrm{d}\mathbf{R}'\,\psi_n^*(\mathbf{R})K(\mathbf{R},\mathbf{R}')\psi_n(\mathbf{R}') \\ &= \int \mathrm{d}\mathbf{R}\,\psi_n^*(\mathbf{R})\left(-\frac{\hbar^2}{2m}\frac{\mathrm{d}^2}{\mathrm{d}\mathbf{R}^2}\right)\psi_n(\mathbf{R}).\end{aligned} \tag{1.58}$$
We note that the eigenvalues of K are *degenerate* in that there are§ $4\pi mp\,\mathrm{d}E/(2\pi\hbar)^3$ different eigenstates having eigenvalues in the range E to $E+\mathrm{d}E$.

The bra associated with the ket $B|n\rangle$ is denoted $\langle n|B^\dagger$, where $B^\dagger$ is the *Hermitian adjoint* of B. Introducing the representation of B, eqn (1.45), and using the definition (1.7), we find that
$$B^\dagger = \sum_{\alpha,\alpha'} |\alpha'\rangle\langle\alpha|B|\alpha'\rangle^*\langle\alpha|. \tag{1.59}$$
Thus
$$(|\alpha\rangle\langle\alpha'|)^\dagger = |\alpha'\rangle\langle\alpha|, \tag{1.60a}$$
$$P_\alpha^\dagger = P_\alpha, \tag{1.60b}$$

§ There are $(2\pi\hbar)^{-3}\,\mathrm{d}\mathbf{p}$ states in the range $\mathbf{p}$ to $\mathbf{p}+\mathrm{d}\mathbf{p}$ (cf. eqn (1.52)). Now $\mathrm{d}\mathbf{p} = \mathbf{p}^2\,\mathrm{d}p\,\mathrm{d}\hat{\mathbf{p}}$ and $\mathrm{d}E = (p/m)\,\mathrm{d}p$ so that $(2\pi\hbar)^{-3}\,\mathrm{d}\mathbf{p} = (2\pi\hbar)^{-3}mp\,\mathrm{d}\hat{\mathbf{p}}\mathrm{d}E$.

STATES AND OBSERVABLES

$$A^\dagger = \sum_\alpha \alpha^* P_\alpha, \tag{1.60c}$$

$$(cA)^\dagger = c^* A^\dagger, \tag{1.60d}$$

and§
$$(AB)^\dagger = B^\dagger A^\dagger. \tag{1.60e}$$

If A is a physical attribute (an *observable*), its spectrum is real, $\alpha = \alpha^*$, and so $A^\dagger = A$. Such an operator is said to be *invariant under Hermitian conjugation* or simply *Hermitian*. Hermitian operators have real eigenvalues. The simplest observable is the operator $P_\alpha = |\alpha\rangle\langle\alpha|$. It has two eigenvalues, one and zero, and two eigenfunctions,

$$P_\alpha(P_\alpha|n\rangle) = 1 \cdot P_\alpha|n\rangle \tag{1.61a}$$

and
$$P_\alpha(I-P_\alpha)|n\rangle = 0(I-P_\alpha)|n\rangle. \tag{1.61b}$$

A Hermitian operator that satisfies eqn (1.61a) is a *projection* operator. ($|\alpha\rangle\langle\alpha'|$, $\alpha \neq \alpha'$ is an example of a non-Hermitian linear operator.)

It is possible to define an operator B^T, the *transpose* of B, by (cf. eqn (1.45))

$$B^T = \sum_{\alpha,\alpha'} |\alpha'\rangle\langle\alpha|B|\alpha'\rangle\langle\alpha| \tag{1.62}$$

so that
$$B^\dagger = (B^T)^* \tag{1.63}$$

or explicitly
$$\langle\alpha|B^\dagger|\alpha'\rangle = \langle\alpha'|B|\alpha\rangle^*. \tag{1.64}$$

When every system in our ensemble is not in the same particular dynamical state, we can no longer represent $\langle A \rangle$ as $\langle n|A|n\rangle$. We can, however, divide the ensemble into sub-ensembles, where in each sub-ensemble all the systems are in the same dynamical state. We then say that the original ensemble corresponds to a 'mixture', and if $p(n)$ is the fraction of the systems in the state $|n\rangle$, we write

$$\langle A \rangle = \sum_n p(n)\langle n|A|n\rangle. \tag{1.65}$$

Clearly
$$\sum_n p(n) = 1, \quad p(n) \geq 0. \tag{1.66}$$

The *statistical* (or *density*) *operator* (ter Haar 1961, Fano 1957, von Neumann 1955) is defined as

$$\rho = \sum_n p(n)|n\rangle\langle n|, \tag{1.67}$$

so that‖
$$\langle A \rangle = \text{Tr}\{\rho A\} = \text{Tr}\left\{\sum_n p(n)|n\rangle\langle n|A\right\}$$
$$= \sum_n p(n)\text{Tr}\{|n\rangle\langle n|A\} = \sum_n p(n)\langle n|A|n\rangle. \tag{1.68}$$

§ The ket $AB|m\rangle$ is defined by $AB|m\rangle = A\sum_n |n\rangle\langle n|B|m\rangle$.

‖ We note that the trace of an operator is independent of the representation. For example, using eqn (1.27),
$$\text{Tr}\{|n\rangle\langle n|\} = \sum_\alpha \langle\alpha|n\rangle\langle n|\alpha\rangle = \sum_\alpha \langle n|\alpha\rangle\langle\alpha|n\rangle = \langle n|n\rangle = 1.$$

Moreover, $$\text{Tr}\{\rho\} = \sum_n p(n)\text{Tr}\{|n\rangle\langle n|\} = \sum_n p(n) = 1. \tag{1.69}$$

ρ is clearly a linear, Hermitian operator. Even though $\text{Tr}\{\rho\} = 1$ we find that, since $p^2(n) \leqslant p(n)$,

$$\text{Tr}\{\rho^2\} = \text{Tr}\Big\{\sum_n |n\rangle p^2(n)\langle n|\Big\} = \sum_n p^2(n) \leqslant 1. \tag{1.70}$$

Thus ρ is not, in general, a projection operator. It is, however, a projection operator when, for a particular value of n, $p(n) = 1$, while $p(n') = 0$ for all $n' \neq n$. This particular case (the *pure state*) corresponds to $\rho = |n\rangle\langle n|$, i.e. to the situation when every system is in the same dynamical state $|n\rangle$. In a given representation ρ is referred to as the *density matrix*. The extension of the concept of 'product space' to density matrices is discussed by Fano (1957).

Finally, we will make a rather brief reference to the theory of measurement (Golden 1968, Furry 1966, Gottfried 1966, Katz 1964, Groenewold 1964, Schwinger 1959). Given an ensemble described by the density operator ρ, let the attribute B be measured. In the first experiment we collect all the systems that yielded the value β for the attribute B. The new ensemble so prepared,§ is taken to have the density operator

$$\rho' = P_\beta \rho P_\beta / \text{Tr}\{P_\beta \rho\}. \tag{1.71}$$

In the second type of experiment we do not sort out the systems according to the different values found for B. In this case the density operator after the experiment is taken to be

$$\rho'' = \sum_\beta P_\beta \rho P_\beta. \tag{1.72 a}$$

If the spectrum of B is non-degenerate we can also write ρ'' as

$$\rho'' = \sum_\beta \text{Tr}\{\rho P_\beta\} P_\beta. \tag{1.72 b}$$

§ If the state $|\beta\rangle$ is not degenerate, so that $\text{Tr}\{P_\beta\} = 1$, ρ' can be simply written as $\rho' = P_\beta$. Thus a pure state can be prepared by a measurement of a non-degenerate eigenvalue.

1.2. The equations of motion

1.2.0. LET $|\psi(t_0)\rangle$ be the state vector at the time t_0. We assume that if the system was isolated during the time interval t_0 to t, its state vector at the time t, $|\psi(t)\rangle$, is determined by the causal law

$$|\psi(t)\rangle = U(t, t_0)|\psi(t_0)\rangle. \tag{2.1}$$

Here $U(t, t_0)$ is the *evolution operator*. We have denoted the state vector $|\psi(t)\rangle$ rather than labelled it by the relevant eigenvalues, since there is no *a priori* reason to expect that if the system was in the state $|\alpha\rangle$ at the time t_0 it will be in the same state at the time t.

The condition that the normalization is conserved in time, i.e. that for all t

$$\langle\psi(t)|\psi(t)\rangle = \langle\psi(t_0)|\psi(t_0)\rangle, \tag{2.2}$$

implies, using eqns (2.1) and (1.60), that

$$\langle\psi(t)|\psi(t)\rangle = \langle\psi(t_0)|U^\dagger(t, t_0)U(t, t_0)|\psi(t_0)\rangle,$$

or
$$U^\dagger(t, t_0)U(t, t_0) = I. \tag{2.3}$$

The equation of motion for U is Schrödinger's time-dependent equation

$$i\hbar\frac{\partial}{\partial t}U(t, t_0) = H(t)U(t, t_0), \tag{2.4}$$

where $H(t)$ is the *Hamiltonian* of the system. For a conservative system, H is time independent and the solution of eqn (2.4), subject to the obvious boundary condition

$$U(t_0, t_0) = I, \tag{2.5}$$

is
$$U(t, t_0) = \exp[-iH(t-t_0)/\hbar]. \tag{2.6}$$

Substituting for $U(t, t_0)$ in eqn (2.1) we obtain

$$i\hbar\frac{d}{dt}|\psi(t)\rangle = H|\psi(t)\rangle, \tag{2.7a}$$

or
$$|\psi(t)\rangle = \exp[-iH(t-t_0)/\hbar]|\psi(t_0)\rangle. \tag{2.7b}$$

The spectrum of H consists of the allowed energy levels of the system. If $|\psi(t_0)\rangle$ is an eigenstate of H with the eigenvalue E,

$$H|\psi(t_0)\rangle = E|\psi(t_0)\rangle, \tag{2.8a}$$

we have, using eqn (2.7b),

$$|\psi(t)\rangle = \exp[-iE(t-t_0)/\hbar]|\psi(t_0)\rangle \tag{2.7c}$$

and
$$H|\psi(t)\rangle = E|\psi(t)\rangle \tag{2.8b}$$
since H commutes with itself.

If the Hamiltonian is time dependent, one can, in general, write the solution of eqn (2.4) only in the form of a series. (See, for example, Dyson (1949), Magnus (1954), and section 3.4.2.)

When the system is not in a pure state we replace eqn (2.1) by a causal law of evolution of the density matrix for an isolated system,

$$\begin{aligned}\rho(t) &= U(t,t_0)\rho(t_0)U^\dagger(t,t_0), \\ \text{or} \quad &= \sum_n p_n |\psi_n(t)\rangle\langle\psi_n(t)| \\ &= \sum_n p_n U(t,t_0)|\psi_n(t_0)\rangle\langle\psi_n(t_0)|U^\dagger(t,t_0),\end{aligned} \tag{2.9}$$

where
$$\rho(t_0) = \sum_n p_n |\psi_n(t_0)\rangle\langle\psi_n(t_0)|.$$

Thus, for a pure state, eqn (2.9) reduced to eqn (2.1).

Differentiating eqn (2.9) with respect to t and using eqn (2.4) and its adjoint,
$$-i\hbar \frac{\partial U^\dagger}{\partial t} = U^\dagger H,$$
we obtain
$$i\hbar \frac{\mathrm{d}}{\mathrm{d}t}\rho(t) = [H, \rho(t)] \tag{2.10}$$
as the equation of motion of the density matrix.

The conservation of normalization can now be written as $\mathrm{Tr}\{\rho(t)\} = 1$ and, from eqn (2.9),
$$\mathrm{Tr}\{\rho(t)\} = \mathrm{Tr}\{\rho(t_0)U^\dagger(t,t_0)U(t,t_0)\}. \tag{2.11}$$
Here we used the cyclic property of the trace
$$\mathrm{Tr}\{ABC\} = \mathrm{Tr}\{BCA\} = \mathrm{Tr}\{CAB\}. \tag{2.12}$$
Since by assumption $\mathrm{Tr}\{\rho(t_0)\} = 1$, we regain eqn (2.3) as a condition on the evolution operator.

Up to this point we followed the view§ that it is the state vectors that evolve with time, and that those observables that are not explicitly time-dependent are stationary. On the other hand, only expectation values can be measured physically, and so only the rate of change of expectation values need be uniquely defined.

In the point of view referred to as 'the Heisenberg representation',

§ This view is often referred to as the 'Schrödinger representation'. The term 'representation' as used in describing a point of view regarding the evolution in time should not be confused with the same term as used in designating an amplitude (cf. eqn (1.8)).

the observables of the system evolve with time. Using eqns (2.12) and (2.9),

$$\begin{aligned}\langle A \rangle &= \mathrm{Tr}\{A\rho(t)\} \\ &= \mathrm{Tr}\{AU(t,t_0)\rho(t_0)U^\dagger(t,t_0)\} \\ &= \mathrm{Tr}\{\rho(t_0)U^\dagger(t,t_0)AU(t,t_0)\} \\ &= \mathrm{Tr}\{\rho(t_0)A_\mathrm{H}(t,t_0)\}.\end{aligned} \quad (2.13)$$

Here we used the cyclic property of the trace and defined

$$A_\mathrm{H}(t,t_0) = U^\dagger(t,t_0)AU(t,t_0) \quad (2.14)$$

as the observable in the Heisenberg representation.

1.2.1. *The Green's operator for the Schrödinger equation*

The concept of a Green's operator (the *resolvent*) is often introduced in applied mathematics in a particular (most often the coordinate) representation, and an extensive discussion can be found, for example, in Morse and Feshbach (1953). The particular application to scattering theory is considered by Mott and Massey (1965, Chapter 4). The present discussion follows more closely the development of the subject in functional analysis. (See, for example, Akhiezer and Glazman (1961) or, in formal scattering theory, Schönberg (1951), Zumino (1956), Friedman (1956), and van Hove (1957).)

Given an operator L, the Green's operator is defined as the inverse operator

$$G(\lambda) = (\lambda - L)^{-1} \quad (2.15)$$

for values of the variable λ (real or complex) which do not belong to the spectrum of L. When a Hermitian operator L has a discrete, non-degenerate, spectrum

$$L|l\rangle = l|l\rangle,$$
$$\langle l'|l\rangle = \delta_{l'l}, \quad (2.16)$$

we can evaluate the matrix elements (cf. eqn (1.28)) as

$$\langle l'|G(\lambda)|l\rangle = (\lambda - l)^{-1}\delta_{l'l}. \quad (2.17\,\mathrm{a})$$

From the definition of the Hermitian adjoint (eqn (1.59))

$$\langle l'|G^\dagger(\lambda)|l\rangle = (\lambda^* - l)^{-1}\delta_{l'l} = \langle l'|G(\lambda^*)|l\rangle, \quad (2.17\,\mathrm{b})$$

or, in operator notation,

$$G^\ddagger(\lambda) = G(\lambda), \quad (2.18\,\mathrm{a})$$

where

$$G^\ddagger(\lambda) \equiv G^\dagger(\lambda^*). \quad (2.18\,\mathrm{b})$$

The conjugation denoted by $\ddagger$ and defined by eqn (2.18 b) was introduced by Lippmann (1957) and is referred to in the book as L- (for lambda or for Lippmann) conjugation. We shall see below (section 2.8.1) that invariance under L-conjugation is closely related to the reciprocity theorem.

The spectral resolution of $G(\lambda)$ can be written as (cf. eqn (1.29))

$$G(\lambda) = \sum_l \frac{P_l}{\lambda - l}, \qquad (2.19)$$

where
$$P_l = |l\rangle\langle l|.$$

Thus, using eqns (2.17) and (2.16),

$$(\lambda - L)G(\lambda) = I = G(\lambda)(\lambda - L). \qquad (2.20\,\text{a})$$

In a representation we write eqn (2.20 a) as

$$\sum_\alpha \langle \alpha'|(\lambda - L)|\alpha\rangle\langle\alpha|G(\lambda)|\alpha''\rangle = \delta_{\alpha',\alpha''}. \qquad (2.20\,\text{b})$$

Here
$$G(\lambda)|\alpha\rangle = \sum_l |l\rangle(\lambda - l)^{-1}\langle l|\alpha\rangle.$$

In the coordinate representation, when the operator L is local, we have in the notation of eqn (1.47),

$$[\lambda - L(\mathbf{R})]\langle \mathbf{R}|G(\lambda)|\mathbf{R}'\rangle = \delta(\mathbf{R} - \mathbf{R}'). \qquad (2.20\,\text{c})$$

The matrix element $\langle l'|G(\lambda)|l\rangle$ is clearly an analytic function of λ in any region in the (complex) λ-plane that excludes the point $\lambda = l$. Thus Cauchy's theorem can be written as

$$\delta_{l',l} = (2\pi i)^{-1} \int_c \langle l'|G(\lambda)|l\rangle \, d\lambda, \qquad (2.21\,\text{a})$$

where c is a closed counter-clockwise contour around the point $\lambda = l$. In an operator notation we can write (cf. eqn (2.19))

$$P_l = (2\pi i)^{-1} \int_{c_l} G(\lambda) \, d\lambda, \qquad (2.21\,\text{b})$$

where c_l is a closed contour on the λ-plane that encloses the point $\lambda = l$ but excludes all points $\lambda = l'$, $l' \neq l$. The spectral resolution of the identity, eqn (1.21), can now be written as

$$I = \sum_l P_l = (2\pi i)^{-1} \sum_l \int_{c_l} G(\lambda) \, d\lambda = (2\pi i)^{-1} \int_{sp} G(\lambda) \, d\lambda. \qquad (2.22)$$

Here sp indicates a closed counter-clockwise contour around the spectrum of L in the complex λ-plane. Moreover, any function $f(L)$ of L can be written, using eqns (2.21 b) and (1.29), as

$$f(L) = (2\pi i)^{-1} \int_{sp} f(\lambda) G(\lambda) \, d\lambda, \qquad (2.23\,\text{a})$$

provided that $f(\lambda)$ is analytic within and on the contour of integration in the λ-plane. Taking matrix elements,

$$\langle l'|f(L)|l\rangle = (2\pi i)^{-1}\int_{sp} f(\lambda)\frac{\delta_{ll'}}{\lambda-1}\,d\lambda = f(l)\delta_{ll'}, \qquad (2.23\,\text{b})$$

we confirm eqn (1.28).

In the case of degeneracy, several eigenfunctions of L correspond to the same eigenvalue l. Denoting these eigenfunctions as $|l,j\rangle$, where $j = 1,\ldots,n$ and enumerates the different degenerate eigenfunctions, we shall put

$$P_l = \sum_{j=1}^{n}|l,j\rangle\langle l,j|.$$

If part of the spectrum is continuous we shall replace (wherever necessary) the summation (over l) by an integration. For example, if the spectrum is purely continuous, eqn (2.19) is replaced by

$$G(\lambda) = \int (\lambda-l)^{-1}P(l)\,dl. \qquad (2.24)$$

Here $P(l)$ is a differential projection operator (eqn (1.36 b))

$$P(l)|l'\rangle = \delta(l-l')|l'\rangle.$$

The continuous spectrum of a Hermitian operator L corresponds to a segment of the real-λ-axis. If L is the Hamiltonian, then most often the continuous spectrum is only bounded from below at some threshold energy. The Green's operator is defined only for values of λ not in the spectrum of L. As λ changes from the upper ($\text{im}\,\lambda > 0$) to the lower ($\text{im}\,\lambda < 0$) parts of the complex λ-plane across the continuous spectrum of L, the Green's operator is discontinuous. To see this we compute $G(l\pm i\epsilon)$, where l is a point in the continuous spectrum of L and ϵ is a small positive number.

The Cauchy principal-value (Pv) of an integral is defined, for the present purpose, as§

$$Pv(l-L)^{-1} = \lim_{\epsilon\to+0}\left\{\int_{-\infty}^{l-\epsilon}(l-l')^{-1}P(l')\,dl' + \int_{l+\epsilon}^{\infty}(l-l')^{-1}P(l')\,dl'\right\}. \qquad (2.25)$$

Using eqn (2.24),

$$G(l\pm i\epsilon) = \int_{-\infty}^{\infty}(l-l'\pm i\epsilon)^{-1}P(l')\,dl'. \qquad (2.26\,\text{a})$$

Indenting the integration path in eqn (2.26 a) at the point l with a half-circle of radius ϵ, centred at l and lying in the upper/lower part of the λ-plane, and using eqn (2.25), we have in the limit $\epsilon \to +0$

$$G(l\pm i\epsilon) = Pv(l-L)^{-1}\mp i\pi P(l). \qquad (2.26\,\text{b})$$

§ The notation $\epsilon \to +0$ means that $\epsilon \to 0$ through positive values of ϵ.

Thus the discontinuity of $G(\lambda)$ as λ crosses the continuous spectrum of L is, in the limit $\epsilon \to +0$,

$$2\pi i P(l) = G(l-i\epsilon) - G(l+i\epsilon), \quad \text{re}\,\lambda = l. \tag{2.27}$$

Indeed, from eqn (2.23a),

$$I = (2\pi i)^{-1} \int_{\text{sp}} G(\lambda)\,d\lambda, \tag{2.28a}$$

and taking the (counter-clockwise) contour around the spectrum as a path just below the real axis ($\lambda = l-i\epsilon$) plus a path just above the real axis ($\lambda = l+i\epsilon$) supplemented by two vertical paths, we have in the limit $\epsilon \to +0$,

$$I = (2\pi i)^{-1} \int dl\,[G(l-i\epsilon) - G(l+i\epsilon)] = \int dl\,P(l), \tag{2.28b}$$

namely the resolution of the identity.

To ensure a unique (single-valued) Green's operator one introduces a cut in the λ-plane along the continuous spectrum of L and defines $G(\lambda)$ everywhere such that

$$\lim_{\substack{\text{im}\,\lambda \to 0 \\ \text{re}\,\lambda = l}} G(\lambda) \to \begin{cases} G(l+i\epsilon), & \text{im}\,\lambda \to +0 \\ G(l-i\epsilon), & \text{im}\,\lambda \to -0. \end{cases} \tag{2.29}$$

As an example of an operator with a purely continuous spectrum we consider the kinetic energy operator K. Using eqns (1.54) and (1.28),

$$(\lambda - K)^{-1}|\mathbf{p}\rangle = [\lambda - (p^2/2m)]^{-1}|\mathbf{p}\rangle, \tag{2.30}$$

where m is the mass of the particle. In the coordinate representation

$$\langle \mathbf{R}|(\lambda-K)^{-1}|\mathbf{R}'\rangle = (2\pi\hbar)^{-3} \int d\mathbf{p}\,\langle \mathbf{R}|\mathbf{p}\rangle[\lambda - (p^2/2m)]^{-1}\langle \mathbf{p}|\mathbf{R}\rangle.$$

Using the transformation function $\langle \mathbf{R}|\mathbf{p}\rangle = \exp(i\mathbf{p}\cdot\mathbf{R}/\hbar)$, we obtain

$$(2\pi\hbar)^{-3}\int p^2\,dp\,d\hat{\mathbf{p}}\,\exp[i\mathbf{p}\cdot(\mathbf{R}-\mathbf{R}')/\hbar][\lambda-(p^2/2m)]^{-1}. \tag{2.31}$$

Introducing the variable $\rho = |\mathbf{R}-\mathbf{R}'|$, we write eqn (2.31) as

$$(2\pi\hbar)^{-3}\int p^2\,dp[\lambda-(p^2/2m)]^{-1}\int \exp[ip\rho\cos\theta/\hbar]\,d\hat{\mathbf{p}},$$

where θ is the angle between $\mathbf{p}$ and $\mathbf{R}-\mathbf{R}'$. Referring all directions to the axis determined by $\mathbf{R}-\mathbf{R}'$ so that $d\hat{\mathbf{p}} = \sin\theta\,d\theta\,d\phi$, and integrating over angles, we obtain

$$(2\pi\hbar)^{-2}\int p^2\,dp[\lambda-(p^2/2m)]^{-1}(ip\rho)^{-1}[\exp(ip\rho/\hbar)-\exp(-ip\rho/\hbar)]$$

$$= \frac{m}{\hbar^2}(8\pi^2\rho)^{-1}\frac{d}{d\rho}\int_{-\infty}^{\infty} \frac{\exp(ik\rho)}{\kappa^2-k^2}\,dk. \tag{2.32}$$

In writing eqn (2.32) we made the change of variables
$$\lambda = \hbar^2\kappa^2/2m \quad \text{and} \quad p = \hbar k.$$
Writing the denominator in eqn (2.32) as $(\kappa+k)(\kappa-k)$, we can evaluate the integral by residues when we close a contour in the complex κ-plane by adding a large semicircle. To ensure a negligible contribution from the integration along the semicircle when its radius tends to infinity, we shall take the semicircle in the upper κ-plane when im $k > 0$ and in the lower κ-plane when im $k < 0$. Closing the contour in the upper κ-plane we obtain (when im $\lambda > 0$) a contribution to the integral from the pole at $k = \kappa$. Then

$$\langle \mathbf{R}|(\lambda-k)^{-1}|\mathbf{R}'\rangle = 2\pi i(m/\kappa 4\pi^2\rho\hbar^2)\frac{\mathrm{d}}{\mathrm{d}\rho}\exp(\mathrm{i}\kappa\rho)$$
$$= -(m/2\pi\hbar^2)\exp(\mathrm{i}\kappa\rho)/\rho \quad (\text{im } \lambda > 0). \qquad (2.33\,\text{a})$$

Similarly, when im $\lambda < 0$ and we close the contour in the lower plane, we obtain
$$-(m/2\pi\hbar^2)\exp(-\mathrm{i}\kappa\rho)/\rho \quad (\text{im } \lambda < 0). \qquad (2.33\,\text{b})$$

In the limit im $\lambda \to \pm 0$ we obtain
$$\langle \mathbf{R}|G(E^\pm)|\mathbf{R}'\rangle = -(m/2\pi\hbar^2)\exp(\pm\mathrm{i}k|\mathbf{R}-\mathbf{R}'|)/|\mathbf{R}-\mathbf{R}'|. \qquad (2.34)$$

Here re $\lambda = \hbar^2k^2/2m$ and the notation $G(E^\pm)$ corresponds to the limits of the Green's function (cf. eqn (2.29)) $E^\pm = E \pm \mathrm{i}\epsilon$.

Using eqn (2.26 b),
$$Pv\langle\mathbf{R}|(E^+-K)^{-1}|\mathbf{R}'\rangle = -(m/2\pi\hbar^2)\cos(k\rho)/\rho \qquad (2.35\,\text{a})$$
and
$$\text{im}\langle\mathbf{R}|(E^+-K)^{-1}|\mathbf{R}'\rangle = -(m/2\pi\hbar^2)\sin(k\rho)/\rho. \qquad (2.35\,\text{b})$$

The Green's operator for the Hamiltonian is closely related to the evolution operator. From the identity (2.B.6) we can write

$$G(E^+) = (-\mathrm{i}/\hbar)\int_0^\infty \exp(-\mathrm{i}Ht/\hbar)\exp(\mathrm{i}Et/\hbar)\,\mathrm{d}E \qquad 2.36$$

or
$$= (-\mathrm{i}/\hbar)\int_{-\infty}^\infty \theta(t)\exp(-\mathrm{i}Ht/\hbar)\exp(\mathrm{i}Et/\hbar)\,\mathrm{d}E,$$

where $\theta(t)$ is the unit step-function,
$$\theta(t) = \begin{cases} 1 & (t > 0) \\ 0 & (t < 0). \end{cases} \qquad (2.37)$$

Conversely, the evolution operator can be expressed in terms of the Green's operator using eqn (2.23).

The operator
$$G^+(t) = -(\mathrm{i}/\hbar)\theta(t)\exp(-\mathrm{i}Ht/\hbar) \qquad (2.38)$$

can be considered as the Green's operator for the equation of motion (2.7 a). Thus, using eqn (2.4) and $\delta(t) = \mathrm{d}\theta(t)/\mathrm{d}t$ we have

$$\left(i\hbar\frac{\partial}{\partial t}-H\right)G^+(t-t') = \delta(t-t'). \tag{2.39}$$

Further discussion of these relations is given in section 2.7.0 and Appendix 2.B.

APPENDIX 1.A

1.A. *The two-body problem and the addition of angular momentum*

THE Hamiltonian of two structureless particles interacting through a two-body local potential V is

$$\mathcal{H} = \frac{P_1^2}{2m_1} + \frac{P_2^2}{2m_2} + V(\mathbf{R}_1 - \mathbf{R}_2), \tag{A.1}$$

where $\mathbf{P}_i$, m_i, and $\mathbf{R}_i$ are the momentum, mass, and position of particle i. The centre of mass position $\mathbf{X}$ and the relative separation $\mathbf{R}$ are defined by

$$\mathbf{X} = (m_1 \mathbf{R}_1 + m_2 \mathbf{R}_2)/(m_1 + m_2),$$
$$\mathbf{R} = \mathbf{R}_1 - \mathbf{R}_2. \tag{A.2}$$

The momenta conjugate to these coordinates are, in the coordinate representation,

$$\mathbf{P} = -i\hbar \frac{\partial}{\partial \mathbf{X}} \quad \text{and} \quad \mathbf{p} = -i\hbar \frac{\partial}{\partial \mathbf{R}}. \tag{A.3}$$

It is easy to show that§

$$\mathbf{P} = -i\hbar\left(\frac{\partial}{\partial \mathbf{R}_1} + \frac{\partial}{\partial \mathbf{R}_2}\right),$$

$$\mathbf{p} = -i\hbar(m_1 + m_2)^{-1}\left(m_2 \frac{\partial}{\partial \mathbf{R}_1} - m_1 \frac{\partial}{\partial \mathbf{R}_2}\right). \tag{A.4}$$

The Hamiltonian $\mathcal{H}$ can now be written

$$\mathcal{H} = \frac{P^2}{2M} + \frac{p^2}{2\mu} + V(\mathbf{R}) = \frac{P^2}{2M} + H, \tag{A.5}$$

where $\mu = m_1 m_2/(m_1 + m_2)$ is the reduced mass for the relative motion. $K_{\text{CM}} = P^2/2M$ is the kinetic energy operator for the motion of the centre of mass and commutes∥ with H, where H can be regarded as a Hamiltonian for a particle of mass μ in the potential $V(\mathbf{R})$. We shall refer to H as the Hamiltonian in the centre of mass system. Eigenfunctions of $\mathcal{H}$ can now be constructed as products of eigenfunctions of K_{CM} and of H. The total angular momentum, $\mathbf{R}_1 \wedge \mathbf{P}_1 + \mathbf{R}_2 \wedge \mathbf{P}_2$, can also be resolved into a sum of the centre of mass angular momentum, $\mathbf{X} \wedge \mathbf{P}$, and the angular momentum for the relative motion, $\hbar \mathbf{L}$,

$$\hbar \mathbf{L} = \mathbf{R} \wedge \mathbf{p}. \tag{A.6}$$

Introducing the spherical coordinates $(R, \theta, \phi) \equiv (R, \hat{\mathbf{R}})$ we obtain the well-known expression

$$\frac{p^2}{2\mu} = -(\hbar^2/2\mu)\frac{\partial^2}{\partial \mathbf{R}^2} = (p_R^2 + \hbar^2 L^2/R^2)/2\mu. \tag{A.7}$$

Here p_R is the 'radial' momentum

$$p_R = -i\hbar\left(R^{-1} + \frac{\partial}{\partial R}\right)$$

so that

$$p_R^2 = -\hbar^2 R^{-2}\frac{\partial}{\partial R}\left(R^2 \frac{\partial}{\partial R}\right). \tag{A.8}$$

§ In classical mechanics, $\mathbf{P} = \mathbf{P}_1 + \mathbf{P}_2$ and $\mathbf{p} = (m_2 \mathbf{P}_1 - m_1 \mathbf{P}_2)/(m_1 + m_2)$.
∥ This is true only if V does not depend on the absolute positions of the particles but only on their relative positions.

The transformation function from the coordinate to the angular momentum representation is

$$\langle \mathbf{R} | R' l m_l \rangle = R^{-1} \delta(R' - |\mathbf{R}|) Y_l^{m_l}(\hat{\mathbf{R}}), \tag{A.9}$$

where $Y_l^{m_l}(\hat{\mathbf{R}})$ is a spherical harmonic,

$$Y_l^{m_l}(\hat{\mathbf{R}}) = \langle \hat{\mathbf{R}} | l m_l \rangle, \tag{A.10}$$

and is an eigenfunction of $\mathbf{L}^2$ and of L_z, where L_z is the projection of $\mathbf{L}$ on a fixed (z) axis,

$$\mathbf{L}^2 Y_l^{m_l}(\hat{\mathbf{R}}) = l(l+1) Y_l^{m_l}(\hat{\mathbf{R}}) \tag{A.11a}$$

and

$$L_z Y_l^{m_l}(\hat{\mathbf{R}}) = m_l Y_l^{m_l}(\hat{\mathbf{R}}). \tag{A.11b}$$

The spherical harmonics are normalized since

$$\begin{aligned}\langle l, m_l | l', m_{l'} \rangle &= \delta_{ll'} \delta_{m_l m_{l'}} \\ &= \int \mathrm{d}\hat{\mathbf{R}} \, \langle l, m_l | \hat{\mathbf{R}} \rangle \langle \hat{\mathbf{R}} | l', m_{l'} \rangle \\ &= \int \mathrm{d}\hat{\mathbf{R}} \, Y_l^{*m_l}(\hat{\mathbf{R}}) Y_{l'}^{m_{l'}}(\hat{\mathbf{R}}).\end{aligned} \tag{A.12}$$

Here $\mathrm{d}\hat{\mathbf{R}} = \sin\theta \, \mathrm{d}\theta \mathrm{d}\phi$, θ and ϕ being the polar angles. Also we can write

$$\begin{aligned}\langle \hat{\mathbf{R}} | \hat{\mathbf{R}}' \rangle &= \delta(\hat{\mathbf{R}} - \hat{\mathbf{R}}') \\ &= \sum_{l,m_l} \langle \hat{\mathbf{R}} | l, m_l \rangle \langle l, m_l | \hat{\mathbf{R}}' \rangle \\ &= \sum_{l,m_l} Y_l^{m_l}(\hat{\mathbf{R}}) Y_l^{*m_l}(\hat{\mathbf{R}}') \\ &= \delta(\theta - \theta') \delta(\phi - \phi') / \sin\theta.\end{aligned} \tag{A.13}$$

Using these relations we can expand an arbitrary function of $\mathbf{R}$ as

$$\psi(\mathbf{R}) = \sum_{l,m_l} \psi_{l,m_l}(R) Y_l^{m_l}(\hat{\mathbf{R}}), \tag{A.14}$$

where

$$\psi_{l,m_l}(R) = \int \psi(\mathbf{R}) Y_l^{*m_l}(\hat{\mathbf{R}}) \, \mathrm{d}\hat{\mathbf{R}}.$$

Another useful relation is the 'addition theorem' for spherical harmonics, where $P_l(\cos\theta)$ is the Legendre polynomial,

$$P_l(\hat{\mathbf{R}} \cdot \hat{\mathbf{R}}') = \frac{4\pi}{2l+1} \sum_{m_l=-l}^{l} Y_l^{*m_l}(\hat{\mathbf{R}}) \cdot Y_l^{m_l}(\hat{\mathbf{R}}'). \tag{A.15}$$

If the mutual interaction, V, is a function of $|\mathbf{R}|$ only, $\mathbf{L}^2$ and L_z commute with H and the Schrödinger equation

$$[(p_R^2/2\mu) + \hbar^2 \mathbf{L}^2 / 2\mu R^2 + V(R) - E] \psi(\mathbf{R}) = 0 \tag{A.16}$$

can be solved by putting

$$\psi(\mathbf{R}) = \psi_l(R) Y_l^m(\hat{\mathbf{R}}), \tag{A.17}$$

so that, with the substitution

$$\psi_l(R) = U_l(R)/R, \tag{A.18}$$

$$\left[-\frac{\hbar^2}{2\mu} \frac{\mathrm{d}^2}{\mathrm{d}R^2} + V(R) + \frac{\hbar^2 l(l+1)}{2\mu R^2} - E \right] U_l(R) = 0. \tag{A.19}$$

Equation (A.19) is the 'radial' Schrödinger equation. If ϵ is a measure of the 'depth' of the interaction $V(R)$, i.e. $V(R) = \epsilon V^*(R)$ where $V^*(R)$ is dimensionless, and σ is a measure of the range of the potential $V(R)$, one can write a reduced form of eqn (A.19) as

$$\left[-\frac{1}{B} \frac{\mathrm{d}^2}{\mathrm{d}z^2} + V^*(z) + \frac{l(l+1)}{z^2} - E^* \right] U_l(z) = 0. \tag{A.20}$$

Here $z = R/\sigma$ and $E^* = E/\epsilon$ are reduced variables and
$$B = 2\mu\epsilon\sigma^2/\hbar^2 \tag{A.21}$$
is the reduced parameter characterizing the system. (For examples of solutions see Harrison and Bernstein (1963).)

In the absence of an interaction eqn (A.16) can be brought into a reduced form using the (reduced) variable $\rho = kR$, where $E = \hbar^2 k^2/2\mu$, so that
$$\left[\frac{1}{\rho^2}\frac{d}{d\rho}\left(\rho^2\frac{d}{d\rho}\right)-\frac{l(l+1)}{\rho^2}+1\right]\psi_l(\rho) = 0. \tag{A.22}$$

Equation (A.22) has a singular point at $\rho = 0$, and of the two independent solutions only one is finite at the origin. The finite ('regular') solution is the spherical Bessel function $j_l(\rho)$,
$$j_l(\rho) = (\pi/2\rho)^{\frac{1}{2}}J_{l+\frac{1}{2}}(\rho),$$
which has the asymptotic behaviour
$$j_l(\rho) \xrightarrow[\rho\to\infty]{} \rho^{-1}\sin(\rho-\tfrac{1}{2}l\pi) \tag{A.23}$$
and is proportional to ρ^l for $\rho \to 0$. The linearly independent solution of eqn (A.22) is $n_l(\rho)$, the spherical Neumann function. Another pair of solutions are the Hankel functions
$$h_l^{\pm}(\rho) \xrightarrow[\rho\to\infty]{} i^{\mp l}\rho^{-1}\exp(\pm i\rho) \tag{A.24}$$
such that
$$j_l(\rho) = \frac{1}{2i}[h_l^+(\rho) - h_l^-(\rho)]. \tag{A.25}$$

The spherical Bessel functions are normalized by
$$\int_0^\infty j_l(kR)j_l(k'R)R^2\,dR = \frac{\pi}{2k^2}\delta(k-k'). \tag{A.26}$$

We shall reserve the notation $|\mathbf{k}\rangle$ for momentum eigenstates normalized by
$$\langle\mathbf{k'}|\mathbf{k}\rangle = (2\pi)^3\delta(\mathbf{k'}-\mathbf{k}) = \int d\mathbf{R}\,\langle\mathbf{k'}|\mathbf{R}\rangle\langle\mathbf{R}|\mathbf{k}\rangle, \tag{A.27}$$
so that
$$\langle\mathbf{R}|\mathbf{k}\rangle = \exp(i\mathbf{k}\cdot\mathbf{R}) \quad\text{and}\quad |\langle\mathbf{R}|\mathbf{k}\rangle|^2 d\mathbf{R} = d\mathbf{R}. \tag{A.28}$$
The probability density for this normalization is therefore 1, and we shall refer to eqn (A.27) as a normalization to unit density. From the equation
$$[H - (\hbar^2 k^2/2\mu)]|\mathbf{k}\rangle = 0,$$
where $H = p^2/2\mu$, it follows, using eqns (A.22), (A.11), and (A.7), that
$$\langle Rlm_l|\mathbf{k}\rangle = c_l j_l(kR)Y_l^{*m_l}(\hat{\mathbf{k}}), \tag{A.29}$$
where c_l is a normalization constant. $|c_l|^2$ is determined using eqns (A.26) and (A.13) and imposing normalization to unit density, so that $|c_l|^2 = (4\pi)^2$, and we adopt the phase convention that $c_l = 4\pi i^l$. Thus
$$\langle\mathbf{R}|\mathbf{k}\rangle = 4\pi\sum_{l,m_l} i^l j_l(kR)Y_l^{*m_l}(\hat{\mathbf{k}})Y_l^{m_l}(\hat{\mathbf{R}}). \tag{A.30}$$

In a similar fashion one can expand the Green's function for the kinetic energy operator K:
$$\langle\mathbf{R}|(E^+ - K)^{-1}|\mathbf{R'}\rangle = \sum_{l,m} Y_l^{m_l}(\hat{\mathbf{R}})Y_l^{*m_l}(\hat{\mathbf{R}'})G_k^{+l}(R,R'). \tag{A.31}$$

Substituting eqn (A.31) in eqn (2.20 c) and using eqns (A.11 a) and (A.7),
$$\left[p_R^2 + \frac{\hbar^2 l(l+1)}{2\mu R^2} - E\right]G_k^+ = \frac{\delta(R-R')}{R^2}. \tag{A.32}$$

Integrating both sides of eqn (A.32) from $r+d$ to $r-d$ we see that the derivative, dG_k^{+l}/dR, has a discontinuity of magnitude $2\mu/(\hbar R)^2$ at $R = R'$. When $R \neq R'$ we can write
$$G_k^{+l}(R, R') = d_l Z_l^{(1)}(R) Z_l^{(2)}(R'), \qquad (A.33)$$
where $Z_l(R)$ is a solution of the homogeneous form of eqn (A.32) and d_l is a constant determined to ensure the required discontinuity of the derivative
$$d_l \left[Z_l^{(1)}(R) \frac{d}{dR} Z_l^{(2)}(R) - Z_l^{(2)}(R) \frac{d}{dR} Z_l^{(1)}(R) \right] = 2\mu/(\hbar R)^2. \qquad (A.34)$$

For $R > R'$, the condition that the Green's function is well-behaved for $R' \to 0$, requires taking $Z_l^{(2)}(R') = j_l(kR')$. The asymptotic form $(R \gg R')$ of the Green's function (eqn (2.2.24)) requires taking $Z_l^{(1)}(R) = h_l^+(kR)$ (cf. eqn (A.24)). From the properties of the spherical Bessel functions $d_l = -2\mu k/\hbar^2$ so that, for $R > R'$,
$$G_k^{+l} = -(2\mu k/\hbar^2) h_l^+(kR) j_l(kR'). \qquad (A.35)$$

A further discussion of the spherical harmonics and of the spherical Bessel functions can be found in most texts of quantum mechanics and applied mathematics, and in particular in that of Morse and Feshbach (1953).

Consider now a situation where one of the particles has an internal angular momentum, $\hbar \mathbf{j}$. (For example, it can be a rigid rotor, cf. section 2.3.1.) For non-interacting particles $\{|jm_j\, lm_l\rangle\}$ is a possible representation where m_j is the eigenvalue of j_z. It is often convenient to couple the internal, $\hbar \mathbf{j}$, and orbital, $\hbar \mathbf{L}$, angular momenta to form a resultant, total angular momentum $\hbar \mathbf{J}$, $\mathbf{J} = \mathbf{L} + \mathbf{j}$. A suitable basis is obtained when we observe that $\mathbf{j}^2$ and $\mathbf{L}^2$ commute with $\mathbf{J}^2$ and J_z, and is thus denoted $\{|jlJM\rangle\}$, where M is the eigenvalue of J_z. Here
$$[\mathbf{j}^2 - j(j+1)]|jlJM\rangle = 0 \qquad (A.36)$$
and similarly for $\mathbf{L}^2$, $\mathbf{J}^2$, and J_z.

The spectrum of $\mathbf{J}^2$ is $J = j+l, j+l-1, \ldots, |j-l|$ and to each value of J there correspond $2J+1$ eigenstates with $M = J, J-1, \ldots, -J$. There is only one state for each value of J and M since
$$\sum_{J=|j-l|}^{j+l} (2J+1) = (2j+1)(2l+1). \qquad (A.37)$$
The restriction on the possible values of J is summarized by the 'triangular' inequality, $|j-l| \leq J \leq j+l$.

The transformation function between the two representations has been extensively discussed (Messiah 1961, Edmonds 1960, Wigner 1959) and is known as the Clebsch–Gordan or vector coupling coefficient. Since $\mathbf{L}^2$ and $\mathbf{j}^2$ have the same eigenvalues in both representations, we can write
$$|jlJM\rangle = \sum_{m_j, m_l} |jm_j\, lm_l\rangle \langle jm_j\, lm_l | jlJM\rangle \qquad (A.38)$$
and regard the transformation as operating in a subspace characterized by given values of j and l (and thus being $(2j+1)(2l+1)$-dimensional). In this subspace
$$\sum_{m_j, m_l} |jm_j\, lm_l\rangle \langle jm_j\, lm_l| = I = \sum_{JM} |jlJM\rangle \langle jlJM|. \qquad (A.39)$$
Here I is the identity operator in the subspace of a given j and l. With the normalization
$$\langle M'J'l'j'|jlJM\rangle = \delta_{jj'} \delta_{ll'} \delta_{JJ'} \delta_{MM'}$$
we obtain from eqns (A.39)
$$\sum_{JM} \langle jm_j\, lm_l | jlJM\rangle \langle jlJM | jm_j'\, lm_l'\rangle = \delta_{m_j m_j'} \delta_{m_l m_l'} \qquad (A.40\,a)$$
and
$$\sum_{m_j, m_l} \langle jlJM | jm_j\, lm_l\rangle \langle jm_j\, lm_l | jlJ'M'\rangle = \delta_{JJ'} \delta_{MM'}. \qquad (A.40\,b)$$

PART 2

Collision theory

Introduction

COLLISION events are characterized experimentally in terms of an initial state in which the colliding molecules are far apart and non-interacting. The relative motion of these molecules may bring them to within the range of their mutual interaction, which will induce transitions to the different possible final states of the collision. The aim of the theory is to specify the final states and to provide an algorithm for the rates (or yields) of their formation, for a given initial state and mutual interaction.

In this part we shall be concerned with isolated binary collision events, where there are two molecules in the initial state and there are no external forces acting on the system during the collision. The Hamiltonian H of the total (isolated) system is resolved as

$$H = H_0 + V,$$

where H_0 is the Hamiltonian for the well-separated molecules in the initial state. The mutual interaction V is that portion of the total Hamiltonian which, in the coordinate representation, tends to zero when the relative separation between the molecules increases. We assume that neither H_0 nor V depends on the absolute position of the molecules in space, or, in other words, that all interactions depend on the relative distances only. We also assume that H does not depend explicitly on the time.

Empirically we observe that under a wide range of experimental conditions it is possible to characterize collision events by parameters (rate constants, cross-sections) that are independent of those features of the initial state that are typical of a particular experimental arrangement (such as the actual numbers of the various molecules, the macroscopic size of the reaction volume, etc). It is clearly advantageous to try to formulate the theory without reference to these experimental details by setting up an idealized conceptual experiment. Provided the relevant expressions are linear, we can then incorporate the actual experimental initial conditions by a superposition of solutions. Deviations from linearity will thus indicate the range of validity of the empirical description.

If we consider the rate of appearance of final states as the primary observable, it is useful to construct a conceptual experiment where this rate is stationary (time-independent), which corresponds to a uniform rate of collisions (namely insignificant depletion of the initial state). This point of view is associated with stationary collision theory. If we are primarily interested in yields (that is probabilities of observation of final states) it is more convenient to follow collision events in time and to compute the yield from the long-time evolution of the system. This is the point of view of time-dependent collision theory.

2.1. Stationary collision theory—the Lippmann–Schwinger equation

THE Hamiltonian H_0 for two non-interacting molecules in the initial state§ is a sum of the Hamiltonians of the individual molecules. For an isolated molecule we can refer all the internal coordinates to the centre of mass, and resolve the molecular Hamiltonian of molecule one, $H_{0,1}$ in the form
$$H_{0,1} = K_1 + h_1, \qquad (1.1)$$
where K_1 is kinetic energy operator for the motion of the centre of mass of molecule one and h_1 is the Hamiltonian for the internal degrees of freedom.

It is often convenient to formulate the problem in the centre of mass system, in which the motion of the total centre of mass is factored out, and the motion of the two molecules is discussed in terms of their relative motion, using the identity (cf. Appendix 1.A)
$$K_1 + K_2 = K + K_{\text{CM}}, \qquad (1.2)$$
where K is the kinetic energy operator for the relative motion of the two individual centres of mass and K_{CM} is the kinetic energy operator for the motion of the total centre of mass. In the centre of mass system
$$H_0 = K + h_1 + h_2 = K + h, \qquad (1.3)$$
and $[K, h] = 0$.

We shall denote by **n** the set of quantum numbers, apart from the total energy, necessary to specify an initial state, and write the Schrödinger equation
$$(E - H_0)|E, \mathbf{n}\rangle = 0. \qquad (1.4)$$
Since both h and K commute with H_0, we consider the state $|E, \mathbf{n}\rangle$ to be an eigenstate of these operators as well,
$$K|E, \mathbf{n}\rangle = (\hbar^2 k^2/2\mu)|E, \mathbf{n}\rangle$$
and
$$h|E, \mathbf{n}\rangle = E_{\text{int}}|E, \mathbf{n}\rangle, \qquad (1.5)$$
where $\hbar \mathbf{k}$ is the relative momentum, μ is the reduced mass, and E_{int} is the sum of the internal energies.

The total energy of the initial state is thus the sum of the internal energies of each molecule, the kinetic energy of their relative motion, and the kinetic energy of the motion of the total centre of mass (in the centre

§ Initial states where the molecules do interact are considered in section 2.4.2.

of mass system the centre of mass is at rest and the last contribution is absent). The total energy can vary continuously as the energy of relative motion is not quantized. The initial state is thus in the continuous spectrum of H_0. We shall arrange the zero of energy so that the continuous spectrum begins at zero. It is important to note that a given total energy does not specify an initial state precisely, since the internal energy can also vary (but only by discrete amounts). Moreover, a specified relative kinetic energy merely determines the magnitude of the relative momentum but not the direction.

Our aim is to determine a solution of the Schrödinger equation $(E-H)|X\rangle = 0$ $(E > 0)$ that will reduce to the state $|E, \mathbf{n}\rangle$ when the mutual interaction, V, vanishes, or

$$(E-H_0)(|X\rangle - |E, \mathbf{n}\rangle) = V|X\rangle. \tag{1.6}$$

To obtain a unique solution the boundary conditions have to be specified. We shall incorporate the boundary conditions in a formal fashion by inverting the operator $(E-H_0)$. The operator $(\lambda - H_0)$, for complex values of the variable λ, possess a unique inverse

$$G_0(\lambda) = (\lambda - H_0)^{-1}. \tag{1.7}$$

We can then define the state vector $|\lambda, \mathbf{n}\rangle$ as the (unique) solution of

$$(\lambda - H)|\lambda, \mathbf{n}\rangle = (\lambda - H_0)|E, \mathbf{n}\rangle \quad (\text{re}\,\lambda = E),$$

namely

$$|\lambda, \mathbf{n}\rangle = |E, \mathbf{n}\rangle + (\lambda - H_0)^{-1}V|\lambda, \mathbf{n}\rangle \tag{1.8a}$$

$$= |E, \mathbf{n}\rangle + (\lambda - H)^{-1}V|E, \mathbf{n}\rangle. \tag{1.8b}$$

A solution with a sharp value for the energy is obtained by the limiting operation $\text{im}\,\lambda \to 0$. As we have seen (section 1.2.1) this limit can be performed in two ways when $\text{im}\,\lambda \to 0$, from above the real axis (denoted by $\text{im}\,\lambda \to +0$) and from below ($\text{im}\,\lambda \to -0$). The two solutions so defined are not equivalent, since the Green's operator $(\lambda - H_0)^{-1}$ has a cut across the continuous spectrum of H_0, and its values on the upper and lower rims of the cut are different (cf. eqn (1.2.29)),

$$\lim_{\substack{\text{im}\,\lambda \to \pm 0 \\ \text{re}\,\lambda = E}} (\lambda - H_0)^{-1} \to Pv(E-H_0)^{-1} \mp i\pi\delta(E-H_0). \tag{1.9}$$

We can now define

$$|E, \mathbf{n}^{\pm}\rangle = \lim_{\text{im}\,\lambda \to \pm 0} |\lambda, \mathbf{n}\rangle \tag{1.10}$$

as two alternative solutions. It is important to remember that the limiting operation $\text{im}\,\lambda \to 0$ must be performed after the explicit evaluation of the Green's function in a specific representation. The designation of a solution as $|E, \mathbf{n}^+\rangle$ is just a reminder of the boundary conditions and of the quantum numbers of the initial state. It does not imply that the quantum numbers of the initial state are conserved during the interaction.

We shall use the notation

$$|E, \mathbf{n}^+\rangle = |E, \mathbf{n}\rangle + (E^+ - H_0)^{-1} V |E, \mathbf{n}^+\rangle \qquad (1.11)$$

for eqn (1.10), where $E^+ = E + i\epsilon$ with ϵ real and positive and the limit $\epsilon \to 0$ is understood at the final stage, so that, for example,

$$G_0(E^+) = Pv(E - H_0)^{-1} - i\pi\delta(E - H_0) \qquad (1.12)$$

is a symbolic instruction for the limiting operation of eqn (1.9) which itself is understood to apply only after taking its matrix elements in some representation.

Equation (1.11) is referred to as the Lippmann–Schwinger (1950) equation. We have derived it here by an explicit introduction of the boundary conditions into the solution of the Schrödinger equation. Since the Green's operator is an integral operator the Lippmann–Schwinger (L.S.) equation is an integral equation.

The first term in eqn (1.11) is the initial state to which $|E, \mathbf{n}^+\rangle$ reduces in the absence of the mutual interaction V. The second term contains the modifications induced by the interaction. Premultiplying both sides of eqn (1.11) by $(E - H_0)$ we see that $|E, \mathbf{n}^+\rangle$ does satisfy§ eqn (1.6) and is the required solution.

The normalization integral for the solutions of the L.S. equation is equal to the normalization integral of their initial states. In the notation of eqn (1.11)

$$\langle E', \mathbf{m}^+ | E, \mathbf{n}^+ \rangle = \langle E', \mathbf{m} | E, \mathbf{n} \rangle. \qquad (1.13)$$

The diagonal elements of this equation equate the norm of the solution of the L.S. equation to the norm of the initial state and thus ensure the conservation of probabilities. A proof of eqn (1.13) is given in Appendix 2.A.

§ We are using the identity
$$(E - H_0) \lim_{\epsilon \to +0} (E + i\epsilon - H_0)^{-1} = I,$$
which can be proved using eqn (1.12) when we recall that $x\delta(x) = 0$.

2.2. The Lippmann–Schwinger equation for 'structureless' molecules

2.2.0. WHEN the colliding molecules have no apparent internal structure, H_0 is simply the sum of the kinetic energy operators. In the centre of mass system (and the coordinate representation),

$$H_0 = -(\hbar^2/2\mu)\nabla_{\mathbf{R}}^2,$$

where μ is the reduced mass and $\mathbf{R}$ the relative separation of the two molecules (cf. Appendix 1.A) and $E = \hbar^2 k^2/2\mu$ where $\hbar k$ is the relative momentum. An initial state of sharp energy and specified direction of the initial relative momentum can be denoted§ $|\mathbf{k}\rangle$.

Equation (1.11) is now specialized to

$$|\mathbf{k}^+\rangle = |\mathbf{k}\rangle + (E^+ - H_0)^{-1} V |\mathbf{k}^+\rangle, \qquad (2.1)$$

where
$$H_0 |\mathbf{k}\rangle = (\hbar^2 k^2/2\mu)|\mathbf{k}\rangle. \qquad (2.2)$$

We shall normalize momentum eigenstates to unit density (cf. eqn (1.A.27))

$$\langle \mathbf{k}' | \mathbf{k} \rangle = (2\pi)^3 \delta(\mathbf{k}' - \mathbf{k}), \qquad (2.3)$$

and the identity operator can then be resolved as

$$I = \int |\mathbf{k}\rangle (2\pi)^{-3} \, d\mathbf{k} \, \langle \mathbf{k}|, \qquad (2.4)$$

so that
$$\langle \mathbf{k}' | (E^+ - H_0)^{-1} I = (2\pi)^{-3} \int d\mathbf{k} \, \frac{\langle \mathbf{k}' | \mathbf{k} \rangle \langle \mathbf{k} |}{E^+ - (\hbar^2 k^2/2\mu)}$$
$$= [E^+ - \hbar^2 k'^2/2\mu]^{-1} \langle \mathbf{k}'|. \qquad (2.5)$$

In future we shall reserve the notation $|E, \mathbf{n}\rangle$ for states normalized on the energy scale

$$\langle E', \mathbf{m} | E, \mathbf{n} \rangle = \delta(E' - E) \delta_{\mathbf{m},\mathbf{n}}. \qquad (2.6)$$

With this normalization

$$\delta(E - H_0) = \sum_{\mathbf{n}} |E, \mathbf{n}\rangle \langle E, \mathbf{n}|, \qquad (2.7)$$

and
$$I = \sum_{\mathbf{n}} \int |E, \mathbf{n}\rangle \, dE \, \langle E, \mathbf{n}|.$$

§ Since a vector is specified by its magnitude and direction. An alternative notation is $|E, \hat{\mathbf{k}}\rangle$, where $\hat{\mathbf{k}}$ is a unit vector in the direction of k.

To determine the transformation between the two normalization conventions we note the identities

$$\delta[E(k)-E(k')] = \delta(k-k')\bigg/\left|\left(\frac{\partial E(k)}{\partial k}\right)_{k=k'}\right| = \delta(k-k')\mu/\hbar^2 k \quad (2.8)$$

and
$$\delta(\mathbf{k}-\mathbf{k}') = \delta(\hat{\mathbf{k}}-\hat{\mathbf{k}}')\delta(k-k')/k^2$$
$$= \delta(\hat{\mathbf{k}}-\hat{\mathbf{k}}')\delta[E(k)-E(k')]\left|\frac{\partial E(k)}{\partial k}\right|\bigg/k^2. \quad (2.9)$$

Comparing eqns (2.3) and (2.6) using eqn (2.9) we find (to within an over-all phase factor)
$$|E,\hat{\mathbf{k}}\rangle = [\rho(k)]^{\frac{1}{2}}|\mathbf{k}\rangle, \quad (2.10)$$
where
$$\rho(k) = k^2\bigg/\left|\frac{\partial E(k)}{\partial k}\right|(2\pi)^3 = \mu k/(2\pi)^3\hbar^2, \quad \text{for } E(k) = \hbar^2 k^2/2\mu.$$

The 'density of states' $\rho(k)$ can also be defined by§

$$I = \int |\mathbf{k}\rangle \rho(k)\, \mathrm{d}E \mathrm{d}\hat{\mathbf{k}}\, \langle\mathbf{k}|, \quad (2.11)$$

or $\rho(k)\, \mathrm{d}E\mathrm{d}\hat{\mathbf{k}}$ is the number of states with momentum in the solid angle in the direction $\hat{\mathbf{k}}$ to $\hat{\mathbf{k}}+\mathrm{d}\hat{\mathbf{k}}$ and whose energy is between E and $E+\mathrm{d}E$. Using eqn (2.4),
$$\rho(k)\, \mathrm{d}E\mathrm{d}\hat{\mathbf{k}} = (2\pi)^{-3}\, \mathrm{d}\mathbf{k}$$
$$= (2\pi)^{-3}k^2\, \mathrm{d}k\mathrm{d}\hat{\mathbf{k}} \quad (2.12)$$
$$= (2\pi\hbar)^{-3}\mu\hbar k\, \mathrm{d}E\mathrm{d}\hat{\mathbf{k}}.$$

The momentum representation of eqn (2.1) is obtained by preoperating with $\langle\mathbf{k}'|$ and, using eqn (2.5),
$$\langle\mathbf{k}'|\mathbf{k}^+\rangle = (2\pi)^3\delta(\mathbf{k}'-\mathbf{k})+[E^+-\hbar^2 k'^2/2\mu]^{-1}\langle\mathbf{k}'|V|\mathbf{k}^+\rangle. \quad (2.13)$$
In a wavefunction notation we can put $\psi_\mathbf{k}^+(\mathbf{k}') = \langle\mathbf{k}'|\mathbf{k}^+\rangle$. An equation for $\psi_\mathbf{k}^+(\mathbf{k}')$ is obtained by using the completeness relation (2.4) in the matrix element of the interaction
$$\langle\mathbf{k}'|V|\mathbf{k}^+\rangle = \int \langle\mathbf{k}'|V|\mathbf{k}''\rangle(2\pi)^{-3}\, \mathrm{d}\mathbf{k}''\, \langle\mathbf{k}''|\mathbf{k}^+\rangle, \quad (2.14)$$
so that
$$\psi_\mathbf{k}^+(\mathbf{k}') = (2\pi)^3\delta(\mathbf{k}'-\mathbf{k})+[E^+-\hbar^2 k'^2/2\mu]^{-1}\int \langle\mathbf{k}'|V|\mathbf{k}''\rangle\psi_\mathbf{k}^+(\mathbf{k}'')(2\pi)^{-3}\, \mathrm{d}\mathbf{k}''. \quad (2.15)$$

This is an integral equation for $\psi_\mathbf{k}^+(\mathbf{k}')$, which is interpreted as the probability amplitude to observe a relative momentum $\mathbf{k}'$ when the initial state had a momentum $\mathbf{k}$. In general, the off-diagonal matrix

§ The general case is discussed in section 2.5.3.

elements of V in the momentum representation will not vanish, and the above amplitude will not vanish when $\mathbf{k}' \neq \mathbf{k}$. If, as expected, the mutual interaction V is a local operator in the coordinate representation, $\langle \mathbf{R}|V|\mathbf{R}'\rangle = V(\mathbf{R})\delta(\mathbf{R}-\mathbf{R}')$,

$$\langle \mathbf{k}'|V|\mathbf{k}''\rangle = \int \langle \mathbf{k}'|\mathbf{R}\rangle V(\mathbf{R})\langle \mathbf{R}|\mathbf{k}''\rangle \, d\mathbf{R}$$
$$= \int \exp[-i(\mathbf{k}'-\mathbf{k}'')\cdot\mathbf{R}]V(\mathbf{R}) \, d\mathbf{R} \quad (2.16)$$
$$= V(\mathbf{k}'-\mathbf{k}''),$$

where we have used the transformation function (eqn 1.1.55)) $\langle \mathbf{k}|\mathbf{R}\rangle = \exp(-i\mathbf{k}\cdot\mathbf{R})$.

The vector $\mathbf{k}$ can change in both magnitude and direction. Since the molecules are structureless we expect that the magnitude of the relative momentum will be conserved because there is no mechanism for the conversion of kinetic energy of relative motion into other forms of energy. That this is indeed the case is shown by eqn (2.19). The role of the interaction V is to change the momentum vector from one direction to another. If the interaction is local, its 'efficiency' for kicking the momentum vector by a given amount is related to its 'shape' (eqn (2.16)).

The wavefunction in the coordinate representation, $\psi_\mathbf{k}^+(\mathbf{R}) = \langle \mathbf{R}|\mathbf{k}^+\rangle$, can be obtained either directly from eqn (1.11) or from the momentum representation and the transformation function $\langle \mathbf{R}|\mathbf{k}\rangle = \exp(i\mathbf{k}\cdot\mathbf{R})$. For a local potential,

$$\psi_\mathbf{k}^+(\mathbf{R}) = \langle \mathbf{R}|\mathbf{k}\rangle + \int d\mathbf{R}' \, \langle \mathbf{R}|(E^+-H_0)^{-1}|\mathbf{R}'\rangle V(\mathbf{R}')\langle \mathbf{R}'|\mathbf{k}^+\rangle. \quad (2.17)$$

If H_0 is the kinetic energy operator for the coordinate R, we have seen (eqn (1.2.34)) that

$$\langle \mathbf{R}|(\lambda-H_0)^{-1}|\mathbf{R}'\rangle = -(\mu/2\pi\hbar^2)\frac{\exp[\pm i\kappa|\mathbf{R}-\mathbf{R}'|]}{|\mathbf{R}-\mathbf{R}'|}, \quad (2.18)$$

where the plus/minus sign corresponds to the sign of im λ and $\lambda = \hbar^2\kappa^2/2\mu$ with re $\kappa > 0$ (so that $\pm i\kappa = \pm ik - \gamma$ with k and γ positive, and as im $\lambda \to 0$ $\gamma \to 0$ while re $\lambda \to \hbar^2 k^2/2\mu$).

In the limit im $\lambda \to 0$

$$\psi_\mathbf{k}^+(\mathbf{R}) \to \exp(i\mathbf{k}\cdot\mathbf{R}) - (\mu/2\pi\hbar^2) \int \frac{\exp[ik|\mathbf{R}-\mathbf{R}'|]}{|\mathbf{R}-\mathbf{R}'|} V(\mathbf{R}')\psi_\mathbf{k}^+(\mathbf{R}') \, d\mathbf{R}', \quad (2.19)$$

where $E = \hbar^2 k^2/2\mu$.

Our physical discussion of the collision process was based on an initial state in which the two molecules are far apart and non-interacting.

Such a state can be described as a linear superposition of states

$$\Phi(\mathbf{R},t) = \int d\mathbf{k}\, \langle \mathbf{R}|\mathbf{k}\rangle A(\mathbf{k},t),$$

where $A(\mathbf{k},t)$ is the weight function which is the probability amplitude of the component $\langle \mathbf{R}|\mathbf{k}\rangle$ at time t. By a suitable selection of $A(\mathbf{k},t)$ the wavefunction $\Phi(\mathbf{R},t)$ can be localized. Such an initial state (a wave packet) cannot in general lead to a uniform collision rate. If $\Psi(\mathbf{R},t)$ is the wavefunction during the collision that evolved from $\Phi(\mathbf{R},t)$ as an initial state, we can put

$$\Psi(\mathbf{R},t) = \int d\mathbf{k}\, \langle \mathbf{R}|\mathbf{k}^+\rangle B(\mathbf{k},t).$$

We shall find it possible, however, to determine $B(\mathbf{k},t)$ (in fact, we shall show that $B(\mathbf{k},t) = A(\mathbf{k},t)$) and thus discuss an arbitrary collision event in terms of the stationary initial state $\langle \mathbf{R}|\mathbf{k}\rangle$ and the corresponding solution of the L.S. equation $\langle \mathbf{R}|\mathbf{k}^+\rangle$.

The collision of wave packets has been discussed by several authors (Low 1959, Ohmura 1964, Goldberger and Watson 1964, Katz 1966, and Hammer and Weber 1967).

2.2.1. *The cross-section*

To discuss the significance of the coordinate representation of the solution of the L.S. equation, we assume that the mutual interaction V is of finite range.§ We can then consider a large sphere, centred at the total centre of mass such that V is non-vanishing inside the sphere only. The first term in eqn (2.19) is the coordinate representation of the initial state. It has a uniform density

$$\rho(\mathbf{R},t)\, d\mathbf{R} = |\langle \mathbf{R}|\mathbf{k}\rangle|^2\, d\mathbf{R} = d\mathbf{R}, \qquad (2.20)$$

so that the probability of finding the two molecules within the sphere of mutual interaction is constant. It is of interest to derive this conclusion from the equation of conservation of probability

$$\int_V \frac{\partial \rho(\mathbf{R},t)}{\partial t}\, d\mathbf{R} = \int_V \mathrm{div}\, \mathbf{j}(\mathbf{R},t)\, d\mathbf{R} = \int_S \mathbf{j}(\mathbf{R},t)\cdot d\mathbf{S}, \qquad (2.21)$$

where $\mathbf{j}(\mathbf{R},t)$ is the quantum-mechanical probability current and the surface integral is over the surface S that encloses the volume V. For a momentum eigenstate normalized to unit density∥

$$\mathbf{j}(\mathbf{R},t) = \hbar \mathbf{k}/\mu = \mathbf{v}. \qquad (2.22)$$

§ We make this assumption for convenience only. It is sufficient that V decreases faster than $1/R$.
∥ Recall that, in general, when a velocity can be defined, the current is the velocity times the density.

The integral of the velocity **v** over the surface of the sphere vanishes by symmetry (since in one half-sphere **v** is in the inward direction and in the other half in the outward direction). The probability density of the initial state is thus time-independent.

The integrand of the second term in eqn (2.19) can be considered as a product of $\psi_{\mathbf{k}}^{+}(\mathbf{R}')$, the amplitude to find the molecules with relative separation $\mathbf{R}'$ times the potential at $\mathbf{R}'$ times the amplitude to propagate as a spherical wave $\exp[ik|\mathbf{R}-\mathbf{R}'|]/|\mathbf{R}-\mathbf{R}'|$ from $\mathbf{R}'$ to $\mathbf{R}$. The total contribution is obtained by summing over all values of $\mathbf{R}'$. It is not implied that necessarily $R > R'$, as the amplitude $\psi_{\mathbf{k}}^{+}(\mathbf{R})$ can itself interact with the potential at the point $\mathbf{R}$ and be scattered into $\mathbf{R}'$:

$$\psi_{\mathbf{k}}^{+}(\mathbf{R}') = \exp(\mathrm{i}\mathbf{k}.\mathbf{R}') - (\mu/2\pi\hbar^2) \int \frac{\exp[\mathrm{i}k|\mathbf{R}'-\mathbf{R}|]}{|\mathbf{R}'-\mathbf{R}|} V(\mathbf{R})\psi_{\mathbf{k}}^{+}(\mathbf{R})\,\mathrm{d}\mathbf{R}. \tag{2.23}$$

The conservation of magnitude of the initial momentum is already evident in the wave number k, since $E = \hbar^2 k^2/2\mu$, where E is the initial energy.

To obtain the form of the wavefunction $\psi_{\mathbf{k}}^{+}(\mathbf{R})$ for values of R outside the range of the potential we consider an expansion of the Green's function, $\exp[\mathrm{i}k|\mathbf{R}-\mathbf{R}'|]/|\mathbf{R}-\mathbf{R}'|$, in terms of the ratio R'/R. We only need an expression correct to order R'/R, since terms of this order will lead to contributions of order $1/R$ for any mutual interaction $V(R')$ that decreases faster than the inverse distance, and will vanish when we let R tend to infinity. Using the expansion $(1-X)^{\frac{1}{2}} = 1 - \tfrac{1}{2}X + O(X^2)$, we find

$$k|\mathbf{R}-\mathbf{R}'| = kR\left[1 - 2\frac{\mathbf{R}.\mathbf{R}'}{R^2} + \frac{(R')^2}{R^2}\right]^{\frac{1}{2}} = kR - k\hat{\mathbf{R}}.\mathbf{R}' + O[k(R')^2/R].$$

Then

$$\frac{\exp[\mathrm{i}k|\mathbf{R}-\mathbf{R}'|]}{|\mathbf{R}-\mathbf{R}'|} \xrightarrow[R\to\infty]{} \frac{\exp(\mathrm{i}kR)}{R}\exp(-\mathrm{i}\mathbf{k}'.\mathbf{R}') + O(R'/R), \tag{2.24}$$

where $\mathbf{k}' = k\hat{\mathbf{R}}$ is a wave vector of magnitude k in the direction of $\mathbf{R}$, so that, substituting eqn (2.24) in eqn (2.19),

$$\psi_{\mathbf{k}}^{+}(\mathbf{R}) \xrightarrow[R\to\infty]{} \exp(\mathrm{i}\mathbf{k}.\mathbf{R}) -$$

$$- \frac{\exp(\mathrm{i}kR)}{R}(\mu/2\pi\hbar^2)\int \exp(-\mathrm{i}\mathbf{k}'.\mathbf{R}')V(\mathbf{R}')\psi_{\mathbf{k}}^{+}(\mathbf{R}')\,\mathrm{d}\mathbf{R}'. \tag{2.25}$$

The second term in eqn (2.25) is a spherical wave $\exp(\mathrm{i}kR)/R$ times the amplitude $f(\mathbf{k}',\mathbf{k})$ (the scattering amplitude),

$$f(\mathbf{k}',\mathbf{k}) = -(\mu/2\pi\hbar^2)\int \exp(-\mathrm{i}\mathbf{k}'.\mathbf{R})V(\mathbf{R})\psi_{\mathbf{k}}^{+}(\mathbf{R})\,\mathrm{d}\mathbf{R}. \tag{2.26}$$

The modifications required if the interaction V is non-local are obvious, by rewriting the L.S. equation for an arbitrary potential
$$V(\mathbf{R}', \mathbf{R}'') = \langle \mathbf{R}'|V|\mathbf{R}''\rangle.$$
Thus for the scattering amplitude we have that
$$f(\mathbf{k}', \mathbf{k}) = -(\mu/2\pi\hbar^2) \iint \exp(-i\mathbf{k}'.\mathbf{R}')V(\mathbf{R}', \mathbf{R}'')\psi_{\mathbf{k}}^+(\mathbf{R}'')\,\mathrm{d}\mathbf{R}'\mathrm{d}\mathbf{R}''$$
$$= -(\mu/2\pi\hbar^2)\langle \mathbf{k}'|V|\mathbf{k}+\rangle. \qquad (2.27)$$

The formal specification of the boundary conditions can now be given a physical interpretation in terms of the asymptotic behaviour of the wavefunction in the coordinate representation. At large relative separations, the scattered wave (the second term in eqn (2.25)) reduces to an outgoing (for E^+) spherical wave,
$$\psi_{\mathbf{k}}^+(\mathbf{R}) \to \exp(i\mathbf{k}.\mathbf{R}) + f(\mathbf{k}', \mathbf{k})\exp(ikR)/R \qquad (2.28)$$
(where $\mathbf{k}' = k\hat{\mathbf{R}}$, $E = \hbar^2 k^2/2\mu$). From eqns (2.25) and (2.18) we see that an E^- specification corresponds to an incoming spherical wave $[\exp(-ikR)/R]$. In both cases energy is conserved.

The probability current associated with the propagation of the scattered wave can be computed by evaluating $\mathbf{j}.\mathrm{d}\mathbf{S})_{\text{sc}}$ on the surface of the sphere that encloses the region of interaction, where the scattered wave has the asymptotic form determined by eqn (2.28). On a sphere§
$$\mathrm{d}\mathbf{S} = \hat{\mathbf{R}} R^2\,\mathrm{d}\omega, \qquad (2.29)$$
where $\mathbf{R}$ is the radius vector and $\mathrm{d}\omega$ is the solid angle subtended by $\mathrm{d}\mathbf{S}$. Since the scattered wave has a well-defined velocity on the sphere we have, from eqn (2.28),
$$\mathbf{j}.\hat{\mathbf{R}})_{\text{sc}} = (\hbar k/\mu)|f(\mathbf{k}', \mathbf{k})/R|^2, \qquad (2.30)$$
where $\hat{\mathbf{k}}' = \hat{\mathbf{R}}$, or
$$\mathbf{j}.\mathrm{d}\mathbf{S})_{\text{sc}} = (\hbar k/\mu)|f(\mathbf{k}', \mathbf{k})|^2\,\mathrm{d}\omega. \qquad (2.31)$$
In any given direction $\hat{\mathbf{R}}$ we thus observe‖ a uniform flux of scattered waves. We expect that the flux of scattered waves is proportional to the incident flux. Indeed we can write eqn (2.31) as
$$\mathbf{j}.\mathrm{d}\mathbf{S})_{\text{sc}} = \frac{\hbar \mathbf{k}}{\mu}.\hat{\mathbf{k}}\,\mathrm{d}\sigma, \qquad (2.32)$$
where $\hat{\mathbf{k}}\,\mathrm{d}\sigma$ is an element of area of magnitude $\mathrm{d}\sigma$ in the direction $\hat{\mathbf{k}}$. In

§ In general $\mathrm{d}\mathbf{S} = \mathbf{n}\,\mathrm{d}S$, where $\mathbf{n}$ is a unit vector in the direction of the outward normal to the surface.
‖ Recall that the origin is at the total centre of mass, and that the observer should be at a distance R, where R is larger than the range of mutual interaction. An alternative derivation is given in the footnote on p. 125.

other words, $d\sigma$ is a magnitude of an area in a plane perpendicular to the direction, $\hat{\mathbf{k}}$, of initial relative motion,

$$d\sigma = |f(\mathbf{k}', \mathbf{k})|^2 \, d\omega. \tag{2.33}$$

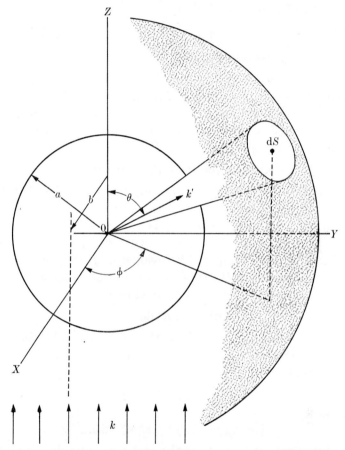

FIG. 2.1. Schematic representation of the collision, with a stationary initial state.

The element of area $d\sigma(\mathbf{k}', \mathbf{k})$ is known as the differential cross-section. The total cross-section σ is defined by§

$$\sigma = \int d\sigma = \int |f(\mathbf{k}', \mathbf{k})|^2 \, d\hat{\mathbf{k}}' \tag{2.34}$$

and has the interpretation of an area in a plane perpendicular to the direction $\hat{\mathbf{k}}$. We also note that by integrating both sides of eqn (2.32)

§ We restate the convention that $d\hat{\mathbf{k}}'$ denotes an element of solid angle in the direction $\hat{\mathbf{k}}'$. Recall that $d\omega$ in eqns (2.29) and (2.31) is in the direction $\hat{\mathbf{R}}$ and that $\hat{\mathbf{k}}' = \hat{\mathbf{R}}$.

over the direction $\hat{\mathbf{k}}'$, of the scattered wave we have

$$(\hbar k/\mu)\sigma = \int \mathbf{j}.\mathrm{d}\mathbf{S})_{\mathrm{sc}}. \tag{2.35}$$

Schematically we can represent our discussion by Fig. 2.1. O is the origin of coordinates. The waves below represent the incident waves of uniform density. Outside the range a of the potential the scattered waves are represented by outgoing spherical waves. Note, however, that their angular distribution need not be isotropic, as it depends on the direction $\hat{\mathbf{k}}'$ (cf. eqn (2.33)). In the absence of the potential a particular component of the incident wave will pass at a distance b from the origin, where all values of b from 0 to infinity are equally likely since the incident wave is of uniform density. In terms of b, the 'impact parameter', we can write the cross-section σ as an area,

$$\sigma = \int_0^\infty \int_0^{2\pi} P(b,\phi) b \, \mathrm{d}b \mathrm{d}\phi, \tag{2.36 a}$$

where ϕ is the azimuthal angle (namely the angular coordinate in a plane in the direction $\hat{\mathbf{k}}$) and $P(b,\phi)$ is defined by this equation.

In a problem with cylindrical symmetry about the direction $\hat{\mathbf{k}}$ we can rewrite eqn (2.36 a) as

$$\sigma = 2\pi \int P(b) b \, \mathrm{d}b. \tag{2.36 b}$$

We can interpret $2\pi b \, \mathrm{d}b$ as an area of a ring of radius b and width $\mathrm{d}b$ in a plane in the direction $\hat{\mathbf{k}}$, and regard $P(b)$ as the probability that a given impact parameter will contribute to the flux of scattered waves.

In classical mechanics one can (Goldstein 1950, Hirschfelder, Curtiss, and Bird 1954) solve the equations of motion for a given central potential $V(R)$ and obtain the deflexion function $\Theta(b)$,

$$\Theta = \pi - 2b \int_{R_\mathrm{M}}^\infty R^{-2}[1 - V(R)/E - b^2/R^2]^{-\frac{1}{2}} \, \mathrm{d}R,$$

defined to be positive for net repulsion and negative for net attraction. Here, R_M is the (classical) distance of closest approach (eqn (2.85)) and b the impact parameter. The observed scattering angle θ is defined to be positive,

$$\theta = |\Theta(b)|, \quad 0 \leqslant \theta \leqslant \pi.$$

For intermolecular potentials containing both attractive and repulsive branches there may be more than one value of b that will lead to a given scattering angle (Fig. 2.2).

In a beam of uniform density the fraction of incident molecules that are scattered into an angle θ to $\theta+d\theta$ is $2\pi b(\theta)\,db(\theta)$. From eqns (2.31)–(2.33) this fraction is equal to the differential cross-section

$$d\sigma_c = \sum d(\pi b^2) = 2\pi \sum b\,db = 2\pi \sum b\left|\frac{d\Theta}{db}\right|^{-1} d\theta, \qquad (2.37\,\text{a})$$

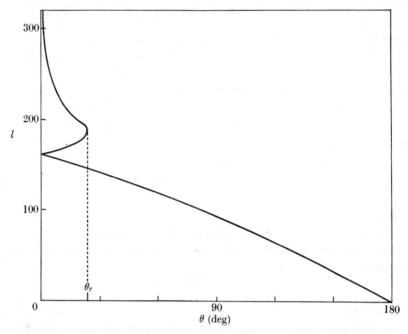

FIG. 2.2. Results of a classical computation of an angle of deflexion (Bernstein and Levine 1968) for a L.-J. (12, 6) potential $V(R) = 4\epsilon[(\sigma/R)^{12}-(\sigma/R)^6]$ when $E = 5\epsilon$ and the reduced parameter (typical for K+HBr) is $B = 2\mu\epsilon\sigma^2/\hbar^2 = 4.10^3$. The impact parameter is given by $(b/\sigma) = 7\cdot 94\times 10^{-2}(l+\tfrac{1}{2})$. θ_r is the 'rainbow' angle for which $(db/d\theta)_{\theta_r} = 0$. Note that large deflexions are produced by essentially 'head on' (i.e. very low b) collisions.

where the summation is over all values of b that can contribute to a scattering in the direction θ. Putting $d\omega = \sin\theta\,d\theta d\phi$, we have

$$\frac{d\sigma_c}{d\omega} = \sum b\Big/\sin\theta\left|\frac{d\Theta}{db}\right|. \qquad (2.37\,\text{b})$$

The subscript c refers to the use of classical theory.

As is clear from Fig. 2.2, there will be only one term in eqn (2.37) for $\theta > \theta_r$ and three terms for $\theta < \theta_r$. At θ_r (the 'rainbow angle') $d\theta/db \to 0$ and the classical cross-section diverges. Fig. 2.3 compares the results of a classical computation, using eqn (2.37) and a quantum-mechanical computation employing eqns (2.74) and (2.76). The quantum-mechanical result remains finite at θ_r.

2.2.2. Separable interactions

The evaluation of a numerical value of the cross-section requires an explicit solution of the L.S. equation (to obtain the scattering amplitude).

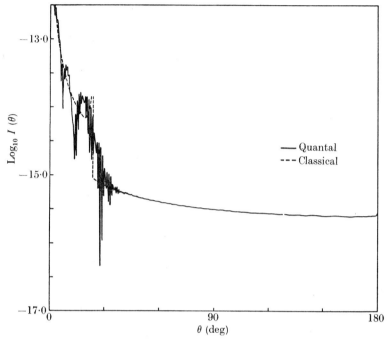

FIG. 2.3. The differential (solid angle) cross-section $I(\theta) = d\sigma(\theta)/d\omega$ computed classically and quantum-mechanically (Bernstein and Levine 1968). The parameters are as in Fig. 2.2 and $\sigma = 3 \cdot 17 \times 10^{-8}$ cm.

In general, a closed exact solution is not possible due to mathematical difficulties. Under certain exceptional circumstances an exact solution is possible, as we demonstrate below. Otherwise we can use approximate solutions, or we can try to rewrite the equation in a manner more suitable for direct numerical methods. (As it stands, the wave function is a function of the three variables, the three components of $\mathbf{R}$, and so cannot be easily handled by numerical techniques.)

Consider a non-local mutual interaction of the form

$$\langle \mathbf{R} | V | \mathbf{R}' \rangle = \eta g(\mathbf{R}) g(\mathbf{R}'), \tag{2.38}$$

where η is independent of $\mathbf{R}$ and $g(\mathbf{R})$ is a given function of $\mathbf{R}$. The L.S. equation, in the coordinate representation for the initial state $|\mathbf{k}\rangle$, is

$$\psi_{\mathbf{k}}^{+}(\mathbf{R}) = \exp(i\mathbf{k}\cdot\mathbf{R}) + \eta \iint G_0^{+}(\mathbf{R},\mathbf{R}') g(\mathbf{R}') g(\mathbf{R}'') \psi_{\mathbf{k}}^{+}(\mathbf{R}'') \, d\mathbf{R}' d\mathbf{R}'', \tag{2.39}$$

where $G_0^+(\mathbf{R}, \mathbf{R}') = \langle \mathbf{R}|(E^+-H_0)^{-1}|\mathbf{R}'\rangle.$

In the second term in eqn (2.39) the integration over $\mathbf{R}''$ is independent of the $\mathbf{R}'$ integration. Thus an interaction of the above type is known as *separable*.§ We can now multiply both sides of eqn (2.39) by $g(\mathbf{R})$ and integrate over $\mathbf{R}$. Due to the separability we have

$$\int g(\mathbf{R})\psi_{\mathbf{k}}^+(\mathbf{R})\,\mathrm{d}\mathbf{R} = \frac{\int g(\mathbf{R})\exp(i\mathbf{k}.\mathbf{R})\,\mathrm{d}\mathbf{R}}{1-\eta\iint \mathrm{d}\mathbf{R}\mathrm{d}\mathbf{R}'g(\mathbf{R})G_0^+(\mathbf{R},\mathbf{R}')g(\mathbf{R}')}. \quad (2.40)$$

Equation (2.40) provides the solution to our problem, when substituted in the second term on the right in eqn (2.39). The scattering amplitude is defined by the asymptotic behaviour of $\psi_{\mathbf{k}}^+(\mathbf{R})$ (cf. eqn (2.28)). Using eqn (2.24),

$$f(\mathbf{k}',\mathbf{k}) = -(\mu/2\pi\hbar^2)\eta\int \exp(-i\mathbf{k}'.\mathbf{R})g(\mathbf{R})\,\mathrm{d}\mathbf{R}\int g(\mathbf{R})\psi_{\mathbf{k}}^+(\mathbf{R})\,\mathrm{d}\mathbf{R}, \quad (2.41)$$

or $\quad f(\mathbf{k}',\mathbf{k}) = -(\mu/2\pi\hbar^2)\langle\mathbf{k}'|V|\mathbf{k}^+\rangle = -(\mu/2\pi\hbar^2)\eta\dfrac{g(\mathbf{k}')g^*(\mathbf{k})}{D(E^+)}, \quad (2.42)$

where $\quad g(\mathbf{k}) = \int \exp(-i\mathbf{k}.\mathbf{R})g(\mathbf{R})\,\mathrm{d}\mathbf{R} \quad (2.43)$

and $\quad D(E^+) = 1-\eta\iint g(\mathbf{R})\langle\mathbf{R}|(E^+-H_0)^{-1}|\mathbf{R}'\rangle g(\mathbf{R}')\,\mathrm{d}\mathbf{R}\mathrm{d}\mathbf{R}' \quad (2.44)$

are assumed to be finite. We note that $D(E^+)$ (and hence $f(\mathbf{k},\mathbf{k})$) possesses an imaginary part. From eqns (1.12), (2.7), and (2.12)

$$-\pi^{-1}\mathrm{im}(E^+-H_0)^{-1} = \int |\mathbf{k}\rangle\rho(E)\,\mathrm{d}\hat{\mathbf{k}}\,\langle\mathbf{k}|, \quad (2.45)$$

so that

$$\mathrm{im}\,D(E^+) = \pi\eta\iiint g(\mathbf{R})\langle\mathbf{R}|\mathbf{k}\rangle\rho(E)\langle\mathbf{k}|\mathbf{R}'\rangle g(\mathbf{R}')\,\mathrm{d}\mathbf{R}\mathrm{d}\mathbf{R}'\mathrm{d}\hat{\mathbf{k}}$$

$$= \pi\eta\rho(E)\int |g(\mathbf{k})|^2\,\mathrm{d}\hat{\mathbf{k}} \quad (2.46)$$

and $\quad \mathrm{im}\,f(\mathbf{k},\mathbf{k}) = (\eta\mu/2\pi\hbar^2)\dfrac{|g(\mathbf{k})|^2\,\mathrm{im}\,D(E^+)}{|D(E^+)|^2}. \quad (2.47)$

It is easy to see from eqns (2.42) and (2.47) that

$$(4\pi/k)\mathrm{im}\,f(\mathbf{k},\mathbf{k}) = \int |f(\mathbf{k}',\mathbf{k})|^2\,\mathrm{d}\hat{\mathbf{k}}' = \sigma(\mathbf{k}). \quad (2.48)$$

Equation (2.48) is of general validity, being a particular case of the optical theorem, which we shall later prove.‖ Equations (2.47) and (2.48) determine the cross-section

$$\sigma(\mathbf{k}) = (2\eta\mu/\hbar^2 k)\frac{|g(\mathbf{k})|^2\,\mathrm{im}\,D(E^+)}{|D(E^+)|^2}. \quad (2.49)$$

§ We shall observe in Chapter 3.1 that an interaction of this type may be due to the presence of internal degrees of freedom. For a discussion of separable interactions see Ghirardi and Rimini (1964) and references therein.

‖ For the general proof see eqn (4.7) or, explicitly, eqn (6.71).

As a function of E, the cross-section will be large at such energies E_0, $E_0 > 0$, that are the simple roots of

$$\text{re } D(E_0) = 0. \tag{2.50}$$

Near such energies we can expand $\text{re } D(E)$ in a Taylor series

$$\text{re } D(E) = \text{re } D(E_0) + (E-E_0)\left[\frac{\partial \text{re } D(E)}{\partial E}\right]_{E=E_0} + \ldots \tag{2.51}$$

Retaining the linear term only, we can write

$$\sigma(\mathbf{k}) \simeq (2\eta\mu/\hbar^2 k)\frac{\Gamma}{(E-E_0)^2+\Gamma^2}\frac{|g(\mathbf{k})|^2}{(\partial \text{re } D/\partial E)_{E=E_0}}, \tag{2.52}$$

where
$$\Gamma = \text{im } D(E^+)\left[\frac{\partial \text{re } D(E)}{\partial E}\right]_{E=E_0}^{-1}. \tag{2.53}$$

Consider in particular a problem with spherical symmetry, so that the scattering amplitude depends on $\hat{\mathbf{k}}.\hat{\mathbf{k}}'$ only. If we regard the direction $\hat{\mathbf{k}}$ as reference and recall $k' = k$, we can take

$$g(\mathbf{k}') = g(k)P_l(\cos\theta), \tag{2.54}$$

where $\cos\theta = \hat{\mathbf{k}}.\hat{\mathbf{k}}'$ and P_l is the Legendre polynomial of degree l, so that§

$$\frac{2l+1}{4\pi}\int |g(\mathbf{k})|^2\,\mathrm{d}\hat{\mathbf{k}} = |g(k)|^2. \tag{2.55}$$

For the scattering amplitude we have

$$f(\mathbf{k}',\mathbf{k}) = -[\eta\mu/2\pi\hbar^2 D(E^+)]|g(k)|^2 P_l(\cos\theta)$$

or, using eqn (2.55) for $|g(k)|^2$ and then eqn (2.46),

$$f(\mathbf{k}',\mathbf{k}) = -\frac{2l+1}{k}P_l(\cos\theta)\,\text{im } D_l(E^+)/D_l(E^+) \tag{2.56a}$$

$$= \frac{2l+1}{k}P_l(\cos\theta)\sin\delta_l\exp(i\delta_l), \tag{2.56b}$$

where $-\delta_l = \arg D_l(E^+)$ and the subscript l is a reminder that D_l is evaluated with the function $g(\mathbf{k}')$ as selected above. Thus

$$\sigma = \frac{4\pi(2l+1)}{k^2}\left|\frac{\text{im } D_l(E^+)}{D_l(E^+)}\right|^2 = \frac{4\pi(2l+1)}{k^2}\sin^2\delta_l. \tag{2.57}$$

Near a root of eqn (2.50) we can write

$$\sigma = \frac{4\pi(2l+1)}{k^2}\frac{\Gamma^2}{(E-E_0)^2+\Gamma^2}, \tag{2.58}$$

and
$$\tan\delta_l(E) = \Gamma/(E_0-E). \tag{2.59}$$

§ $\iint P_{l'}(\cos\theta)P_l(\cos\theta)\,\mathrm{d}\cos\theta\,\mathrm{d}\phi = \frac{4\pi}{2l+1}\delta_{ll'}.$

It is common (but by no means always correct; see the example following eqn (5.100)) to refer to the situation where the cross-section becomes large in the vicinity of a certain energy as a 'resonance'. Our present example refers to a resonance due to the mutual interaction between two structureless particles and is thus classified as a single degree of freedom resonance (namely the degree of freedom or relative motion). A more reasonable definition of resonance is as the situation where, for certain values of the energy, $|\psi_{\mathbf{k}}^+(\mathbf{R})|^2$ is significantly larger for R inside the interaction range than for most other values of the energy. Since the magnitude of $V\psi_{\mathbf{k}}^+(\mathbf{R})$ determines the magnitude of the scattering amplitude, the cross-section may§ indeed be large at a resonance. As is clear from eqns (2.36), (2.40), and (2.44), $\psi_{\mathbf{k}}^+(\mathbf{R})$ does increase in magnitude around the roots of eqn (2.50).

In a similar fashion we can evaluate the coordinate representation of the Green's operator for H, using the operator identity

$$G(E^+) = (E^+ - H)^{-1} = G_0(E^+)[1 + V G(E^+)], \qquad (2.60)$$

to get

$$\langle \mathbf{R}'|G(E^+)|\mathbf{R}\rangle = \langle \mathbf{R}'|G_0(E^+)|\mathbf{R}\rangle +$$
$$+ D^{-1}(E^+)\eta \int G_0^+(\mathbf{R}', \mathbf{R}'') g(\mathbf{R}'') \, d\mathbf{R}'' \int g(\mathbf{R}'') G_0^+(\mathbf{R}'', \mathbf{R}) \, d\mathbf{R}'', \qquad (2.61)$$

and obtain the solution of the L.S. equation from the coordinate representation of the formal solution, eqn (1.8b).

To consider possible generalizations of our method of solution we write the L.S. equation as an integral equation of the second kind,

$$\psi^+(\mathbf{R}) = \phi(\mathbf{R}) + \int K(\mathbf{R}, \mathbf{R}') \psi^+(\mathbf{R}') \, d\mathbf{R}', \qquad (2.62)$$

where
$$K(\mathbf{R}, \mathbf{R}') = \langle \mathbf{R}|(E^+ - H_0)^{-1} V|\mathbf{R}'\rangle \qquad (2.63)$$

is known as the kernel of the equation. Provided the kernel is square integrable, namely, $\text{Tr}\, KK^\dagger$ is finite,

$$\text{Tr}\, KK^\dagger = \iint d\mathbf{R} d\mathbf{R}' |K(\mathbf{R}, \mathbf{R}')|^2, \qquad (2.64)$$

it can be approximated by a kernel K_0 of finite rank, for which solving the integral equation (2.62) can be reduced to solving a finite set of

---

§ It is generally possible to resolve $\psi_{\mathbf{k}}^+(\mathbf{R})$ into symmetry components and associate a cross-section with each component (for example, eqn (2.77)) while only the sum of these cross-sections is observable. A resonance in one of these cross-sections may well be masked by the rest. Also if $V = V_1 + V_2$ and V_2 causes a resonance, its effect may be masked by the scattering due to V_1.

linear equations (Smithies 1958). A kernel of finite rank can be written as

$$K_0(\mathbf{R}, \mathbf{R}') = \sum_{n=1}^{N} a_n(\mathbf{R}) b_n^*(\mathbf{R}'), \qquad (2.65)$$

where N is finite, while $a_n(\mathbf{R})$ and $b_n(\mathbf{R})$ are square integrable.

For the separable interaction of eqn (2.38) we can write K itself as a kernel of rank one. However,

$$a(\mathbf{R}) = \int G_0^+(\mathbf{R}, \mathbf{R}') g(\mathbf{R}') \, d\mathbf{R}'$$

is not square integrable. We can however multiply eqn (2.62) by $g(\mathbf{R})$, thus obtaining a square integrable kernel, and a soluble equation. The explicit construction of the inversion of the set of linear equations for an arbitrary square integrable kernel is known as the Fredholm method (Smithies 1958; for applications to the L.S. equation see Khuri 1957, and Baker 1958).

The resolvent kernel J is defined to satisfy the equations

$$J - K = KJ = JK,$$

so that
$$J = K(I-K)^{-1} \qquad (2.66)$$

and the solution of eqn (2.62) can be written as (cf. eqn (1.8b))

$$\psi^+(\mathbf{R}) = \phi(\mathbf{R}) + \int J(\mathbf{R}, \mathbf{R}') \phi(\mathbf{R}') \, d\mathbf{R}',$$

where
$$J(\mathbf{R}, \mathbf{R}') = \langle \mathbf{R} | (E^+ - H)^{-1} V | \mathbf{R}' \rangle.$$

The power series expansion of eqn (2.66)

$$J = K + K^2 + K^3 + \dots \qquad (2.67)$$

is known as the Born–Neumann series (cf. section 2.2.4).

2.2.3. *Partial wave analysis*

Our discussion of separable interactions with spherical symmetry is a particular example of the method of partial wave analysis. The method is based on the existence of a complete set of functions on the surface of the unit sphere, namely the spherical harmonics (Appendix 1.A). Using these functions we can write

$$f(\mathbf{k}) = \sum_{l,m} f_{l,m}(k) Y_l^m(\hat{\mathbf{k}})$$

and thus simplify the angular integrations necessary for the evaluation of the scattering amplitude and the cross-section. We shall put

$$\psi_\mathbf{k}^+(\mathbf{R}) = \sum_{l,m} \psi_{klm}^+(R) Y_l^{m*}(\hat{\mathbf{k}}) Y_l^m(\hat{\mathbf{R}}) \qquad (2.68)$$

and (cf. eqn (1.A.30))

$$\exp(-i\mathbf{k}.\mathbf{R}) = 4\pi \sum_{l,m} i^{-l} j_l(kR) Y_l^m(\hat{\mathbf{k}}) Y_l^{m*}(\hat{\mathbf{R}}), \qquad (2.69)$$

where $j_l(kR)$ is the regular spherical Bessel function of order l and $R\psi_{klm}(R)$ is a solution of the radial Schrödinger equation (eqn (1.A.19)).

The scattering amplitude for a local potential can be written from eqn (2.27) as

$$f(\mathbf{k'},\mathbf{k}) = \sum_{l,m,l',m'} i^{l'} Y_{l'}^{m'}(\hat{\mathbf{k}}') Y_l^{m*}(\hat{\mathbf{k}}) \times$$

$$\times \iint Y_{l'}^{m'*}(\hat{\mathbf{R}}) Y_l^m(\hat{\mathbf{R}}) j_l(kR) \psi_{klm}^{\pm}(R) U(\mathbf{R}) R^2 \, \mathrm{d}R \mathrm{d}\hat{\mathbf{R}}, \quad (2.70)$$

where $U(\mathbf{R}) = (2\mu/\hbar^2) V(\mathbf{R})$ and $\mathrm{d}\mathbf{R} = R^2 \, \mathrm{d}R \mathrm{d}\hat{\mathbf{R}}$. If the potential is spherically symmetric $V = V(R)$, the integration over $\hat{\mathbf{R}}$ can be performed, and by the orthogonality relation (cf. eqn (1.A.12)) yields $\delta_{ll'}\delta_{mm'}$. A central potential cannot induce l or m transitions.§ Using the addition theorem of spherical harmonics (eqn (1.A.15)), we can sum eqn (2.70) over m to get

$$f(\mathbf{k'},\mathbf{k}) = \sum_{l=0}^{\infty} \frac{2l+1}{k} \beta_{k,l} P_l(\cos\theta), \quad (2.71)$$

where $\cos\theta = \hat{\mathbf{k}}' \cdot \hat{\mathbf{k}}$ and we put

$$\psi_{\mathbf{k}}^{\pm}(\mathbf{R}) = \sum_l i^l (2l+1) \psi_{kl}^{\pm}(R) P_l(\cos\theta), \quad (2.72)$$

$$\beta_{kl} = k \int_0^{\infty} \psi_{kl}^{\pm}(R) U(R) j_l(kR) R^2 \, \mathrm{d}R. \quad (2.73)$$

The angular distribution of the scattered waves depends on the angle θ only. From eqn (2.33)

$$\mathrm{d}\sigma(\theta) = k^{-2} \sum_{l,l'} (2l+1)(2l'+1) P_l(\cos\theta) P_{l'}(\cos\theta) \beta_{kl} \beta_{kl'}^* \, \mathrm{d}\omega. \quad (2.74)$$

Terms with different l values can interfere in their contribution to the differential cross-section. To determine $\mathrm{d}\sigma$ we observe the flux of scattered waves in a particular direction θ, and so expect to observe interference between the different angular momentum components scattered into a specified direction. We can always put

$$\beta_{kl} = |\beta_{kl}| \exp[i\delta_l(E)],$$

so that the interference pattern can be studied in terms of the dependence of $\delta_l(E)$ on l. An exhaustive study of these phenomena has been recently reviewed (Bernstein 1965; see also Ford and Wheeler 1959).

For a central potential the different l components scatter independently of one another, so that the optical theorem (eqn (2.48); for a

§ If the potential is central the angular momentum operator for the relative motion, **l**, commutes with the total Hamiltonian, and so its eigenvalues are good quantum numbers.

proof see section 2.5.4), can be applied to each term in the sum over l in eqn (2.71) or, since $P_l(1) = 1$,

$$(2l+1)\operatorname{im}\beta_{kl} = \frac{(2l+1)^2}{4\pi}\iint |P_l(\cos\theta)|^2\,\mathrm{d}\cos\theta\,\mathrm{d}\phi|\beta_{kl}|^2$$

so that
$$\operatorname{im}\beta_{kl} = |\beta_{kl}|^2, \tag{2.75}$$

which we can write as

$$\beta_{kl} = \sin\delta_l\exp(\mathrm{i}\delta_l) = (2\mathrm{i})^{-1}[\exp(2\mathrm{i}\delta_l)-1], \tag{2.76}$$

where δ_l, the *phase shift*, is a function of E.

The total cross-section is given by

$$\sigma = \int \frac{\mathrm{d}\sigma}{\mathrm{d}\omega}\,\mathrm{d}\omega$$

$$= \frac{4\pi}{k^2}\sum_{l=0}^{\infty}(2l+1)|\beta_{kl}|^2 \tag{2.77}$$

$$= \frac{4\pi}{k^2}\sum_{l=0}^{\infty}(2l+1)\sin^2\delta_l,$$

where we have used the orthogonality relation of the Legendre polynomials (footnote to eqn (2.55)) in integrating eqn (2.74) over ω. The different angular momentum components are now distinct.

From a computational point of view we have reduced the three-dimensional integrations necessary to evaluate the scattering amplitude and cross-section to a sum over one-dimensional integrals. If $\mathbf{p}$ is the momentum operator for the relative motion we can write $\mathbf{l}$, the angular momentum operator for relative motion as $\mathbf{l} = \mathbf{R}\wedge\mathbf{p}$, and the eigenvalues of $\mathbf{l}^2$ are $\hbar^2 l(l+1)$. We can therefore consider l as the discrete quantum-mechanical analogue of the impact parameter b, and note the correspondence $\hbar l \leftrightarrow \hbar bk$. (More accurately, $l(l+1) \leftrightarrow (bk)^2$ or, since $l(l+1) = (l+\tfrac{1}{2})^2 - \tfrac{1}{4}$, we have $l+\tfrac{1}{2} \sim bk$.) As is clear from Fig. 2.1, when b is larger than a, the range of the potential, the incident wave is not modified by the potential. This classical argument suggests (and eqn (2.87) below confirms) that the infinite sum in eqn (2.77) can in practice be truncated when $l \geqslant 2ka$. Molecular interactions are often of considerable range and so we require to retain a large number of terms. We can then try to approximate the summation over l by an integration over b, with the replacement $l+\tfrac{1}{2} = bk$,

$$\sigma = 2\pi\int bP(b)\,\mathrm{d}b. \tag{2.78}$$

A further discussion and derivation of eqn (2.78) is given in section 3.4.0.

The close formal similarity between eqns (2.77) and (2.57) suggests that it may be possible to introduce a non-local interaction V of the type

$$V(\mathbf{R}, \mathbf{R}') = \sum_l \eta_l V_l(\mathbf{R}) V_l(\mathbf{R}'), \qquad (2.79)$$

with $V_l(\mathbf{R}) = V(\mathbf{R})P_l(\cos\theta)$, that will lead to the same cross-section as a given local interaction. This is indeed the case, and the relative ease of manipulating separable interactions lends many uses to this idea. (For the construction of V see, for example, Lloyd 1965.)

The asymptotic form of $\psi_{kl}^+(R)$ is obtained by performing an angular-momentum resolution of the general result:

$$\psi_{\mathbf{k}}^+(\mathbf{R}) \to \exp(i\mathbf{k}\cdot\mathbf{R}) + f(\mathbf{k}', \mathbf{k})\exp(ikR)/R.$$

Using eqns (2.69), (2.71), and (2.72),

$$\psi_{kl}^+(R) \to j_l(kR) + i^{-l}k^{-1}\beta_{kl}\exp(ikR)/R. \qquad (2.80)$$

The asymptotic behaviour of $j_l(kR)$ is

$$j_l(kR) \to (kR)^{-1}\sin(kR - \tfrac{1}{2}l\pi), \qquad (2.81)$$

so that using eqn (2.76) and $i^{-l} = \exp(-\tfrac{1}{2}il\pi)$, we have

$$\psi_{kl}^+(R) \to (kR)^{-1}\exp(i\delta_l)\sin(kR - \tfrac{1}{2}l\pi + \delta_l). \qquad (2.82)$$

The significance of the label 'phase shift' for δ_l is apparent on comparing eqns (2.81) and (2.82). $j_l(kR)$ is the solution of the Schrödinger equation in the absence of the potential, while ψ_{kl}^+ is the solution of the same equation in the presence of the interaction. Outside the range of the interaction the two functions differ by a phase shift.

In practice, if an analytic solution for the problem is unknown the phase shift is often determined by a numerical integration of the one-dimensional equation that determines ψ_{kl}^+. To avoid complex arithmetic we put

$$\psi_{kl}^+(R) = \frac{U_{kl}^+(R)}{R}\exp(i\delta_l) \qquad (2.83)$$

and since by construction $\psi_{kl}^+(R)$ satisfies the Schrödinger equation, $U_{kl}^+(R)$ satisfies the radial equation (1.A.19),

$$\left[-\frac{d^2}{dR^2} + U(R) + \frac{l(l+1)}{R^2} - k^2\right]U_{kl}^+(R) = 0, \qquad (2.84)$$

with $U(R) = (2\mu/\hbar^2)V(R)$, and the boundary conditions that are implied by eqn (2.82), $U_{kl}^+(0) = 0$

and $\qquad U_{kl}^+(R) \to k^{-1}\sin(kR - \tfrac{1}{2}l\pi + \delta_l).$

The effective potential for the functions U_{kl}^+ is not only the mutual interaction $U(R)$ but also the centrifugal term $l(l+1)/R^2$, which acts as

a barrier (see Fig. 2.4). The classical turning-point R_M is defined as the root of
$$k^2 = U(R_M) + l(l+1)/R_M^2. \qquad (2.85)$$

For molecular problems $U(R)$ is repulsive at a very short range, so that eqn (2.85) has at least one root (it may, however, have more than one

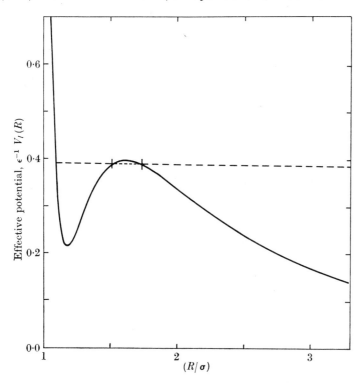

FIG. 2.4. The effective potential $V_l(R)$ for a given partial wave (eqn (2.84)). $V_l(R)$ is a sum of the mutual interaction $V(R)$ and the centrifugal barrier $\hbar^2 l(l+1)/2\mu R^2$. $V(R)$ is taken as a L.-J. (12, 6) potential and the value of l is determined by $l(l+1) = 1 \cdot 6B$, where B is the reduced parameter. Shown also is the energy $E = 0 \cdot 39\epsilon$, and the three classical turning-points at this energy (eqn (2.85)).

root). The shape of the wavefunction is determined mainly by the form of the potential to the right of the smallest classical turning-point. If the potential has a finite range a, the functions ψ_{kl}^+ and j_l would not differ much when $k^2 < l(l+1)/a^2$ or
$$l > ka, \qquad (2.86)$$
so that δ_l is small when $l > ka$ and
$$\lim_{l \to \infty} \delta_l \to 0. \qquad (2.87)$$

The low-energy value of the phase shift is known to satisfy Levinson's theorem (eqn (5.99), Fig. 2.8)

$$\pi n_l = \lim_{E \to 0} \delta_l(E),$$

where n_l is the number of bound states of the Hamiltonian H, of angular momentum l (each of which is $(2l+1)$-fold degenerate.) A qualitative understanding of this relation can be based on the orthogonality of the bound and continuous eigenfunctions of H. Thus $U_{kl}^+(R)$ has to be orthogonal to n bound eigenfunctions, in contrast to $j_l(kR)$. In the limit $k \to 0$, $U_{kl}^+(R)$ will necessarily have n more nodes than $j_l(kR)$. (Recall that in one dimension the $(n+1)$th excited state has n nodes.) Consider now the low-energy scattering of an electron by a helium atom. When the electron is very near the atom its wavefunction should resemble that of a bound $2s$ electron, since it is excluded from the $1s$ orbitals by the Pauli principle. Thus the wavefunction should have a node even though there are no physically bound states, and this is indeed observed in that the experimentally deduced s-wave ($l = 0$) phase shift tends to π at zero energy (Massey and Burhop 1952).

To compute a phase shift we consider the collision of two rigid spheres of radii a_1 and a_2 ($a_1+a_2 = a$), so that the mutual interaction is infinite for $R < a$ and vanishes for $R > a$. Thus $U_{kl}^+(R)$ should vanish at $R = a$, while for $R > a$ it is a solution of the radial equation in the absence of a potential. For $l = 0$ we have

$$U_{k0}^+(R > a) = k^{-1}\sin(kR+\delta_0),$$

since $j_0(kR) = (kR)^{-1}\sin(kR)$. To satisfy the boundary conditions we must have

$$\delta_0 = -ka \qquad (2.88)$$

and the $l = 0$ component of the cross-section is given by

$$\sigma_0 = \frac{4\pi}{k^2}\sin^2 ka.$$

In the limit $ka \ll 1$ we can expand the sine function to obtain

$$\sigma_0 \to 4\pi a^2.$$

In this limit $\sigma \to \sigma_0$ by virtue of the criteria (2.86) and (2.87), so that the cross-section does appear to be an effective area in a classical sense, although it is not equal to the expected geometrical cross-section (πa^2), and the approximation is restricted to very low velocities.

Our example does suggest, however, that a parameter a can be defined by

$$a = -(d\delta_0/dk)_{k=0} = -\lim_{k \to 0} f, \qquad (2.89)$$

where f is the scattering amplitude (eqn (2.71)), so that

$$\lim_{k \to 0} \sigma(k) = 4\pi a^2. \tag{2.90}$$

The problem of the next term in the expansion of δ_l in powers of k has received considerable theoretical attention (Bethe 1949, Blatt and

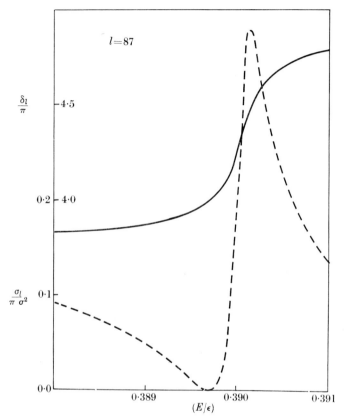

FIG. 2.5. The phase shift and cross-section for the lth partial wave, computed by Bernstein *et al.* (1966) for the potential of Fig. 2.4. The cross-section does not have the symmetric Lorentzian shape of eqn (2.58) since δ_l does not tend to zero (or a multiple of π) away from the resonance.

Jackson 1949, Baker 1958, and also section 2.6.4). Unfortunately the results are only simple if $V(R)$ vanishes identically for $R > a$. For potentials with a long-range tail considerable ingenuity is required (Levy and Keller 1963, O'Malley 1964, Berger and Spruch 1965).

The effective potential in the Schrödinger equation (2.84) for the radial function $U_{kl}^{+}(R)$ can lead to so-called 'orbiting' resonances. From a qualitative point of view we have seen that a single degree of freedom resonance was associated with a large density of the scattered wave in the region of mutual interaction. The superposition of a centrifugal barrier on an attractive potential can, for energies below the top of the barrier, provide a region where the density of the wavefunction $U_{kl}^{+}(R)$ is high (Fig. 2.4). The results of numerical integrations of eqn (2.84) by Bernstein, Curtiss, Imam-Rahajoe, and Wood (1966) are shown in Fig. 2.5. We expect a similar behaviour whenever the potential $V(R)$ possess more than one classical turning-point.

2.2.4. *The Born approximation*

From a computational point of view, the (first) Born approximation replaces the matrix element $\langle \mathbf{k}'|V|\mathbf{k}^{+}\rangle$ by $\langle \mathbf{k}'|V|\mathbf{k}\rangle$ in the evaluation of the scattering amplitude (eqn (2.27)). It is therefore formally equivalent to replacing the exact L.S. equation by

$$|\mathbf{k}^{+}\rangle_{B} = |\mathbf{k}\rangle + G_{0}(E^{+})V|\mathbf{k}\rangle, \qquad (2.91)$$

where the subscript B refers to the Born approximation. Equation (2.91) can be regarded as the first approximation to an iterative solution of the L.S. equation which is generated by substituting the nth order approximation in the second term of the L.S. equation, thus generating the $(n+1)$th term. Starting with $|\mathbf{k}\rangle$ as the zero order approximation we have eqn (2.91). Using $|\mathbf{k}^{+}\rangle_{B}$ as the first approximation we have

$$|\mathbf{k}^{+}\rangle_{2ndB} = |\mathbf{k}\rangle + G_{0}(E^{+})V|\mathbf{k}^{+}\rangle_{B}$$
$$= |\mathbf{k}\rangle + G_{0}(E^{+})V|\mathbf{k}\rangle + G_{0}(E^{+})VG_{0}(E^{+})V|\mathbf{k}\rangle. \qquad (2.92)$$

Thus in the second Born approximation the transition amplitude is given by
$$\langle \mathbf{k}'|V + VG_{0}(E^{+})V|\mathbf{k}\rangle \qquad (2.93)$$

and in general in the nth order, by

$$\langle \mathbf{k}'|V \sum_{m=0}^{n-1} [G_{0}(E^{+})V]^{m}|\mathbf{k}\rangle. \qquad (2.94)$$

From a practical point of view one rarely goes beyond the first order approximation in this series since a gross error in the first Born approximation is usually an indication of some features that are best handled by special methods, and for this reason the first Born approximation is called 'the' Born approximation. We derive below some criteria for the numerical validity of the Born approximation. However, it is easy to see in a qualitative manner that the approximation cannot be quantitatively accurate for large scattering amplitudes since $\langle \mathbf{k}|V|\mathbf{k}\rangle$ and hence $f_{B}(\mathbf{k}, \mathbf{k})$ are necessarily real numbers, and so the optical theorem, which is, as we shall show, the statement of conservation of probabilities, is not satisfied. Even outside its range of validity the Born approximation often provides reliable information on angular distributions and other attributes which depend on the ratios of transition amplitudes. As an example we can consider the scattering by a separable interaction, when

$$f_B(\mathbf{k'},\mathbf{k}) = -(\mu/2\pi\hbar^2) \iint \langle \mathbf{k'}|\mathbf{R'}\rangle\langle \mathbf{R'}|V|\mathbf{R}\rangle\langle \mathbf{R}|\mathbf{k}\rangle \, d\mathbf{R'}d\mathbf{R}$$
$$= -(\mu/2\pi\hbar^2)\eta g(\mathbf{k})g^*(\mathbf{k}), \quad (2.95)$$

where $g(\mathbf{k})$ is defined by eqn (2.43). In this example (which must not be taken to be truly representative) the angular distribution is predicted exactly, as is seen on comparing the Born approximation with the exact result, eqn (2.42). In fact

$$f(\mathbf{k'},\mathbf{k}) = f_B(\mathbf{k'},\mathbf{k})/D(E^+), \quad (2.96)$$

so that ratios are also predicted correctly. Equation (2.96) is not, however, of universal validity, even though for short-range potentials and a suitable extension of the definition of $D(E^+)$ it can be used as an approximation.§ A less satisfactory aspect is the convergence of the Born series for this case. As is clear the Born iteration corresponds to expanding the denominator of eqn (2.42) in powers of η,

$$(1-\eta c)^{-1} = 1+\eta c+\eta^2 c^2+\ldots$$

where $\eta c = 1-D(E^+)$. The series thus fails to converge at the zeros of $D(E)$. It is, therefore, not meaningful to go to higher Born approximations in that case; rather one has to express the denominator exactly.

For a local potential the Born approximation is

$$f_B(\mathbf{k'},\mathbf{k}) = -(\mu/2\pi\hbar^2) \int \langle \mathbf{k'}|\mathbf{R}\rangle V(\mathbf{R})\langle \mathbf{R}|\mathbf{k}\rangle \, d\mathbf{R} = -(\mu/2\pi\hbar^2)V(\mathbf{q}), \quad (2.97)$$

where $\mathbf{q} = \mathbf{k'}-\mathbf{k}$, $\mathbf{q}$ being the momentum transfer during the collision. Since $k' = k$, this momentum is used to change the direction of the relative motion by an angle θ, such that

$$q^2 = |\mathbf{k'}-\mathbf{k}|^2 = 2k^2(1-\cos\theta).$$

From the point of view of the momentum representation the potential induces a single transition from the initial momentum $\mathbf{k}$ to the final momentum $\mathbf{k'}$. To produce this deflexion $V(\mathbf{k'}-\mathbf{k})$ must be significant.

§ The function $D(E^+)$ is really the Fredholm determinant, eqn (5.94). Using eqn (5.96) the reader can show that the scattering amplitude for the lth partial wave can be written as $N_l(E^+)/D_l(E^+)$. For separable interactions this leads to eqn (2.96). In other cases there are higher-order contributions to the numerator. For further details see DeWitt (1956) and Chew and Mandelstam (1960). For inelastic collisions see Sugar and Blankenbecler (1964).

From the general theory of Fourier transforms we know that the long-range behaviour of $V(\mathbf{R})$ is reflected by the short-range behaviour of $V(\mathbf{q})$, and vice versa. So that, for example, low-energy collisions (small $\mathbf{k}$ hence small $\mathbf{q}$) can only explore the smooth part of the long-range tail of the potential. To see this more explicitly we consider a central potential where§

$$f_B(\mathbf{q}) = -(\mu/2\pi\hbar^2) \iint \exp(-i\mathbf{q}\cdot\mathbf{R})V(R)R^2\,\mathrm{d}R\,\mathrm{d}\hat{\mathbf{R}}$$
$$= -q^{-1}\int_0^\infty \sin(qR)U(R)R\,\mathrm{d}R. \quad (2.98)$$

By performing a partial wave analysis of eqn (2.98) or from eqn (2.73) we can show that the Born approximation is equivalent to replacing β_{kl} by $-\delta_l$, or in other words

$$-(\beta_{kl})_B = (\delta_l)_B = -k\int_0^\infty [j_l(kR)]^2 U(R)R^2\,\mathrm{d}R. \quad (2.99)$$

The numerical validity of the Born approximation is best judged from the requirement that the violation of the optical theorem is minimal, or from eqn (2.75) that $\beta_{kl} \ll 1$. (In other words, that σ be small compared to its maximum possible value.) For a square well of depth V_0 and range a we can provide an over-estimate of β_{kl} by replacing $j_l(kR)$ by $(kR)^{-1}$ or

$$(\beta_{kl})_B \leqslant \frac{2\mu V_0 a}{\hbar^2 k} = \frac{2V_0 a}{\hbar v} \ll 1. \quad (2.100)$$

We expect this result to be representative‖ of potentials of range a and average strength V_0. However, intermolecular potentials typically possess a strongly repulsive part which cannot be treated in this fashion.

It can be shown (for example, Messiah (1961)) that our condition is equivalent to requiring that the amplitude of the scattered wave be small. In the region of strong repulsion the exact wavefunction $\psi_{kl}^+(R)$ has an exponentially small amplitude to the left of the classical turning-point (cf. Fig. 3.1), while the spherical Bessel function has a large amplitude up to the origin. In this case the naïve Born approximation

---

§ If x is the angle between $\mathbf{q}$ and $\mathbf{R}$, then $\mathrm{d}\hat{\mathbf{R}} = \mathrm{d}\cos x\,\mathrm{d}\phi$,

$$\int_0^{2\pi}\int_{-1}^{1} \exp(-iqR\cos x)\,\mathrm{d}\cos x\,\mathrm{d}\phi = \frac{4\pi}{qR}\sin qR$$

where we noted that only the even part of the integrand can contribute. We also assume that the integral in eqn (2.98) converges.

‖ A more precise criterion is available, however, from eqns (2.86) and (2.87). Equation (2.99) is thus applicable as soon as $l \geqslant 2ka$. We can then also take note that

$$j_l(kR) \underset{l \gg kR}{=} (kR)^l/1.3.5\ldots(2l+1).$$

will be very misleading. In eqn (2.73) one can, however, replace $\psi_{kl}^+(R)$ by $j_l(kR)$ to the right of the turning-point only and neglect the contribution from the left of the turning-point. Equation (2.99) is thus modified by changing the lower limit to R_M, as determined by eqn (2.85). An estimate of the integral is

$$-\delta_l \sim 2\mu V(R_M) R_M / \hbar^2 k. \qquad (2.101)$$

Equation (2.101) does predict correctly the k-dependence of the rigid sphere phase-shift ($R_M = a$, $V(R_M) = E$).

We can also consider the condition from a physical point of view. A potential of range a will have a range $\hbar/a$ in momentum space. To ensure a small cross-section the actual momentum transfer Δp must be small in this range, and $\Delta p \sim V_0/v$, since V_0/v is the average force per unit time. This criterion can be re-expressed by noting that $h/v\Delta p$ is the 'duration' of the action of the potential leading to the momentum transfer Δp. To ensure a small cross-section this has to be longer compared with the duration of an average collision, a/v. These qualitative arguments are clearly only valid when $E \gg V_0$, since only then can we regard v, the asymptotic velocity, as unchanged during the interaction.

It may seem that the discussion of 'structureless' molecules cannot be relevant to any realistic situation. As we shall see it is generally possible, when the energy is low enough and internal excitation cannot occur, to reduce the L.S. equation to an equation for the relative motion only. The complexity of the problem is then contained in a suitable definition of the mutual interaction, which may be non-local and energy dependent. As a preliminary step in this direction we consider in the next section the collision of molecules that possess internal degrees of freedom. The final reduction of the problem is carried out in section 3.2.0.

2.3. Internal excitation in collisions

2.3.0. WHEN the colliding molecules possess internal degrees of freedom it is convenient to consider an initial state that is a product of the internal state vectors of the two molecules and a state vector of their relative motion,
$$|\mathbf{k}, n\rangle = |\mathbf{k}\rangle|n\rangle, \tag{3.1}$$
where
$$|n\rangle = |n_\alpha\rangle|n_\beta\rangle.$$

In general, such an initial state is not an eigenstate of all the operators that commute with H. In particular, such a state will normally not possess the required symmetry properties. For a collision of two hydrogen atoms in their ground state, for example, such a state is a product of a spin-orbital on each atom and the state vector $|\mathbf{k}\rangle$. It clearly is not antisymmetric with respect to the permutation of the electrons (or of the protons). A properly symmetrized initial state can be written as a linear combination
$$|\Phi\rangle = \sum_n C_n |\mathbf{k}, n\rangle, \tag{3.2}$$
where the coefficients C_n are determined to make $|\Phi\rangle$ an eigenstate of all the operators that commute with H.§

Let $\mathbf{r}$ be the set of internal coordinates of the two molecules, so that
$$\langle \mathbf{r}, \mathbf{R} | \mathbf{k}, m \rangle = \langle \mathbf{R} | \mathbf{k}\rangle\langle \mathbf{r} | m\rangle = \exp(i\mathbf{k}\cdot\mathbf{R})\varphi_m(\mathbf{r}), \tag{3.3}$$
where
$$\varphi_m(\mathbf{r}) = \langle \mathbf{r}_1 | m_\alpha\rangle\langle \mathbf{r}_2 | m_\beta\rangle = \phi_{m_\alpha}(\mathbf{r}_1)\phi_{m_\beta}(\mathbf{r}_2) \tag{3.4}$$
is a product of the internal wavefunctions of the two molecules, with $\mathbf{r}_1$ denoting the internal coordinates of molecule one and $\mathbf{r}_2$ those of molecule two. As eigenfunctions of the internal Hamiltonians,
$$h_1 \phi_{m_\alpha}(\mathbf{r}_1) = E_{m_\alpha} \phi_{m_\alpha}(\mathbf{r}_1),$$
$$h_2 \phi_{m_\beta}(\mathbf{r}_2) = E_{m_\beta} \phi_{m_\beta}(\mathbf{r}_2), \tag{3.5}$$
they are assumed orthonormal:
$$\langle m | n\rangle = \langle m_\alpha | n_\alpha\rangle\langle m_\beta | n_\beta\rangle = \delta_{m_\alpha n_\alpha}\delta_{m_\beta n_\beta} = \delta_{mn} \tag{3.6}$$
and
$$h|m\rangle = (h_1 + h_2)|m_\alpha\rangle|m_\beta\rangle = (E_{m_\alpha} + E_{m_\beta})|m\rangle,$$
so that
$$\langle n|(E^+ - H_0)^{-1}|m\rangle = (E^+ - E_m - K)^{-1}\delta_{mn}, \tag{3.7}$$
where $E_m = E_{m_\alpha} + E_{m_\beta}$, $H_0 = h + K$, and K is the kinetic energy operator for the relative motion.

§ From a computational point of view it is often advantageous to consider such an initial state. See section 2.3.1.

Consider the matrix element $\langle m, \mathbf{R} | \mathbf{k}, n^+ \rangle$. It is the amplitude to find the interacting molecules at a relative separation $\mathbf{R}$, with the internal state m, when the initial state of the collision was $|\mathbf{k}, n\rangle$. If $\psi^+_{\mathbf{k},n}(\mathbf{R}, \mathbf{r})$ is the total wavefunction for the interacting molecules in the coordinate representation, we can expand it in the form

$$\psi^+_{\mathbf{k},n}(\mathbf{R}, \mathbf{r}) = \sum_m \varphi_m(\mathbf{r}) F^m_{\mathbf{k},n}(\mathbf{R}) \tag{3.8}$$

and make the identification

$$F^m_{\mathbf{k},n}(\mathbf{R}) = \langle m, \mathbf{R} | \mathbf{k}, n^+ \rangle = \langle \mathbf{R} | \langle m | \mathbf{k}, n^+ \rangle \tag{3.9}$$

by writing

$$\psi^+_{\mathbf{k},n}(\mathbf{R}, \mathbf{r}) = \langle \mathbf{r}, \mathbf{R} | \mathbf{k}, n^+ \rangle = \sum_m \langle \mathbf{r} | m \rangle \langle m, \mathbf{R} | \mathbf{k}, n^+ \rangle, \tag{3.10}$$

where we have assumed§ that the states $|m\rangle$ form a complete set in the space of internal coordinates.

The orthogonality of the internal state vectors ensures the mutual orthogonality of the terms in eqn (3.8) and implies that the events 'the molecules are in the internal state m' are mutually exclusive in that the probability density of finding the two molecules at a relative separation R can be written as a sum over internal states of the probability densities of finding the molecules, in a given internal state, at a relative separation R, namely,

$$\int |\psi^+_{\mathbf{k},n}(\mathbf{R}, \mathbf{r})|^2 \, d\mathbf{r} = \sum_m |F^m_{\mathbf{k},n}(\mathbf{R})|^2. \tag{3.11}$$

The integral equations for $F^m_{\mathbf{k},n}(\mathbf{R})$ can be obtained by writing down the L.S. equation for $\psi^+_{\mathbf{k},n}(\mathbf{R}, \mathbf{r})$, premultiplying by $\varphi^*_m(\mathbf{r})$ and integrating over $\mathbf{r}$. We can, however, consider $F^m_{\mathbf{k},n}(\mathbf{R})$ as the coordinate representation of $\langle m | \mathbf{k}, n^+ \rangle$. From the L.S. equation

$$|\mathbf{k}, n^+\rangle = |\mathbf{k}, n\rangle + (E^+ - H_0)^{-1} V | \mathbf{k}, n^+\rangle \tag{3.12a}$$

and eqn (3.7) we have

$$\langle m | \mathbf{k}, n^+ \rangle = |\mathbf{k}\rangle \delta_{mn} + (E^+ - E_m - K)^{-1} \langle m | V | \mathbf{k}, n^+ \rangle \tag{3.12b}$$

$$= |\mathbf{k}\rangle \delta_{mn} + (E^+ - E_m - K)^{-1} \sum_{m'} \langle m | V | m' \rangle \langle m' | \mathbf{k}, n^+ \rangle. \tag{3.12c}$$

Note that $\langle m | V | m' \rangle$ is an operator in the space of relative motion of the

§ If, as we have implied, the states $|m\rangle$ are the bound internal states of the molecules then this assumption is in error, and eqn (3.8) is only approximate. When the total energy is well below the threshold of dissociation of the isolated molecules we expect the approximation of restricting m to be discrete, to be reasonable. To keep the formulation exact one can, however, understand the summation in eqn (3.8) to include integration.

two molecules such that, in the coordinate representation,§

$$V_{m,m'}(\mathbf{R}) = \int \varphi_m^*(\mathbf{r}) V(\mathbf{r},\mathbf{R}) \varphi_{m'}(\mathbf{r}) \, d\mathbf{r}. \tag{3.13}$$

The second term on the right in eqn (3.12c) depends in general on all the amplitudes $\langle m'|\mathbf{k},n^+\rangle$. Equation (3.12c) is just one member of a set of coupled integral equations. In the coordinate representation

$$F_{\mathbf{k},n}^m(\mathbf{R}) = \exp(i\mathbf{k}\cdot\mathbf{R})\delta_{mn} +$$
$$+ \sum_{m'} \int \langle \mathbf{R}|(E^+ - E_m - K)^{-1}|\mathbf{R}'\rangle V_{m,m'}(\mathbf{R}') F_{\mathbf{k},n}^{m'}(\mathbf{R}') \, d\mathbf{R}'. \tag{3.14}$$

The advantage of the set of integral equations (3.14) over the single integral equation (3.12a) is now apparent. Equation (3.12a) is an integral equation in the space of internal and relative coordinates, while eqn (3.14) is an equation for the relative motion only and can be handled by extending the approach of section 2.2.0. An example is discussed below.

In considering the asymptotic behaviour of $F_{\mathbf{k},n}^m(\mathbf{R})$ as a function of R we distinguish two cases. When $E > E_m$ we intuitively expect that as a result of the collision it should be possible, in principle, to observe the excitation of the state m. From eqn (3.14) and the asymptotic form of the translational Green's function at the kinetic energy $E - E_m$ (cf. eqn (2.24)),

$$F_{\mathbf{k},n}^m(\mathbf{R}) \to \exp(i\mathbf{k}\cdot\mathbf{R})\delta_{mn} -$$
$$- (\mu/2\pi\hbar^2)\frac{\exp(ik_m R)}{R} \sum_{m'} \int \exp(-i\mathbf{k}'\cdot\mathbf{R}') V_{m,m'}(\mathbf{R}') F_{\mathbf{k},n}^{m'}(\mathbf{R}') \, d\mathbf{R}', \tag{3.15a}$$

where $\mathbf{k}' = k_m \hat{\mathbf{R}}$ and $E = E_m + \hbar^2 k_m^2/2\mu$ defines k_m and ensures the conservation of energy. The scattering amplitude $f_{m,n}(\mathbf{k}',\mathbf{k})$ is determined by the asymptotic form (3.15a) to be

$$f_{m,n}(\mathbf{k}',\mathbf{k}) = -(\mu/2\pi\hbar^2) \sum_{m'} \int \exp(-i\mathbf{k}'\cdot\mathbf{R}) V_{m,m'}(\mathbf{R}) F_{\mathbf{k},n}^{m'}(\mathbf{R}) \, d\mathbf{R}$$
$$= -(\mu/2\pi\hbar^2)\langle m,\mathbf{k}'|V|\mathbf{k},n^+\rangle, \tag{3.16}$$

where in the second line we have used eqn (3.9). If we define the direction of the relative motion $\hat{\mathbf{R}}$ as being from molecule two to molecule one, then the scattering amplitude above corresponds to observing molecule one in the internal state m_α in the direction $\hat{\mathbf{k}}'$ and molecule two in the internal state m_β in the direction $-\hat{\mathbf{k}}'$. In practice, one seldom performs the coincidence experiment and simply assumes that the other molecule

§ We have anticipated in our notation the fact that $V(\mathbf{r},\mathbf{R})$ is usually a local operator in R.

does appear in the reverse direction. Its internal state can, in principle, be determined from conservation of energy.

When $E < E_m$ the asymptotic behaviour of $F_{\mathbf{k},n}^m(\mathbf{R})$ is that of a decaying exponential $\exp(-\kappa R)$ with $\hbar^2\kappa^2 = (E_m-E)2\mu$. Asymptotically such states cannot be occupied. It is important to note that they can, however, be occupied at a finite relative distance. We shall return to their role in collision problems in section 3.2.2.

The asymptotic form of the wavefunction $\psi_{\mathbf{k},n}^+(\mathbf{R}, \mathbf{r})$ is given from eqns (3.8) and (3.15) by

$$\psi_{\mathbf{k},n}^+(\mathbf{R}, \mathbf{r}) \xrightarrow[R\to\infty]{} \sum_m \varphi_m(\mathbf{r})\left[\delta_{m,n}\exp(i\mathbf{k}\cdot\mathbf{R}) + \frac{\exp(ik_m R)}{R} f_{m,n}(\mathbf{k}', \mathbf{k})\right].$$

(3.15b)

Each term in the sum above is a product of an internal state (an eigenfunction of h) and a wavefunction for relative motion (an eigenfunction of K), such that the sum of the internal energy and the energy of relative motion is the same for all terms and is equal to the total energy of the initial state. At large relative separation $\psi_{\mathbf{k},n}^+(\mathbf{R},\mathbf{r})$ is an eigenfunction of H_0.§ Thus our discussion is, as yet, too restrictive to include true chemical rearrangements, where the well-separated final states are characterized by a different Hamiltonian than that of the initial state.

We can now consider an arbitrary initial state $|\Phi\rangle = \sum_n C_n|\mathbf{k}_n, n\rangle$. If we write the L.S. equation in the form

$$(E^+ - H)|\Psi\rangle = (E^+ - H_0)|\Phi\rangle,$$

we see that, if $|\Psi_1\rangle$ is a solution with $|\Phi_1\rangle$ as the initial state and $|\Psi_2\rangle$ is a different solution with $|\Phi_2\rangle$ as the initial state (of the same total energy as $|\Phi_1\rangle$), then $|\Psi_1\rangle + |\Psi_2\rangle$ satisfies the L.S. equation with the initial state $|\Phi_1\rangle + |\Phi_2\rangle$. We can thus take in general

$$\langle m|\Psi\rangle = \sum_n C_n\langle m|\mathbf{k}_n, n^+\rangle$$

and

$$\Psi^+(\mathbf{R}, \mathbf{r}) = \sum_m \varphi_m(\mathbf{r}) \sum_n F_{\mathbf{k}_n,n}^m(\mathbf{R}) C_n.$$

By interchanging the order of summation we also have

$$\Psi^+(\mathbf{R},\mathbf{r}) = \sum_n C_n \sum_m F_{\mathbf{k}_n,n}^m(\mathbf{R})\varphi_m(\mathbf{r}) = \sum_n C_n \psi_{\mathbf{k}_n,n}^+(\mathbf{R},\mathbf{r})$$

§ At large relative separation the molecules are no longer interacting and the Hamiltonian H can be replaced by H_0 so that the conclusion above is not unexpected. Equation (3.15a) provides more than a qualitative conclusion, since the wavefunction for the relative motion is specified and an explicit algorithm of the scattering amplitude was given in eqn (3.16).

as the solution of the L.S. equation with an initial state
$$\Phi = \sum_n C_n \phi_{\mathbf{k}_n,n}(\mathbf{R}, \mathbf{r}),$$
where
$$\phi_{\mathbf{k}_n,n}(\mathbf{R}, r) = \varphi_n(\mathbf{r})\exp(i\mathbf{k}_n \cdot \mathbf{R})$$
is a set of initial states of the same total energy E, $E = E_n + \hbar^2 k_n^2/2\mu$.

A formal proof of our considerations is readily available by writing the L.S. equation as
$$|\Psi\rangle = |\Phi\rangle + (E^+ - H)^{-1}(H - E)|\Phi\rangle. \tag{3.17}$$
We also see explicitly that if Λ is a constant of motion, $[\Lambda, H] = 0$, we can ensure that Ψ is an eigenfunction of Λ by selecting Φ as an eigenfunction
$$\Lambda|\Phi\rangle = \lambda_j|\Phi\rangle \tag{3.18}$$
so that
$$\Lambda|\Psi\rangle = \lambda_j|\Phi\rangle + (E^+ - H)^{-1}(H - E)\Lambda|\Phi\rangle = \lambda_j|\Psi\rangle. \tag{3.19}$$

As the first application of eqn (3.19) we consider the collision of two identical molecules.§ If Q is the operator for the exchange of the two molecules, the total wavefunction should satisfy
$$Q\Psi = \delta_Q \Psi, \tag{3.20}$$
where‖
$$\delta_Q = \begin{cases} +1 \text{ (for bosons)} \\ -1 \text{ (for fermions)}. \end{cases}$$
Since $Q^2 = I$ and $\delta_Q^2 = 1$ we can define an operator O by
$$O = \tfrac{1}{2}[I + \delta_Q Q] \tag{3.21}$$
such that
$$QO = \delta_Q O \quad \text{and} \quad O^2 = O. \tag{3.22}$$
O projects out of any function the proper symmetry component that satisfies eqn (3.20). When the two molecules are identical, $[Q, H] = 0$, since H must be invariant under the exchange of identical molecules, or
$$H = QHQ^{-1}. \tag{3.23}$$
Thus O is a constant of motion.

Consider an unsymmetrized initial state
$$\Phi = \varphi_{n_\alpha}(\mathbf{r}_1)\varphi_{n_\beta}(\mathbf{r}_2)\exp(i\mathbf{k} \cdot \mathbf{R}).$$
Under the operation Q, $r_1 \to r_2$, $r_2 \to r_1$, but also $\mathbf{R} \to -\mathbf{R}$ since $\mathbf{R}$ is the position vector from molecule two to molecule one. We define
$$|O\Phi\rangle = MO|\Phi\rangle \tag{3.24}$$

§ For further discussion see Kerner 1953, Gioumousis and Curtiss 1958, Davison 1962, and Waldmann 1964.

‖ The molecules are, of course, composite structures, so that the labels 'bosons' or 'fermions' are just labels for the sign of δ_Q. The sign is determined from first principles by the consideration of section 2.8.4, in particular eqn (8.66).

and determine the normalization constant M by requiring that
$$\langle O\Phi | O\Phi \rangle = \langle \Phi | \Phi \rangle. \tag{3.25}$$
From our previous considerations it follows that
$$\langle O\Psi | O\Psi \rangle = \langle \Phi | \Phi \rangle,$$
and, since $O^2 = O$ and $O = O^\dagger$,
$$\langle O\Phi | O\Psi \rangle = M \langle \Phi | O\Psi \rangle. \tag{3.26}$$
In the present case $M = \sqrt{2}$ and the coordinate representation of $|O\Phi\rangle$ is
$$\Phi_o = \frac{1}{\sqrt{2}}[\varphi_{n_\alpha}(\mathbf{r}_1)\varphi_{n_\beta}(\mathbf{r}_2)\exp(i\mathbf{k}\cdot\mathbf{R}) + \delta_Q\,\varphi_{n_\alpha}(\mathbf{r}_2)\varphi_{n_\beta}(\mathbf{r}_1)\exp(i\mathbf{k}\cdot\mathbf{R})]. \tag{3.27}$$
From eqns (3.16), (3.19), and (3.27)
$$\Psi_o^* \to \Phi_o + (R\sqrt{2})^{-1} \sum_{m_\alpha m_\beta} \exp(ik_m R)[f_{m_\alpha m_\beta, n}(\mathbf{k}_m', \mathbf{k})\varphi_{m_\alpha}(\mathbf{r}_1)\varphi_{m_\beta}(\mathbf{r}_2) +$$
$$+ \delta_Q f_{m_\alpha m_\beta, n}(-\mathbf{k}_m', \mathbf{k})\varphi_{m_\alpha}(\mathbf{r}_2)\varphi_{m_\beta}(\mathbf{r}_1)]. \tag{3.28}$$
In view of eqn (3.26) we can write the scattering amplitude to observe one of the identical molecules in the internal state m_α in the direction of $\hat{\mathbf{k}}_m'$ and the other molecule in the internal state m_β in the opposite direction, as
$$MR\exp(-ik_m R)\langle \varphi_{m_\alpha}\varphi_{m_\beta}|O\Psi\rangle. \tag{3.29}$$
Two terms in the summation in eqn (3.28) contribute once to give
$$f_{m_\alpha m_\beta, n}(\mathbf{k}_m', \mathbf{k}) + \delta_Q f_{m_\beta m_\alpha, n}(-\mathbf{k}_m', \mathbf{k}). \tag{3.30}$$
We note that in accordance with our interpretation of eqn (3.16), both these amplitudes refer to the physical event, one molecule in the state m_α in the direction $\hat{\mathbf{k}}_m'$ and the other in the state m_β, in the direction $-\hat{\mathbf{k}}_m'$.

Equation (3.30) corresponds to the final state
$$MO\varphi_{m_\alpha}(\mathbf{r}_1)\varphi_{m_\beta}(\mathbf{r}_2)\exp(i\mathbf{k}_m\cdot\mathbf{R}).$$
Instead we could have considered the final state
$$MO\varphi_{m_\alpha}(\mathbf{r}_2)\varphi_{m_\beta}(\mathbf{r}_1)\exp(i\mathbf{k}_m\cdot\mathbf{R}),$$
which corresponds to a molecule in the internal state m_β being in the direction $\hat{\mathbf{R}}$, and we have obtained the amplitude
$$f_{m_\beta m_\alpha, n}(\mathbf{k}_m', \mathbf{k}) + \delta_Q f_{m_\alpha m_\beta, n}(-\mathbf{k}_m', \mathbf{k}). \tag{3.31}$$
In general, the two amplitudes, (3.30) and (3.31), are different; however, the integral of their square over all directions $\hat{\mathbf{k}}_m'$ is necessarily the same, since both results determine the amplitude to observe the scattering leading to one molecule in the state m_α and one in the state m_β.

2.3.1. *The theory of rotational excitation*

When we consider the collision of two molecules we usually assume a description of the system that is invariant under over-all rotation in space, where by an over-all rotation we mean rotating the whole system as a rigid body. If **j** is the internal angular momentum of the molecules and **l** is the angular momentum of their relative motion, our description implies that the total angular momentum, **J** is a constant of motion,

$$\mathbf{J} = \mathbf{j} + \mathbf{l}. \tag{3.32}$$

The general theory using the invariance of **J** has been extensively discussed (Blatt and Biedenharn 1952, Lane and Thomas 1958, Newton 1958, 1960), and several applications to molecular collisions have been presented (Arthurs and Dalgarno 1960, Gioumousis and Curtiss 1961, Davison 1962, Takayanagi 1963, 1965).

To simplify the notation we shall consider explicitly the collision of a structureless atom and a rigid rotor. Since the projection of **J** on a fixed axis is also conserved it is useful to introduce the $(jlJM)$ representation where the states $|jlJM\rangle$ are eigenstates of $\mathbf{J}^2$, $\mathbf{J}_z$, $\mathbf{l}^2$, and $\mathbf{j}^2$. For example,

$$\mathbf{J}^2|jlJM\rangle = \hbar^2 J(J+1)|jlJM\rangle, \tag{3.33a}$$

or
$$J_z|jlJM\rangle = \hbar M|jlJM\rangle. \tag{3.33b}$$

Let **R** be the relative separation and **r** the internal coordinate of the rotor, so that the functions

$$\mathscr{Y}^M_{Jlj}(\hat{\mathbf{R}},\hat{\mathbf{r}}) = \langle \hat{\mathbf{r}},\hat{\mathbf{R}}|jlJM\rangle \tag{3.34}$$

are the coordinate representation of our basis and are eigenfunctions of $\mathbf{J}^2$ and J_z. Using the expansion (cf. Appendix 1.A),

$$|jlJM\rangle = \sum_{m_j}\sum_{m_l}|jm_j\, lm_l\rangle\langle jm_j\, lm_l|jlJM\rangle, \tag{3.35a}$$

we can express the functions $\mathscr{Y}^M_{Jlj}$ as§

$$\mathscr{Y}^M_{Jlj}(\hat{\mathbf{R}},\hat{\mathbf{r}}) = \sum_{m_j}\sum_{m_l} Y_l^{m_l}(\hat{\mathbf{R}})Y_j^{m_j}(\hat{\mathbf{r}})\langle jm_j\, lm_l|jlJM\rangle. \tag{3.35b}$$

To specify the direction of the initial relative momentum it is convenient to introduce the functions

$$D^{Mm_j}_{Jlj}(\hat{\mathbf{k}}) = \sum_{m_l} i^{-l}\langle jm_j\, lm_l|jlJM\rangle Y_l^{m_l}(\hat{\mathbf{k}}). \tag{3.36}$$

Using the unitarity of the vector coupling coefficients (cf. eqn (1.A.40)) $\langle jm_j\, lm_l|jlJM\rangle$, we can resolve the initial state

$$\exp(i\mathbf{k}_j\cdot\mathbf{R})Y_j^{m_j}(\hat{\mathbf{r}}) = 4\pi\sum_{l,m_l} i^l j_l(kR)Y_l^{m_l}(\hat{\mathbf{R}})Y_l^{m_l*}(\hat{\mathbf{k}})Y_j^{m_j}(\hat{\mathbf{r}}), \tag{3.37a}$$

---

§ The Clebsch-Gordan coefficients vanish unless $m_j+m_l = M$; however, the redundant sum in eqns (35) is kept to show the completeness relations.

using the definitions (3.35) and (3.36), as

$$= 4\pi \sum_{JMl} j_l(kR) \mathcal{Y}_{Jlj}^M(\hat{\mathbf{R}}, \hat{\mathbf{r}}) D_{Jlj}^{Mm_j^*}(\hat{\mathbf{k}}_j), \quad (3.37\,\text{b})$$

where $E = E_j + \hbar^2 k_j^2/2\mu$.

Each component in the sum above is an eigenfunction of $\mathbf{J}^2$ and J_z and can be taken as an initial state in an L.S. equation.§ Such an initial state corresponds to an initial relative momentum of magnitude k_j in the direction $\hat{\mathbf{k}}_j$ to the rotor being in the state j and to the lth partial wave of relative motion. Since only J and M are conserved during the collision we expect transitions out of this state to all other j and l values consistent with the given total J.

We denote a particular j,l combination by γ, and expand the solution of the L.S. equation as (cf. eqns (3.37) and (3.8))

$$\psi_{\mathbf{k},\gamma}^{+,J}(\mathbf{R}, \hat{\mathbf{r}}) = 4\pi \sum_{\gamma'} R^{-1} F_\gamma^{\gamma'}(R) \mathcal{Y}_{J,\gamma'}^M(\hat{\mathbf{R}}, \hat{\mathbf{r}}) D_{J,\gamma}^{Mm_j^*}(\hat{\mathbf{k}}). \quad (3.38)$$

To write the L.S. equation we consider the Green's function for the relative motion of the non-interacting rotor and atom

$$\langle \hat{\mathbf{r}}, \mathbf{R} | (E^+ - h - K)^{-1} | \hat{\mathbf{r}}', \mathbf{R}' \rangle = \sum_{j,m_j} \langle \hat{\mathbf{r}} | jm_j \rangle \langle \mathbf{R} | (E^+ - E_j - K)^{-1} | \mathbf{R}' \rangle \langle jm_j | \hat{\mathbf{r}}' \rangle, \quad (3.39)$$

where h is the internal Hamiltonian

$$h|jm_j\rangle = E_j|jm_j\rangle. \quad (3.40)$$

Introducing the partial wave expansion of the Green's function for K and using the unitarity relation we can transform the Green's function to

$$\sum_{JMlj} \mathcal{Y}_{Jlj}^M(\hat{\mathbf{R}}, \hat{\mathbf{r}}) G_{k_j}^{+l}(R, R') \mathcal{Y}_{Jlj}^{m^*}(\hat{\mathbf{R}}', \hat{\mathbf{r}}'), \quad (3.41)$$

where $G_{k_j}^{+l}$ is the outgoing Green's function for the lth partial wave at the energy $E - E_j$, $E - E_j = \hbar^2 k_j^2/2\mu$. For $R > R'$ (cf. eqn (1.A.35)),

$$G_{k_j}^{+l}(R, R') = -(2\mu/\hbar^2) k_j h_l^+(k_j R) j_l(k_j R'), \quad (3.42)$$

where $h_l^+(kR)$ is the outgoing solution of the radial Schrödinger equation

$$h_l^+(kR) \to i^{-l} \exp(ikR)/kR. \quad (3.43)$$

The orthogonality relations of the spherical harmonics and the unitarity relations of the vector coupling coefficients imply that

$$\iint \mathcal{Y}_{J,\gamma'}^{M*}(\hat{\mathbf{R}}, \hat{\mathbf{r}}) \mathcal{Y}_{J,\gamma}^M(\hat{\mathbf{R}}, \hat{\mathbf{r}}) \, d\hat{\mathbf{R}} d\hat{\mathbf{r}} = \delta_{\gamma\gamma'}. \quad (3.44)$$

Thus substituting eqns (3.37), (3.38), and (3.41) into the L.S. equation, premultiplying by $\mathcal{Y}_{J,\gamma'}^{M*}$, and integrating over the angles we obtain an

§ The factor $D_{Jlj}^{Mm_j^*}(\hat{\mathbf{k}}_j)$ is irrelevant for this discussion. It is essentially an expansion coefficient that ensures that the separate partial waves combine linearly to yield a plane wave, in the direction $\hat{\mathbf{k}}_j$ and the rotor in an orientation specified by m_j.

L.S. equation for an initial state of a specified value of γ

$$F_\gamma^{\gamma'}(R) = Rj_l(kR)\delta_{\gamma\gamma'} + \sum_{\gamma''} \int g_{k_{j'}}^{+l'}(R, R')V_{\gamma',\gamma''}^J(R')F_\gamma^{\gamma''}(R')\,dR', \quad (3.45)$$

where
$$V_{\gamma',\gamma''}^J(R) = \int \mathscr{Y}_{J,\gamma'}^{M*} V(\mathbf{R}, \hat{\mathbf{r}}) \mathscr{Y}_{J,\gamma''}^M \, d\hat{\mathbf{R}} d\hat{\mathbf{r}} \quad (3.46)$$

and
$$g_{k_j}^{+l}(R, R') = RR'G_{k_j}^{+l}(R, R'), \quad (3.47)$$

so that
$$\left[E - E_{j'} + \frac{\hbar^2}{2\mu}\frac{d^2}{dR^2} - \frac{\hbar^2 l'(l'+1)}{2\mu R^2}\right]g_{k_{j'}}^{+l'} = \delta(R-R'). \quad (3.48)$$

A method for the numerical solution of such a set of coupled integral equations has recently been discussed by Johnson (1967). (See also Johnson and Secrest (1966), Secrest and Johnson (1966), and Calogero (1967), Chapter 19 and references therein.) Most other studies have obtained numerical solutions by converting the integral equations to differential equations with specified boundary conditions. To obtain the differential equations we use eqn (3.48) in eqn (3.45) to get

$$\left[E - E_{j'} + \frac{\hbar^2}{2\mu}\frac{d^2}{dR^2} - \frac{\hbar^2 l'(l'+1)}{2\mu R^2}\right]F_\gamma^{\gamma'}(R) = \sum_{\gamma''} V_{\gamma',\gamma''}^J(R)F_\gamma^{\gamma''}(R). \quad (3.49)$$

The necessary boundary conditions are determined by the integral equation (3.45). As $R \to 0$ both terms in eqn (3.45) tend to zero since

$$\lim_{R \to 0} Rj_l(kR) \to 0.$$

(The limit of g_k^{+l} is determined from eqns (3.47) and (3.43).) As $R \to \infty$ we find, using eqns (3.42), (3.43), and (3.47),

$$F_\gamma^{\gamma'}(R) \to k_j^{-1}\sin(k_j R - \tfrac{1}{2}l\pi)\delta_{\gamma\gamma'} -$$
$$- \sum_{\gamma''} (2\mu/\hbar^2)i^{-l}\exp(ik_{j'} R) \int R'j_{l'}(k_{j'} R')V_{\gamma',\gamma''}^J(R')F_\gamma^{\gamma''}(R')\,dR', \quad (3.50)$$

which we rewrite as

$$(i/2k_j)\{\delta_{\gamma\gamma'}\exp[-i(k_j R - \tfrac{1}{2}l\pi)] -$$
$$- (k_j/k_{j'})^{\frac{1}{2}} S^J(\gamma', \gamma)\exp[i(k_{j'} R - \tfrac{1}{2}l\pi)]\}, \quad (3.51)$$

where $S^J(\gamma', \gamma)$ is defined by

$$S^J(\gamma', \gamma) = \delta_{\gamma'\gamma} + i(k_j k_{j'})^{\frac{1}{2}}(4\mu/\hbar^2)\sum_{\gamma''}\int Rj_{l'}(k_{j'} R)V_{\gamma',\gamma''}^J(R)F_\gamma^{\gamma''}(R)\,dR. \quad (3.52)$$

If we regard the functions $F_\gamma^{\gamma'}$ as forming an array $\mathbf{F}$, with different columns corresponding to different values of γ (which specifies the initial state), we can write the set of coupled equations (3.49) as a matrix equation,

$$(\mathbf{E}-\mathbf{H}_0)\mathbf{F} = \mathbf{VF}, \quad (3.53)$$

where $\mathbf{E}-\mathbf{H}_0$ is a diagonal matrix of elements

$$\left\{E-E_j+(\hbar^2/2\mu)\left[\frac{d^2}{dR^2}-\frac{l(l+1)}{R^2}\right]\right\}\delta_{\gamma\gamma'}$$

and
$$(\mathbf{V})_{\gamma',\gamma''} = V^J_{\gamma',\gamma''}.$$

A separate matrix equation holds for each J value.

For the purpose of numerical integration (for example, Allison and Dalgarno 1967) one is forced to truncate the matrix equation to a matrix of small dimension (see, however, Lester and Bernstein 1967). It is also possible to avoid the use of complex quantities by the introduction of 'standing wave' boundary conditions (section 2.5.1).

An alternative approach to the present problem (Jacob and Wick 1959, Lawley and Ross 1965) can be based on the fact that the component of J along the direction of relative motion is also conserved. Since $\mathbf{l}$ has no component in this direction ($\mathbf{R}.\mathbf{R}\wedge\mathbf{p} = 0$) one can transform to this representation by rotating the orientation of the diatomic molecule only, so that its axis of quantization coincides with $\mathbf{R}$.

The scattering amplitude is defined as usual by

$$f_{j'm_{j'},jm_j}(\mathbf{k}_{j'}\mathbf{k}_j) = -(\mu/2\pi\hbar^2)\langle\mathbf{k}_{j'}j'm_{j'}|V|\mathbf{k}_j jm_j^+\rangle. \quad (3.54)$$

Introducing the expansions (3.37) and (3.38) we obtain for the scattering amplitude

$$\sum_{JM}\sum_{ll'} -(2\pi)^2(k_j k_{j'})^{-\frac{1}{2}}T^J(\gamma',\gamma)D^{Mm_{j'}}_{J,\gamma'}(\hat{\mathbf{k}}_{j'})D^{Mm_j^*}_{J,\gamma}(\hat{\mathbf{k}}_j), \quad (3.55)$$

where $T^J(\gamma',\gamma)$ is defined by

$$S^J(\gamma',\gamma) = \delta_{\gamma'\gamma}+2\pi i T^J(\gamma',\gamma). \quad (3.56)$$

We shall find it possible to interpret $-(2\pi)^2(k_j k_{j'})^{-\frac{1}{2}}T^J(\gamma',\gamma)$ as the scattering amplitude from the state labelled by γ to the state labelled by γ'. Introducing the definition of the D functions we can rewrite the scattering amplitude as

$$\sum_{lm_l}\sum_{l'm_{l'}} -\mathrm{i}^{(l-l')}(2\pi)^2(k_j k_{j'})^{-\frac{1}{2}}T(l'm_{l'}j'm_{j'},lm_l jm_j)Y^{m_{l'}}_{l'}(\hat{\mathbf{k}}_{j'})Y^{m_l^*}_l(\hat{\mathbf{k}}_j), \quad (3.57)$$

where

$$T(l'm_{l'}j'm_{j'},lm_l jm_j) = \sum_{JM}\langle j'm_{j'},l'm_{l'}|j'l'JM\rangle T^J(\gamma',\gamma)\langle jlJM|jm_j lm_l\rangle.$$

To extend the present discussion to the collision of two diatomic molecules we assume that initially each molecule has a definite internal angular momentum so that we can put

$$\varphi_{n_\alpha}(\mathbf{r}_1) = \varphi_{n_\alpha}(r_1)Y^{m_{j_1}}_{j_1}(\hat{\mathbf{r}}_1), \quad (3.58)$$

and similarly for the second molecule. We then combine the two spherical harmonics to form
$$\mathscr{Y}^{m_j}_{jj_1j_2}(\hat{\mathbf{r}}_1, \hat{\mathbf{r}}_2),$$
as in eqn (3.35). j is now the internal angular momentum and the initial state is taken to be
$$\exp(i\mathbf{k}_n\cdot\mathbf{R})\varphi_{n_\alpha}(r_1)\varphi_{n_\beta}(r_2)\mathscr{Y}^{m_j}_{jj_1j_2}(\hat{\mathbf{r}}_1, \hat{\mathbf{r}}_2). \tag{3.59}$$
The remainder of the discussion is unchanged except of course that the evaluation of $V^J(R)$ involves integrating over the variables $\mathbf{r}_1$ and $\mathbf{r}_2$ with respect to both magnitude and direction. (See, for example, Davison 1962, Takayanagi 1965.)

2.3.2. *The Born approximation*

In the Born approximation the exact scattering amplitude (3.16) is replaced by
$$f_{m,n}(\mathbf{k}',\mathbf{k})_B = -(\mu/2\pi\hbar^2)\langle m,\mathbf{k}'|V|\mathbf{k},n\rangle, \tag{3.60}$$
where $E_m + \hbar^2 k'^2/2\mu = E_n + \hbar^2 k^2/2\mu$. In terms of the matrix elements of the potential
$$f_{m,n}(\mathbf{k}',\mathbf{k})_B = -(\mu/2\pi\hbar^2)\int \exp(-i\mathbf{q}\cdot\mathbf{R})V_{m,n}(\mathbf{R})\,d\mathbf{R}, \tag{3.61}$$
where $\mathbf{q}$ is the momentum transfer. The form of the result is an obvious extension of the purely elastic ($m = n$) case
$$f_{m,n}(\mathbf{k}',\mathbf{k})_B = -(\mu/2\pi\hbar^2)V_{m,n}(\mathbf{q}). \tag{3.62}$$
Our previous limits on the strength of the mutual interaction are equally applicable here.

As an example we consider a simple model for the collision of an atom A with a diatomic molecule BC, where we regard the atom as structureless and the molecule as two structureless atoms bound by a mutual interaction. We approximate the atom–molecule interaction as a sum of two-body terms,
$$V = V_C(\mathbf{R}_{AB}) + V_B(\mathbf{R}_{AC}), \tag{3.63}$$
where the coordinates are labelled in Fig. 2.6 and the potential is labelled by the 'absent' particle, so that the diatomic is bound together by $V_A(R_{BC})$. This model has been discussed by Kerner (1953), Brout (1954), and Tobocman (1961), and is sometimes referred to as a 'dumb-bell' model.

The Born scattering amplitude is now given as a sum of two terms
$$f_{m,n}(\mathbf{k}',\mathbf{k}) = -(\mu/2\pi\hbar^2)[U^{m,n}_C(\mathbf{q}) + U^{m,n}_B(\mathbf{q})]. \tag{3.64}$$
In this model we can separate the integrations necessary to evaluate $U(\mathbf{q})$ by the change of variables
$$\mathbf{R} = \mathbf{R}_{AB} - \mathbf{R}_B = \mathbf{R}_{AB} - \alpha\mathbf{R}_{BC} = \mathbf{R}_{AC} - \mathbf{R}_C = \mathbf{R}_{AC} + (1-\alpha)\mathbf{R}_{BC}, \tag{3.65}$$
with
$$\alpha = M_C/(M_B + M_C), \quad \mathbf{R}_{BC} = \mathbf{R}_B - \mathbf{R}_C.$$

Then
$$U_C^{m,n}(\mathbf{q}) = \iint d\mathbf{R} d\mathbf{R}_{BC} \exp(-i\mathbf{q}\cdot\mathbf{R})V_C(\mathbf{R}_{AB})\varphi_m^*(\mathbf{R}_{BC})\varphi_n(\mathbf{R}_{BC})$$
$$= \int d\mathbf{R}_{AB} \exp(-i\mathbf{q}\cdot\mathbf{R}_{AB})V_C(\mathbf{R}_{AB}) \times$$
$$\times \int d\mathbf{R}_{BC} \exp(i\alpha\mathbf{q}\cdot\mathbf{R}_{BC})\varphi_m^*(\mathbf{R}_{BC})\varphi_n(\mathbf{R}_{BC})$$
$$= V_C(\mathbf{q}) \cdot g_{m,n}(-\alpha\mathbf{q}) \qquad (3.66)$$

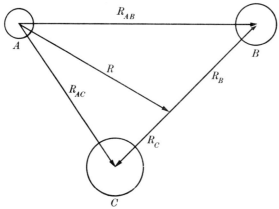

FIG. 2.6

and in the same fashion
$$U_B^{m,n}(\mathbf{q}) = V_B(\mathbf{q}) \cdot g_{m,n}[(1-\alpha)\mathbf{q}]. \qquad (3.67)$$
$-(\mu/2\pi\hbar^2)U_B^{m,n}$ can be interpreted as due to the collision of the atom A with the atom C of the diatomic. The Born scattering amplitude for the collision of the (structureless) atoms A and C is given by
$$-(\mu/2\pi\hbar^2)V_B(\mathbf{q}) \qquad (3.68)$$
and the effects of the molecular binding are only present in the function $g_{m,n}$, the form factor, which is the Fourier transform of the transition density $\varphi_m^*(\mathbf{R}_{BC})\varphi_n(\mathbf{R}_{BC})$. The orthonormality of the molecular states implies that
$$\lim_{q\to 0} g_{m,n}(\mathbf{q}) \to \delta_{mn} \qquad (3.69)$$
and similarly
$$\lim_{\alpha\to 0} g_{m,n}(\alpha\mathbf{q}) \to \delta_{mn}. \qquad (3.70)$$
The limit $\alpha \to 0$ corresponds physically to the situation $M_C \ll M_B$. In this limit the interaction of the atom A with the (very heavy) atom B of the diatomic will not contribute much to the inelastic transitions.

From a physical point of view the most severe limitation on the Born approximation in the present model is the neglect of amplitudes for the events where the atom A interacts first with atom B and then with atom C and then leaves. It is important to distinguish these events (whose amplitudes are at least quadratic in V) from the interference between the amplitudes U_C and U_B, in that

$$|f_{m,n}|^2 = (\mu/2\pi\hbar^2)^2[(U_C^{m,n})^2 + (U_B^{m,n})^2 + 2\,\text{re}\,U_C^{m,n}\,U_B^{m,n*}]. \quad (3.71)$$

The last term in the square bracket is the interference between the amplitudes that refer to the scattering by atoms B and C. The two atoms are at a finite distance, and the observer who is far removed cannot consider the two alternatives (scattering by either B or C) to be mutually exclusive. For example, in a homonuclear diatomic $V_B = V_C$ and $\alpha = \frac{1}{2}$. If the transition density is real

$$|f_{m,n}|^2 = (\mu/2\pi\hbar^2)^2|V(\mathbf{q})|^2 4|g(\mathbf{q}/2)|^2, \quad (3.72)$$

or twice the value expected for mutually exclusive alternatives.

To evaluate the form factor $g_{m,n}$ we can perform a partial wave expansion of the transition density

$$\varphi_m^*(\mathbf{r})\varphi_n(\mathbf{r}) = \sum_{l=0}^{\infty} \sum_{m_l=-l}^{l} Y_l^{m_l*}(\hat{\mathbf{r}}) g_{m,n}^{l,m_l}(r),$$

$$g_{m,n}^{l,m_l}(r) = \int d\hat{\mathbf{r}}\, \varphi_m^*(\mathbf{r})\varphi_n(\mathbf{r}) Y_l^{m_l}(\hat{\mathbf{r}}). \quad (3.73)$$

Using the partial wave expansion for a plane wave, eqn (2.69),

$$g_{m,n}(-\alpha\mathbf{q}) = \sum_l \sum_{m_l} 4\pi i^l Y_l^{m_l*}(\hat{\mathbf{q}}) \int_0^{\infty} dR_{BC}\, R_{BC}^2\, j_l(\alpha q R_{BC}) g_{m,n}^{l,m_l}(R_{BC}). \quad (3.74)$$

In the limit $q \to 0$, $j_l(\alpha q R_{BC}) \to (\alpha q R_{BC})^l/(2l+1)!!$ and so the lth term in the above expansion is (for $m \neq n$) proportional to $(\alpha q)^l$. Thus the contribution of the lth wave to the inelastic scattering amplitude vanishes for zero momentum transfer, and is proportional, for small q, to $(\alpha q)^l$.

For a rigid rotor the internal quantum number n specifies the rotational angular momentum j and its projection on a fixed axis m_j. Thus with $m = j, m_j$

$$\varphi_m(\mathbf{r}) = \varphi(r) Y_j^{m_j}(\hat{\mathbf{r}}),$$

$$|\varphi(r)|^2 = r^{-2}\delta(r - r_e), \quad (3.75)$$

where r_e is the equilibrium separation. For rotational excitation from the ground state $(n = 0, 0)$ to some level m

$$g_{m,n}^{l,m_l} = r^{-2}\delta(r - r_e)(4\pi)^{-\frac{1}{2}} \int Y_j^m(\hat{\mathbf{r}}) Y_l^{m_l}(\hat{\mathbf{r}})\, d\hat{\mathbf{r}}$$

$$= r^{-2}\delta(r - r_e)(4\pi)^{-\frac{1}{2}} \delta_{jl}\delta_{m_j m_l}, \quad (3.76)$$

since $Y_0^0 = (4\pi)^{-\frac{1}{2}}$. For the form factor we obtain from eqn (3.74)

$$g_{m,n}(\alpha\mathbf{q}) = (4\pi)^{\frac{1}{2}}i^j Y_j^{m_j^*}(\hat{\mathbf{q}})j_j(\alpha q_e). \tag{3.77}$$

In the following section we prove that the differential cross-section for the inelastic transition $n \to m$ is given by§

$$(k'/k)|f_{m,n}(\mathbf{k'},\mathbf{k})|^2 \tag{3.78}$$

and that the cross-section for a transition into a group of final states is obtained by summing over the respective cross-sections into individual final states.

Thus the differential cross-section for the excitation of a homonuclear diatomic from the ground state to the rotational state j is obtained from eqns (2.78), (3.77), and (3.72), as

$$(k'/k)(\mu/\hbar^2)^2 V^2(\mathbf{q})(2j+1)4j_j^2(\tfrac{1}{2}qr_e), \tag{3.79}$$

where we summed over the quantum number m_j using the addition theorem

$$\sum_{m_j} Y_j^{m_j^*}(\hat{\mathbf{q}})Y_j^{m_j}(\hat{\mathbf{q}}) = (2j+1)/4\pi. \tag{3.80}$$

When the mutual interaction between the atoms is a function of their relative distance only $V(\mathbf{q})$ is a function of q and the total cross-section is easily computed by relating q to the angle between $\mathbf{k}$ and $\mathbf{k'}$ by

$$q^2 = |\mathbf{k}-\mathbf{k'}|^2 = k^2+k'^2-2kk'\cos\theta, \tag{3.81}$$

so that

$$\frac{d\cos\theta}{dq} = q/kk'$$

and the limits $\theta = 0$ and $\theta = \pi$ correspond to $q = k-k'$ and $q = k+k'$ respectively. We thus have for the total cross-section, eqn (2.34),

$$4(2j+1)(\mu/\hbar^2)^2\frac{2\pi}{k^2}\int_{k-k'}^{k+k'} V^2(q)j_j^2(\tfrac{1}{2}qr_e)q\,dq, \tag{3.82}$$

where we have integrated taking k as the polar axis and have put $d\hat{\mathbf{k}}' = d\cos\theta\,d\phi$.

Kerner (1953) has performed the q integration for the excitation of N_2 by protons using for each term in eqn (3.63) the interaction between a point charge and the Thomas–Fermi field of the isolated atom. His

---

§ Equation (3.78) also follows from the definition of $d\sigma(\mathbf{k'},\mathbf{k})$ as the ratio of the scattered flux (in the direction $\mathbf{k'}$) to the incident current (in the direction $\mathbf{k}$), eqn (2.32),

$$\frac{\hbar k}{\mu}d\sigma = \mathbf{j}.d\mathbf{S})_{sc} = \frac{\hbar k'}{\mu}|f|^2,$$

when we recall that due to the internal excitation the final relative momentum is k' $\hbar^2 k'^2/2\mu = E-E_m$ (cf. eqns (3.15)).

results are summarized in Fig. 2.7. As in the elastic case the high-energy cross-section is proportional to k^{-2}.

The position of the maximum value of the Born cross-section is seen to shift to higher relative momenta as the energy transfer increases. We

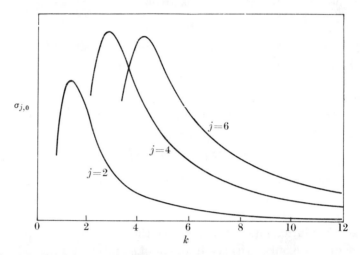

FIG. 2.7. Total cross-section for rotational excitation of N_2 by protons, computed by Kerner (1953), as function of the relative momentum in atomic units. The cross-sections to the $j = 4$ and 6 levels are magnified 20 and 100 times respectively in relation to the excitation to the $j = 2$ level. Vibrational excitation cross-sections were found to be 10^{-4}–10^{-5} times smaller, even at their maximal value. This is in qualitative accord with experiments on diatomics at moderate temperatures (10^3 °K).

can obtain a qualitative understanding of this phenomena when we recall that in the Born approximation the relative motion is not distorted by the potential so that the energy transferred by the potential is ΔE, where $\Delta E = E_m - E_n$. A transfer of this magnitude is most likely when the collision duration, a/v, is equal to $\hbar/\Delta E$ (where a is the range of the interaction and v is the relative velocity), or

$$v = a\Delta E/\hbar. \tag{3.83}$$

This criterion (Massey 1949) with $\Delta E = E_m - E_n$ is only valid when the Born approximation is valid. In molecular problems there is often considerable distortion of the relative motion (cf. section 3.2.0) and the simple estimate for ΔE is no longer valid. The cross-section can thus attain significant values also for values of v other than those predicted by eqn (3.83). The average relative velocity at translational temperature T is

$$v = (8\mathbf{k}T/\pi\mu)^{\frac{1}{2}} \simeq 2\cdot 8 \cdot 10^{-3}(T/\mu)^{\frac{1}{2}} \text{ a.u.},$$

where μ is the reduced mass in atomic units and k is Boltzmann's constant. From eqn (3.83),

$$\Delta E = \hbar v/a \simeq 2\cdot 8.10^{-3}(T/\mu)^{\frac{1}{2}}a^{-1} \text{ a.u.}$$

We expect a to be of the order of a few atomic units so that, even at fairly high temperatures, ΔE (for large cross-sections) is restricted to the range of rotational energy differences (about 10^{-4} a.u. Vibrational-energy differences are typically of the order of 10^{-2} a.u.).

Introducing the initial relative kinetic energy $E = \frac{1}{2}\mu v^2$,

$$E/\Delta E \leqslant \tfrac{1}{4}B(\Delta E/\epsilon). \tag{3.84}$$

Here $B = 2\mu\epsilon a^2/\hbar^2$, where ϵ is the depth of the mutual interaction (cf. eqn. 1.A.21). The inequality refers to the low velocity (the so-called 'adiabatic') regime. When the inequality sign in eqn. (3.84) is reversed, one is in the 'sudden' regime, where ΔE is within the energy uncertainty, $\hbar v/a$, during the collision.

A rough estimate of the importance of distortion is obtained by comparing the relative velocity with the velocity of the internal motions of the isolated molecules. We expect the Born approximation to be valid only when the relative velocity is higher than the internal velocities. For a rigid rotor the internal velocities are quite low§ and the Born approximation should be valid to fairly low velocities. The situation is less satisfactory for vibrational excitation where the internal velocities are of the order of 10^{-3} a.u. (Recall also that 1 a.u. of velocity is the velocity of an electron in the first Bohr orbit.) In such cases we expect eqn (3.83) to predict only one possible range of velocities where an efficient transfer can take place. As we have seen, apart from rotational excitation (or cases where $\mu \sim 1$ rather than $\mu \sim 10^4$, as is the case for molecule–molecule collisions) the velocities predicted by eqn (3.83) with $\Delta E = E_n - E_m$ are outside the range of thermal velocities. Stueckelberg (1932), using a semi-classical analysis, has modified the definition of ΔE to account for the distortion of the relative motion. When the velocity does satisfy the criterion (3.83) we expect a cross-section of the order of magnitude of πa^2 (Zener (1931), London (1932)). We can see this in a qualitative manner from eqn (2.78) by putting

$$P(b) = \begin{cases} 1 & (b \leqslant a) \\ 0 & (b > a), \end{cases}$$

since $b > a$ corresponds to passage outside the range of the potential (cf. Fig. 2.1 and section 3.4.1).

§ For N_2 the internal velocity is about $2.10^{-5}[j(j+1)]$ a.u.

2.4. The rate of change of observables

2.4.0. THE net rate of change of some observable is usually expressed in terms of some relaxation equation. As an example we can think of a small density of diatomic molecules in a monoatomic gas and consider as an observable the internal energy (whether vibrational or rotational or both) of the diatomic. Let the index n label the states of the diatomic so that A_n is the internal energy and C_n is the density of the state n. If $k_{m,n}$ is the rate of transitions from the state n to the state m, we expect the rate of change of internal energy per unit volume due to collisions of the diatomic in the state n with the monoatomic host to be given by

$$C_n \sum_m (A_m - A_n) k_{m,n},$$

and the total rate of change of internal energy by

$$\sum_n \frac{\mathrm{d}}{\mathrm{d}t}(A_n C_n) = \sum_n C_n \sum_m (A_m - A_n) k_{m,n}. \tag{4.1}$$

The total internal energy (per unit volume) $\sum_n A_n C_n$ is not conserved in time due to the changes in the densities C_n induced by collisions with the host gas. In fact the operational approach to the derivation of eqn (4.1) starts from the rate equations for the densities

$$\frac{\mathrm{d}C_n}{\mathrm{d}t} = \sum_m k_{n,m} C_m - \sum_m k_{m,n} C_n. \tag{4.2}$$

Equation (4.1) is obtained on multiplying by A_n and summing over n.

In this section we begin the quantum-mechanical discussion of such relations by deriving exact expressions for the rate of change of the expectation value of an observable A that can be measured both before and after the collision. Our main result will be the derivation of eqn (4.1), when before the collision the relative momentum of the collision partners is well defined. The generalization to ensembles can then be made by making suitable assumptions about the density matrix of the system in the absence of collisions.

We consider an operator A that commutes with the Hamiltonian H_0, $[A, H_0] = 0$, and has no explicit time-dependence, so that the only cause of time dependence of $\langle A \rangle$ is the mutual interaction V. In order to avoid any explicit normalization convention we shall label the eigenstates of H_0 as $|f\rangle$, where f stands both for the total energy and any other quantum numbers, $H_0|f\rangle = E_f|f\rangle$, and since A and H_0 commute these states can be arranged to be eigenvectors of A, $A|f\rangle = A_f|f\rangle$. The spectral resolution of A is given formally by

$$A = \sum_f |f\rangle A_f \langle f|, \qquad (4.3)$$

and the L.S. equation is written as

$$|i^+\rangle = |i\rangle + (E_i^+ - H_0)^{-1} V |i^+\rangle,$$

where $|i\rangle$ is the initial state. We shall use the following matrix element derived from this equation by taking its scalar product with the state $\langle f|$, and using the basic property of the scalar product, $\langle i^+|f\rangle = \langle f|i^+\rangle^*$, so that

$$\langle i^+|f\rangle = \delta_{f,i} + (E_i^- - E_f)^{-1} \langle f|V|i^+\rangle^*. \qquad (4.4)$$

The expectation value of the rate of change of A during the stationary collision process is obtained by taking the expectation value of Heisenberg's equation of motion of A,

$$\frac{d}{dt}\langle i^+|A|i^+\rangle = (i\hbar)^{-1} \langle i^+|[A,H]|i^+\rangle$$
$$= (i\hbar)^{-1}[\langle i^+|AV|i^+\rangle - \langle i^+|VA|i^+\rangle]$$
$$= (2/\hbar)\mathrm{im}\langle i^+|AV|i^+\rangle. \qquad (4.5)$$

Here, as A is measurable in the absence of the interaction,

$$[A, H_0 + V] = [A, V].$$

Using eqn (4.3),

$$\frac{d}{dt}\langle i^+|A|i^+\rangle = (2/\hbar)\mathrm{im} \sum_f \langle i^+|f\rangle A_f \langle f|V|i^+\rangle$$
$$= (2/\hbar) \sum_f A_f \mathrm{im}\langle f|V|i^+\rangle[\delta_{fi} + (E_i^- - E_f)^{-1}\langle f|V|i^+\rangle^*]$$
$$= (2/\hbar) A_i \mathrm{im}\langle i|V|i^+\rangle + (2\pi/\hbar) \sum_f A_f |\langle f|V|i^+\rangle|^2 \delta(E_i - E_f), \qquad (4.6)$$

where in the last expression we have used eqn (4.4), recalling that summation over the index f involves integration over the continuous energy E_f. It is important to note that we have taken the limit $\epsilon \to 0$ only in the last stage of the above derivation, after the evaluation of the commutator and after taking the matrix elements of the Green's operator. This ensures that there are no surface terms that contribute to the commutator in the coordinate representation, so that all operators are Hermitian. It is also possible to proceed by working explicitly in the coordinate representation (section 2.6.4). In this case one goes to the limit $\epsilon \to 0$ during the evaluation of $\langle \mathbf{r}|i^+\rangle = \psi_i^+(\mathbf{r})$, and the commutator has to be evaluated using Green's theorem since $\psi_i^+(\mathbf{r})$ does not vanish asymptotically when $\epsilon \to 0$. (The final results are, of course, identical.)

If A is the identity operator, the left-hand side is the rate of change of the normalization integral $\langle i^+|i^+\rangle$. Since the identity commutes with the total Hamiltonian, the normalization integral is time independent and eqn (4.6) reduces to

$$-(2/\hbar)\mathrm{im}\langle i|V|i^+\rangle = (2\pi/\hbar)\sum_f |\langle f|V|i^+\rangle|^2 \delta(E_i - E_f). \qquad (4.7)$$

Equation (4.7) is thus equivalent to the statement of conservation of probabilities, and is referred to as *the optical theorem*. Using the optical theorem in eqn (4.6),

$$\frac{\mathrm{d}}{\mathrm{d}t}\langle i^+|A|i^+\rangle = \frac{2\pi}{\hbar}\sum_f (A_f - A_i)|\langle f|V|i^+\rangle|^2 \delta(E_i - E_f). \qquad (4.8)$$

If we put $A = |f\rangle\langle f|$ (or equivalently from eqn (4.3) $A_j = \delta_{jf}$) we have that for $f \neq i$

$$R_{f,i} = \frac{\mathrm{d}}{\mathrm{d}t}\langle i^+|f\rangle\langle f|i^+\rangle = \frac{2\pi}{\hbar}|\langle f|V|i^+\rangle|^2 \delta(E_f - E_i). \qquad (4.9)$$

$R_{f,i}$, defined by the first equation, is the rate of change of the probability to observe the state $|f\rangle$ due to a collision with the initial state $|i\rangle$. The final form of eqn (4.8) is

$$\frac{\mathrm{d}}{\mathrm{d}t}\langle i^+|A|i^+\rangle = (i\hbar)^{-1}\langle i^+|[A,V]|i^+\rangle = \sum_f (A_f - A_i) R_{f,i}. \qquad (4.10)$$

Equation (4.10) was derived by Lippmann (1965a, 1966) who named it 'the generalized Ehrenfest theorem'. The second form of eqn (4.10) is indeed the expression that one would tend to write for the rate of change of $\langle A\rangle$ on physical grounds. The generalized Ehrenfest theorem establishes the connection between this expression and the commutator $[A, V]$.

The rate of transitions into a group of states f, in some class c, is obtained by taking A to be the projection onto the group of states

$$A_c = \sum_{f \in c} |f\rangle\langle f|. \qquad (4.11)$$

$$\frac{\mathrm{d}\langle A_c\rangle}{\mathrm{d}t} = \frac{2\pi}{\hbar}\sum_{f \in c} |\langle f|V|i^+\rangle|^2 \delta(E_f - E_i) = \sum_{f \in c} R_{f,i}. \qquad (4.12)$$

In particular we can, using the optical theorem, eqn (4.7), interpret $-(2/\hbar)\mathrm{im}\langle i|V|i^+\rangle$ as the total rate of transitions out of the state i.

These exact results for the transition rates bear considerable similarity to the celebrated 'Golden rule' of time-dependent perturbation theory. The essential difference is that the Golden rule replaces the matrix elements $\langle f|V|i^+\rangle$ by their Born approximation $\langle f|V|i\rangle$, and is thus the lowest (Born) approximation to the rate.

2.4.1. *Collision rates and cross-sections*

Quantitative characterization of collision processes can often be achieved in terms of concepts which can be interpreted in a classical fashion, even though their numerical magnitude must be determined in a quantum-mechanical fashion. Rates of collision events can be expressed in terms of an effective area, the cross-section, which is characteristic of the state of the colliding molecules and is introduced by analogy to the kinetic theory of gases. For an observer located on one molecule, the cross-section σ is the area presented by the collision partner during the collision. Let the relative velocity of the collision partners be v_r, so that $v_r \sigma$ is the effective volume swept by a collision partner per unit time. If the collision partners are randomly distributed with a density n_2, the observer will express the rate as $n_2 v_r \sigma$. If there are n_1 observers per unit volume the total rate per unit volume is $n_1 n_2 v_r \sigma$. The average distance between successive collisions of the same observer is $l = v_r/n_2 v_r \sigma = 1/n_2 \sigma$. Our definition of the cross-section referred to an arbitrary distribution of the relative directions of the observer and the collision partners. If all the collision partners move in the same direction towards the observer, with a flux of I molecules per unit area, one can rewrite the collision rate as $\sigma \, \mathrm{d}I/\mathrm{d}t$ (since $\mathrm{d}I/\mathrm{d}t = v_r n_2$ in this case). From the definition it is clear that one can introduce a cross-section for any type of rate measured by the observer, in that the observer may refer only to a particular subset of events in his definition of 'collision': for example, those events in which the relative kinetic energy does not change (elastic collisions) or only those events in which the molecules emerge from the collision in a specified set of internal states, etc.

Consider a single stationary observer (the 'target') in a volume V ($n_1 = \mathsf{V}^{-1}$). In the initial state its collision partners (the 'beam') have a well-defined momentum $\mathbf{k}$. To show explicitly the volume dependence of various quantities, we shall, in this section only, normalize states to a finite volume V, so that the coordinate representation of the initial state is

$$\phi_{\mathbf{k}}(\mathbf{R}) = \mathsf{V}^{-\frac{1}{2}} \exp(i\mathbf{k}.\mathbf{R}),$$

and
$$n_2 = \mathsf{V}^{-1} \int_{\mathsf{V}} \phi_{\mathbf{k}}^{*}(\mathbf{R})\phi_{\mathbf{k}}(\mathbf{R}) \, \mathrm{d}\mathbf{R} = \mathsf{V}^{-1}. \tag{4.13}$$

The relative velocity is $v_r = \hbar k/m$, where m is the mass of a beam molecule. The total transition rate $\mathrm{d}\dot{N}/\mathrm{d}\omega$, out of the beam state $|\mathbf{k}_i\rangle$ into a solid angle $\mathrm{d}\omega$, is given by eqn (4.14), where V is the interaction between the target and the beam:

$$\frac{d\dot{N}(\mathbf{k}_f \leftarrow \mathbf{k}_i)}{d\omega} = \frac{d}{d\omega}\left\{\frac{2\pi}{\hbar}\sum_{\mathbf{k}_f}\delta[E(k_f)-E(k_i)]|\langle\phi_{\mathbf{k}_f}|V|\psi_{\mathbf{k}_i}^+\rangle|^2\right\}. \quad (4.14)$$

The summation over $\mathbf{k}_f$ is replaced by an integration over $\mathbf{k}_f$ using the substitution, valid for a large volume

$$\sum_{\mathbf{k}_f} = (2\pi)^{-3}\mathsf{V}\int d\mathbf{k}_f \quad (4.15)$$

and the delta function in energy is replaced by a delta function of conservation of the magnitude of the relative momentum (eqn (2.8)) to give

$$\frac{d\dot{N}(\mathbf{k}_f \leftarrow \mathbf{k}_i)}{d\omega} = \mathsf{V}\frac{2\pi}{\hbar}\int \delta(k_f-k_i)|\langle\phi_{\mathbf{k}_f}|V|\psi_{\mathbf{k}_i}^+\rangle|^2 \frac{m\, d\mathbf{k}_f}{(2\pi)^3\hbar^2 k_f\, d\omega}$$

$$= \frac{m^2 v_\mathrm{r}}{(2\pi)^2\hbar^4}\mathsf{V}|\langle\phi_{\mathbf{k}_f}|V|\psi_{\mathbf{k}_i}^+\rangle|^2, \quad (4.16)$$

where $d\mathbf{k}_f = k_f^2\, dk_f\, d\omega$ and in the matrix element $k_f = k_i$ after the integration.

The differential cross-section $d\sigma/d\omega$ is defined by

$$\frac{d\sigma}{d\omega} = \mathsf{V}^{-1}\frac{d\dot{N}(\mathbf{k}_f \leftarrow \mathbf{k}_i)}{d\omega}\bigg/n_1 n_2 v_\mathrm{r} = \frac{m^2\mathsf{V}^2}{(2\pi\hbar)^2}|\langle\phi_{\mathbf{k}_f}|V|\psi_{\mathbf{k}_i}^+\rangle|^2. \quad (4.17)$$

In the present normalization (eqn (4.13)) the scattering amplitude $f(\mathbf{k}_f, \mathbf{k}_i)$ is given by (eqn (2.27)),

$$f(\mathbf{k}_f, \mathbf{k}_i) = -\frac{m\mathsf{V}}{2\pi\hbar}\langle\phi_{\mathbf{k}_f}|V|\psi_{\mathbf{k}_i}^+\rangle, \quad (4.18)$$

so that

$$\frac{d\sigma}{d\omega} = |f(\mathbf{k}_f, \mathbf{k}_i)|^2. \quad (4.19)$$

The scattering amplitude and the differential cross-section do not depend on the normalization volume, since the explicitly indicated volume-dependence in eqn (4.18) exactly cancels the implicit volume-dependence of the initial and final states.

We turn now to the collision of two molecules. In a space fixed system of coordinates the wavefunction for the initial state ϕ_I is a product of the wavefunction in the centre of mass system ϕ_i, times a translational wavefunction for the motion of the total centre of mass

$$\phi_I = \phi_i\, \mathsf{V}^{-\frac{1}{2}}\exp(i\boldsymbol{\varkappa}_i\cdot\boldsymbol{\rho}_i),$$

$$\phi_i = \varphi_{n_\alpha}(\mathbf{r}_1)\varphi_{n_\beta}(\mathbf{r}_2)\mathsf{V}^{-\frac{1}{2}}\exp(i\mathbf{k}_n\cdot\mathbf{R}),$$

where $\boldsymbol{\rho}$ is the position and $\boldsymbol{\varkappa}$ the momentum of the total centre of mass. The internal coordinates $\mathbf{r}_1$ and $\mathbf{r}_2$ refer to the centre of mass of molecules

1 and 2 respectively. The wavefunction ϕ_i can be interpreted as holding one molecule fixed at the origin and directing a beam with momentum $\hbar k_n$ and density $n_2 = V^{-1}$ towards it. Thus we see, without any further computation, that the collision rate computed in the centre of mass system is actually n_1 times the rate and $n_1 = V^{-1}$. In other words, the rate computed in the centre of mass system is the rate per unit volume. The situation is very familiar in chemical kinetics, where it is empirically well known that in collisions between two molecules the invariant quantity is the collision rate per unit volume (when we hold the number of molecules constant and change the volume). It is important to realize that our discussion depends on the assumption that the mutual interaction does not depend on the absolute position of the total centre of mass, so that

$$\langle \phi_F | V | \phi_I \rangle = \langle \phi_f | V | \phi_i \rangle V^{-1} \langle \varkappa_f | \varkappa_i \rangle, \tag{4.20}$$

since $|\phi_I\rangle = |\phi_i\rangle V^{-\frac{1}{2}} |\varkappa_i\rangle$. We can write

$$V^{-1}\langle \varkappa_f | \varkappa_i \rangle = V^{-1} \int_V d\rho \exp[i\rho(\varkappa_i - \varkappa_f)] = \delta_{\varkappa_f \varkappa_i} = \frac{(2\pi)^3}{V} \delta(\varkappa_f - \varkappa_i). \tag{4.21}$$

Thus, with $E_I = E_i + \hbar^2 \kappa_i^2 / 2M$,

$$\sum_{\varkappa_f} \delta(E_i + \hbar^2 \kappa_i^2/2M - E_f - \hbar^2 \kappa_f^2/2M) |\langle \phi_F | V | \psi_I^+ \rangle|^2$$

$$= \int d\varkappa_f \delta(\varkappa_f - \varkappa_i) V^{-1} \langle \varkappa_f | \varkappa_i \rangle |\langle \phi_f | V | \psi_i^+ \rangle|^2 \delta(E_i + \hbar^2 \kappa_i^2/2M - E_f - \hbar^2 \kappa_f^2/2M)$$

$$= \delta(E_i - E_f) |\langle \phi_f | V | \psi_i^+ \rangle|^2, \tag{4.22}$$

where we have used the fact that the normalization volume is V so that

$$V^{-1}\langle \varkappa_i | \varkappa_i \rangle = V^{-1} \int_V d\rho = 1. \tag{4.23}$$

The transition rate into the internal state m is given by the rate of increase of the probability of observing the internal state m occupied, if in the initial state the state n is occupied. Using Ehrenfest's theorem with $A = |\varphi_m\rangle\langle\varphi_m|$ we have

$$\dot{N}(m \leftarrow \mathbf{n}) = \frac{2\pi}{\hbar} \sum_{\mathbf{k}_m} \sum_{\varkappa_f} \delta(E_F - E_I) |\langle \phi_F | V | \psi_I^+ \rangle|^2, \tag{4.24}$$

where
$$\phi_I = \phi_i \, V^{-\frac{1}{2}} \exp(i\rho \cdot \varkappa_i)$$

and
$$\phi_i = \varphi_n(\mathbf{r}) V^{-\frac{1}{2}} \exp(i\mathbf{k}_n \cdot \mathbf{R}), \qquad E_i = E_n + \hbar^2 k_n^2/2\mu, \tag{4.25}$$

with similar definition for the final state. From eqn (4.22)

$$\mathsf{V}^{-1}\dot{N}(m \leftarrow \mathbf{n}) = \frac{2\pi}{\hbar \mathsf{V}} \sum_{\mathbf{k}_m} \delta(E_i - E_f) |\langle \phi_f | V | \psi_i^+ \rangle|^2$$

$$= \frac{2\pi}{\hbar} \int (2\pi)^{-3} \, d\mathbf{k}_m \frac{\mu}{\hbar^2 k_m} \delta\{k_m - [(E_i - E_m) 2\mu/\hbar]^{\frac{1}{2}}\} |\langle \phi_f | V | \psi_i^+ \rangle|^2$$

$$= \left(\frac{2\pi}{\hbar}\right) \frac{\mu \hbar k_m}{(2\pi\hbar)^3} \int d\omega_m |\langle \phi_f | V | \psi_i^+ \rangle|^2, \tag{4.26}$$

where $E_i = E_m + \hbar^2 k_m^2 / 2\mu$.

Equation (4.26) determines the rate of transitions per unit volume for a pair of reactants ($n_1 = n_2 = \mathsf{V}^{-1}$). The rate for an arbitrary density of reactants ($n_1 = N_1/\mathsf{V}$, $n_2 = N_2/\mathsf{V}$) for those events where one molecule of type one collides with one molecule of type two is expected to be $N_1 N_2$ times the rate for a single-pair collision. To see this we note that the observable A can now be resolved as

$$A = \sum_\alpha A_\alpha, \tag{4.27}$$

where α refers to a specific pair of molecules, so that

$$\frac{d\langle A \rangle}{dt} = \sum_\alpha \frac{d\langle A_\alpha \rangle}{dt}. \tag{4.28}$$

If only pairwise collisions can take place, $\langle A_\alpha \rangle$ changes with time due to the collision of the pair α only. When all the pairs are equivalent

$$\frac{d\langle A \rangle}{dt} = N_1 N_2 \frac{d\langle A_\alpha \rangle}{dt}. \tag{4.29}$$

We thus obtain for the transition rate§ per unit volume

$$\mathsf{V}^{-1}\dot{N}(m \leftarrow \mathbf{n}) = (N_1/\mathsf{V})(N_2/\mathsf{V}) \frac{\mu \hbar k_m}{(2\pi\hbar)^3} \left(\frac{2\pi}{\hbar}\right) \times$$

$$\times \int d\hat{\mathbf{k}}_m |\langle \mathbf{k}_m, m | V | \mathbf{k}_n, n^+ \rangle|^2, \tag{4.30}$$

where we have removed a factor V^{-1} from each matrix element (cf. eqn (4.25))

$$\mathsf{V}^{-1}\langle \mathbf{k}_m, m | V | \mathbf{k}_n, n^+ \rangle = \langle \phi_f | V | \psi_i^+ \rangle.$$

The cross-section for the transition $n \to m$, defined by

$$n_1 n_2 v_n \sigma_{m,n} = \dot{N}(m \leftarrow \mathbf{n})/\mathsf{V}, \quad v_n = \hbar k_n / \mu, \tag{4.31}$$

§ The rate we have computed refers to the situation where the transition $n \to m$ occurs in a single-pair collision (say, in crossed molecular beams). When the densities are high and one cannot neglect triple (and higher) collisions one should consider the proper L.S. equation for $N_1 + N_2$ molecules (section 2.5.5). One then finds eqn (4.30) as the first contribution plus additional corrections which are of higher order in the density.

does have the dimensions of area and is given by

$$\sigma_{m,n} = \frac{\mu^2 v_m}{(2\pi\hbar^2)^2 v_n} \int d\hat{\mathbf{k}}_m |\langle \mathbf{k}_m, m|V|\mathbf{k}_n, n^+\rangle|^2. \tag{4.32}$$

The differential cross-section can now be written

$$d\sigma_{m,n} = \frac{v_m}{v_n} |f_{m,n}(\mathbf{k}_m, \mathbf{k}_n)|^2 \, d\hat{\mathbf{k}}_m, \tag{4.33}$$

where (eqn (3.16))

$$\langle \mathbf{k}_m, m|V|\mathbf{k}_n, n^+\rangle = -(2\pi\hbar^2/\mu) f_{m,n}(\mathbf{k}_m, \mathbf{k}_n). \tag{4.34}$$

Using the optical theorem (eqn (4.7)) we also note that

$$v_n \sum_m \sigma_{m,n} = -(2/\hbar)\text{im}\langle \mathbf{k}_n, n|V|\mathbf{k}_n, n^+\rangle \tag{4.35 a}$$

or, using eqn (4.34),

$$\sum_m \sigma_{m,n} = (4\pi/k_n)\text{im}\, f_{n,n}(\mathbf{k}_n, \mathbf{k}_n). \tag{4.35 b}$$

From now on we shall denote the rate per unit density of reactants by R. According to eqn (4.30) we can compute $V^{-1}R$ directly from Ehrenfest's theorem using states normalized to unit density.

Equation (4.33) can also be obtained directly from Ehrenfest's theorem by evaluating the rate of transitions to the internal state m and a relative momentum in the direction $\hat{\mathbf{k}}_m$. The relevant observable is $A = |m, \mathbf{k}_m\rangle\langle \mathbf{k}_m, m|$.

$$V^{-1}R(\mathbf{m} \leftarrow \mathbf{n}) = \frac{2\pi}{\hbar} \int \delta(E_i - E_f)|\langle \mathbf{k}_m, m|V|\mathbf{k}_n, n^+\rangle|^2 (2\pi)^{-3} k_m^2 \, dk_m, \tag{4.36}$$

where we have put $d\mathbf{k}_m = k_m^2 \, dk_m \, d\hat{\mathbf{k}}_m$ and $E_f = E_m + \hbar^2 k_m^2/2\mu$. We can perform the k_m integration by introducing the density of final translational states by $\rho(E)\, dE\, d\hat{\mathbf{k}} = (2\pi)^{-3} \, d\mathbf{k}$. Thus

$$V^{-1}R(\mathbf{m} \leftarrow \mathbf{n}) = \frac{2\pi}{\hbar} |\langle \mathbf{k}_m, m|V|\mathbf{k}_n, n^+\rangle|^2 \rho(E - E_m) = v_n \, d\sigma_{m,n}, \tag{4.37}$$

where the magnitude of k_m is now restricted (by the delta function in eqn (4.36) to conserve energy, $E_i = E_m + \hbar^2 k_m^2/2\mu$).

In the same fashion we have that

$$V^{-1}R(m \leftarrow \mathbf{n}) = \frac{2\pi}{\hbar} \int \delta(E_i - E_f)|\langle \mathbf{k}_m, m|V|\mathbf{k}_n, n^+\rangle|^2 (2\pi)^{-3} \, d\mathbf{k}_m = v_n \sigma_{m,n}, \tag{4.38}$$

where $R(m \leftarrow \mathbf{n})$ is the rate of the transitions $n \to m$ from an initial state, with a relative momentum $\hbar \mathbf{k}_n$, and of unit density. $R(m \leftarrow \mathbf{n})$ is the most general rate that we can compute for an initial state with a well-defined relative momentum. To compute the observed rate in an actual experiment we must have explicit information about the actual initial state. We consider this problem in section 2.4.2 and again in section 2.6.3.

We shall show in section 2.6.0 that a matrix element of the type $\langle f|V|i^+\rangle$ determines also the rate of reactive collisions. In this paragraph we consider the volume dependence of such a collision with p molecules in the initial state and q molecules in the final state. The matrix element $\langle f|V|i^+\rangle$ has a volume dependence of $\mathsf{V}^{-(p+q)/2}$. In evaluating the rate, dN/dt, we have to square the matrix element and integrate over q final momenta. Thus
$$dN/dt \sim \mathsf{V}^{-(p+q)} \cdot \mathsf{V}^q \cdot \mathsf{V}$$
$$\sim \mathsf{V}^{-(p-1)},$$
where the factor V^q comes from the density of states factor (eqn (4.15)), and an additional factor of V is due to conservation of the total centre of mass momentum (eqn (4.23)). As expected, we find that
$$d(N/\mathsf{V})/dt \sim \mathsf{V}^{-p}, \tag{4.39}$$
a well-known empirical result. Moreover, it is also clear that
$$d(N/\mathsf{V})/dt \sim \prod_{i=1}^{p} (N_i/\mathsf{V}), \tag{4.40}$$
where N_i is the number of molecules of type i in the initial state. As we have shown, the proportionality factor is the rate per unit volume computed with states normalized to unit density. The dimensions of eqns (4.37) and (4.38) are both (length)^3(time)$^{-1}$.

As an example, consider the rate when the density of the initial molecules is specified in terms of a density function f such that
$$C_{n_\alpha} f_\alpha(\mathbf{k}_1) \, d\mathbf{k}_1 \tag{4.41}$$
is the density of molecules with the internal state n_α occupied and with momentum between $\hbar \mathbf{k}_1$ and $\hbar(\mathbf{k}_1 + d\mathbf{k}_1)$. The function $f(\mathbf{k})$ is normalized to unit volume so that C_{n_α} is the density of molecules in the state n_α. From eqn (4.30) the number of transition $\mathbf{n} \to \mathbf{m}$ per unit time and unit volume is
$$\mathsf{V}^{-1}\dot{N}(\mathbf{m} \leftarrow \mathbf{n}) = C_{n_\alpha} C_{n_\beta} v_n \, d\sigma_{m,n} \, f_\alpha(\mathbf{k}_1) f_\beta(\mathbf{k}_2) \, d\mathbf{k}_1 \, d\mathbf{k}_2$$
$$= C_{n_\alpha} C_{n_\beta} \mathsf{V}^{-1} R_{f_\alpha f_\beta}(\mathbf{m} \leftarrow \mathbf{n}), \tag{4.42}$$
where (cf. Appendix 1.A) $d\mathbf{k}_1 \, d\mathbf{k}_2 = d\mathbf{k}_n \, d\mathbf{\varkappa}_n$ and the second line defines the rate per unit density.

To illustrate the use of Ehrenfest's theorem for other observables we consider the rate of change of the relative momentum of two molecules due to collisions. We take A as the momentum operator for the relative motion, $\mathbf{A} = -i\hbar\nabla_\mathbf{R}$, and work in the centre of mass system. From eqns (4.10) and (4.15)

$$-\hat{\mathbf{k}}_i.\langle\dot{\mathbf{A}}\rangle = \hat{\mathbf{k}}_i.\langle i^+|[\nabla_\mathbf{R}, V]|i^+\rangle = \langle i^+|\hat{\mathbf{k}}_i.\mathbf{grad}\,V|i^+\rangle$$
$$= \frac{2\pi}{\hbar}\int (2\pi)^{-3}\,\mathrm{d}\mathbf{k}_f\,\hbar(k_i-k_f\hat{\mathbf{k}}_i.\hat{\mathbf{k}}_f)|\langle f|V|i^+\rangle|^2\delta(E_i-E_f), \quad (4.43)$$

where $\hat{\mathbf{k}}_i$ and $\hat{\mathbf{k}}_f$ are the directions of the initial and final momenta. For elastic collisions the magnitude of the momentum is conserved, so that replacing the energy-conserving delta function by the momentum conservation (eqn (2.8)),

$$-\hat{\mathbf{k}}_i.\langle\dot{\mathbf{A}}\rangle = (2\pi)^{-2}\int k_f^2\,\mathrm{d}k_f\,\mathrm{d}\hat{\mathbf{k}}_f\,(k_i-k_f\hat{\mathbf{k}}_i.\hat{\mathbf{k}}_f)|\langle f|V|i^+\rangle|^2\frac{\mu\delta(k_i-k_f)}{\hbar^2 k_f}$$
$$= \mu k_i^2(2\pi\hbar)^{-2}\int \mathrm{d}\hat{\mathbf{k}}_f\,(1-\hat{\mathbf{k}}_i.\hat{\mathbf{k}}_f)|\langle f|V|i^+\rangle|^2, \quad (4.44)$$

where in the matrix element $E_f = E_i$. We can introduce the scattering amplitude as usual by

$$f(\mathbf{k}_f, \mathbf{k}_i) = -(\mu/2\pi\hbar^2)\langle f|V|i^+\rangle,$$

where the states are now assumed normalized to unit density, so that with $E_i = \hbar^2 k_i^2/2\mu$,

$$-\hat{\mathbf{k}}_i.\langle\dot{\mathbf{A}}\rangle = 2E_i\sigma_d, \quad (4.45)$$

where σ_d is the momentum transfer cross-section§

$$\sigma_d = \int \mathrm{d}\hat{\mathbf{k}}_f\,(1-\hat{\mathbf{k}}_i.\hat{\mathbf{k}}_f)|f(\mathbf{k}_f, \mathbf{k}_i)|^2$$
$$= (2E_i)^{-1}\langle i^+|\hat{\mathbf{k}}_i.\mathbf{grad}\,V|i^+\rangle. \quad (4.46)$$

An alternative derivation has been discussed by Gerjuoy (1965). It is clear that the present method can be generalized to include inelastic processes, when we obtain

$$\sigma_d = \frac{1}{k_n}\sum_m \frac{k_m}{k_n}\int \mathrm{d}\hat{\mathbf{k}}_m\,(k_n-k_m\hat{\mathbf{k}}_m.\hat{\mathbf{k}}_n)|f_{m,n}(\mathbf{k}_m, \mathbf{k}_n)|^2, \quad (4.47)$$

where m labels the different internal states of the molecules (cf. section 2.3.0)).

A rate equation in the form of eqn (4.10) can thus be associated with any measurable attribute of the initial and final states, where by 'measurable' we mean an operator A such that $[A, H_0] = 0$. If we plan to measure several observables simultaneously, we can expect eqn (4.8) to be valid only if all our observables commute. Otherwise they cannot

§ The factor E_i is introduced to ensure that σ_d has the dimension of area.

be diagonalized by the same basis. For example, for a pair of non-commuting observables, A and B, if we write A as in eqn (4.3), B will be non-diagonal,

$$B = \sum_{f,f'} |f\rangle B_{f,f'} \langle f'|,$$

and eqn (4.8) for B will involve summations over f and f'. We have seen one example already, in eqn (2.74). The relevant operator is $|\mathbf{k}\rangle\langle\mathbf{k}|$, which is not diagonal in the $|klm\rangle$ representation.

2.4.2. *Collision rates in ensembles*

Our previous discussion of transition rates was restricted by the assumption of a sharply defined initial state. If the initial state is a mixture characterized by the density matrix ρ_0,

$$\rho_0 = \sum_i p_i P_i, \tag{4.48}$$

where $P_i = |i\rangle\langle i|$ and p_i is the statistical weight ('the mole fraction') of the state i,

$$\sum_i p_i = 1,$$

then, anticipating eqn (5.16), the density matrix for the interacting system is

$$\rho_+ = \sum_i p_i P_i^+, \quad P_i^+ = |i{+}\rangle\langle i{+}|, \tag{4.49}$$

and

$$\operatorname{Tr}(A\rho_+) = \sum_i p_i \operatorname{Tr}(A P_i^+)$$

$$= \sum_i p_i \langle i{+}|A|i{+}\rangle. \tag{4.50}$$

Under our assumption about ρ_0 we now have

$$\frac{d}{dt}\langle A\rangle = \frac{d}{dt}\operatorname{Tr}(A\rho_+) = \sum_i p_i \sum_f (A_f - A_i) R_{f,i}. \tag{4.51}$$

Equation (4.51) is often summarized by the rule 'average over initial states, sum over final states'.

We consider two particular applications of this rule. The first is the so-called 'degeneracy averaged' cross-section, relevant to the case where the initial state is degenerate. For example, the eigenstates of a rigid rotor with the same value of angular momentum j, but with different values of m_j, the projection of j on some fixed axis, are degenerate in the absence of a field.

An unpolarized (namely an equilibrium) rotor, in a state j, is thus characterized by the density matrix (cf. eqn (1.1.72 b))

$$\rho_0^j = (2j+1)^{-1} \sum_{m_j=-j}^{j} |jm_j\rangle\langle jm_j|, \tag{4.52}$$

so that the cross-section for the transition $j \to j'$ is

$$\sigma_{j',j} = (2j+1)^{-1} \sum_{m_{j'} m_j} \sigma_{j'm_{j'},jm_j}, \qquad (4.53)$$

where we averaged over the initial states and summed over the final states. It should be stressed that classical averaging over the initial states is the correct procedure only when the initial state is a mixture, eqn (4.48), where the weights p_j are purely statistical in character and refer to our experimental failure to specify the initial state more precisely. This is in contrast to an initial state of the type $|\Phi_i\rangle = \sum_m C_{im}|m\rangle$, with the initial density operator $\rho_0^i = |\Phi_i\rangle\langle\Phi_i|$,

$$\rho_0^i = \sum_{m,m'} C_{im} C_{im'}^* |m\rangle\langle m'|. \qquad (4.54)$$

In this case, due to the cross terms in eqn (4.54), which correspond to definite phase relations between the elements C_{im}, we are unable to write $\mathrm{Tr}(A\rho_+^i)$ as a sum over the index m. We recall that such cross terms will be present whenever we are unable in principle to distinguish the states m before the collision.

As another example we consider the collision of two atoms with spin angular momentum S. The degeneracy (or number of spin states of the same energy) is $2S+1$. For half-integer S these states are divided in the ratio $S(S+1)^{-1}$ between the space symmetric ($|S\rangle$) and space antisymmetric $|A\rangle$ states, and so

$$\rho_0 = S(2S+1)^{-1}|S\rangle\langle S| + (S+1)(2S+1)^{-1}|A\rangle\langle A|$$

or
$$\sigma = S(2S+1)^{-1}\sigma_S + (S+1)(2S+1)^{-1}\sigma_A, \qquad (4.55)$$

where σ_S is the cross-section for the space-symmetric states and σ_A for the antisymmetric states.

In the theory of collisions of systems with internal angular momentum (whether spin or orbital) the selection of a suitable representation can often simplify the problem considerably. Interesting comments on this problem were made by Lippmann (1953) and Smith (1967).

Another case of a degenerate initial state occurs when the experimental conditions§ are such that all the directions of the initial relative momentum $\hbar\mathbf{k}_n$ are equally likely. Under these conditions we can explicitly evaluate $\mathrm{d}\dot{N}(m \leftarrow n)/\mathrm{d}E$, namely, the rate of the transitions $n \to m$ when the total energy of the system is between E and $E+\mathrm{d}E$ and the internal energy of the initial state is E_n. We can consider the initial state as a microcanonical ensemble with the density matrix

$$\rho_0 = \delta(E - E_n - K), \qquad (4.56)$$

§ Cf. eqns (4.41) and (4.42). An example of this case is a translational equilibrium.

where K is the kinetic energy operator for the relative motion. The spectral resolution of ρ_0 is obtained by multiplying both sides by the identity operator for states normalized to unit density,

$$\rho_0 = \int |\mathbf{k}_n, n\rangle \mathsf{V}\rho(E-E_n)\,\mathrm{d}\hat{\mathbf{k}}_n \langle n, \mathbf{k}_n|, \tag{4.57}$$

where $\rho(E)$ is the density of translational states per unit volume. The transition rate is thus given by

$$\frac{\mathrm{d}\dot{N}(m \leftarrow n)}{\mathrm{d}E} = \int \mathsf{V}^{-1} R(m \leftarrow \mathbf{n})\rho(E-E_n)\,\mathrm{d}\hat{\mathbf{k}}_n$$

$$= \int v_n\, \sigma_{m,n}\, \rho(E-E_n)\,\mathrm{d}\hat{\mathbf{k}}_n. \tag{4.58}$$

If the cross-section depends on k_n only, as is the case when there is no preferred space direction, the angular integration can be carried out to give

$$\frac{\mathrm{d}\dot{N}(m \leftarrow n)}{\mathrm{d}E} = (2\pi\hbar)^{-3} 8\pi\mu(E-E_n)\sigma_{m,n}$$

$$= (2\pi\hbar)^{-1} \pi^{-1} k_n^2\, \sigma_{m,n}. \tag{4.59}$$

Equation (4.59) determines the rate of transitions $n \to m$ in a microcanonical ensemble. Using the optical theorem in the form of eqn (4.35) we can sum over the index m, to obtain§

$$\sum_m \frac{\mathrm{d}\dot{N}(m \leftarrow n)}{\mathrm{d}E} = 4\pi(-2/\hbar)\mathrm{im}\langle E, \mathbf{n}|V|E, \mathbf{n}^+\rangle. \tag{4.60}$$

In situations where the distribution of the initial momenta are anisotropic (say in crossed molecular beams experiments) the present discussion does not apply. Rather one has to use eqn (4.42), sum over the final states (by integrating over $\hat{\mathbf{k}}_m$), and average over the initial states by integrating the given experimental densities. In this case we obtain

$$\mathsf{V}^{-1} R_{f_\alpha f_\beta}(m \leftarrow n) = \int \mathrm{d}\mathbf{k}_n\, \mathrm{d}\mathbf{\varkappa}_n\, f_\alpha(\mathbf{k}_1) f_\beta(\mathbf{k}_2) v_n\, \sigma_{m,n}. \tag{4.61}$$

When the translational motion of the molecules is in equilibrium at a temperature T the distribution of momenta is isotropic and

$$\mathrm{f}(\mathbf{k}_1) = (\beta m_1/2\pi)^{\frac{3}{2}} \exp\left(-\frac{\beta \hbar^2 k_1^2}{2m_1}\right), \tag{4.62}$$

where m_1 is the mass of molecule one and $\beta = (\mathbf{k}T)^{-1}$ where $\mathbf{k}$ is Boltzmann's constant. We can now perform the integration in eqn (4.61)

§ The factor 4π is due to the integration of the matrix element $\langle E, \mathbf{n}|V|E, \mathbf{n}^+\rangle$ over $\hat{\mathbf{k}}_n$. By our previous assumption the matrix element is independent of $\hat{\mathbf{k}}_n$.

when we note that the total translational energy is independent of the coordinate system,

$$k_1^2/2m_1 + k_2^2/2m_2 = k_n^2/2\mu + \kappa_n^2/2(m_1+m_2),$$

where μ is the reduced mass, $\mu = m_1 m_2/(m_1+m_2)$. The result of integration over κ can be written as

$$V^{-1}R_T(m \leftarrow n) = Z_{tr}^{-1}(2\pi\hbar)^{-1} \int \exp(-\beta e)\pi^{-1}k_n^2 \sigma_{m,n}\, de$$

$$= Z_{tr}^{-1} \int \exp(-\beta e) \frac{d\dot{N}(m \leftarrow n)}{de}\, de, \qquad (4.63)$$

where e is the relative kinetic energy, $e = \hbar^2 k_n^2/2\mu$ and Z_{tr} is the translational partition function (per unit volume) for the relative motion,

$$Z_{tr} = (\mu/2\pi\hbar^2\beta)^{\frac{3}{2}}. \qquad (4.64)$$

The second line indicates that the same result can be obtained by averaging over the distribution of the total energy. To see this explicitly we consider the density matrix $\rho_0^n(T)$ for a non-interacting pair of molecules in the centre of mass system, when the translational motion is in thermal equilibrium and the internal state n is occupied.

$$\rho_0^n(T) = Z^{-1} \int \exp(-\beta E)\delta(E-E_n-K)P_n\, dE, \qquad (4.65)$$

where Z is the normalization factor,

$$Z = \mathrm{Tr} \int \exp(-\beta E)\delta(E-E_n-K)P_n\, dE = \exp(-\beta E_n)VZ_{tr}. \qquad (4.66)$$

Introducing the relative translational energy e, $E = E_n + e$, using eqn (4.51) with $A = P_m$ recalling our discussion of eqn (4.39), and using eqn (4.58) we obtain eqn (4.63).

Equation (4.63) is appropriate to the situation where we know that initially the system is in the internal state n. In other situations the initial state is a mixture of internal states, so that the density matrix is

$$\rho_0(T) = \sum_n p_n \rho_0^n(T), \qquad (4.67)$$

where p_n is the probability of finding the internal state n occupied.

2.4.3. *The relaxation equation*

As the second application of eqn (4.51) we consider the derivation§ of the relaxation equation for the system discussed in the introduction, namely a small density of diatomic molecules in a monoatomic host gas.

§ The present formal similarity between the two equations is slightly misleading. The quantum numbers n and m in our relaxation equation refer to internal states only, while the quantum numbers i and f in eqn (4.51) refer to all the relevant quantum numbers. Of course, in a generalized sense, eqn (4.51) is itself a relaxation equation.

We assume that the system has reached translational equilibrium and describe it by the density matrix (4.67), where n is an index of the internal levels of the diatomic.

The observable A that determines the internal energy of the diatomic is simply the Hamiltonian h for the internal degrees of freedom

$$A = \sum_n A_n P_n, \qquad (4.68)$$

where A_n is the internal energy of the nth level and $P_n = |n\rangle\langle n|$. It is obvious that $[A, H_0] = 0$, and from eqn (4.67)

$$\mathrm{Tr}[A\rho_0(T)] = \sum_n A_n p_n. \qquad (4.69)$$

Thus A is the average internal energy per molecule. From eqns (4.67), (4.57), and (4.51),

$$\frac{\mathrm{d}}{\mathrm{d}t}\langle A\rangle = \sum_n p_n \sum_m (A_m - A_n) \mathsf{V}^{-2} Z_{\mathrm{tr}}^{-1} \times$$
$$\times \iiint \exp(-\beta e) R(\mathbf{m} \leftarrow \mathbf{n}) \rho(e) \, \mathrm{d}\hat{\mathbf{k}}_n \, \mathrm{d}\hat{\mathbf{k}}_m \, \mathrm{d}e. \qquad (4.70)$$

The integrations over $\hat{\mathbf{k}}_m$, $\hat{\mathbf{k}}_n$, and e are performed using eqns (4.38), (4.58), and (4.63) respectively, when we find that

$$\frac{\mathrm{d}}{\mathrm{d}t}\langle A\rangle = \sum_n p_n \sum_m (A_m - A_n) \mathsf{V}^{-2} R_\mathrm{T}(m \leftarrow n). \qquad (4.71)$$

Equation (4.71) is the relaxation equation. We have not used the fact that A_n is an energy in the derivation and so eqn (4.71) holds for any observable A that can be written in the form (4.68). In particular we want to consider the observable A such that $A_m = \delta_{m,n}$. In this case $\langle A\rangle = \langle P_n\rangle$, and the relaxation equation reads

$$\frac{\mathrm{d}}{\mathrm{d}t}\langle P_n\rangle = \sum_m p_m \mathsf{V}^{-2} R_\mathrm{T}(n \leftarrow m) - \sum_m p_n \mathsf{V}^{-2} R_\mathrm{T}(m \leftarrow n). \qquad (4.72)$$

Equations (4.71) and (4.72) refer to a single atom and a single diatomic molecule. If there are N_g host gas atoms and N diatomic molecules, we should modify eqn (4.68) to

$$A_N = \sum_{r=1}^{N} \sum_n A_n P_{n,r}, \qquad (4.73)$$

where $P_{n,r}$ is the projection on the nth internal state of the rth diatomic, and so

$$\mathrm{Tr}[A_N \rho_0(T)] = N \sum_n p_n A_n = \mathsf{V} \sum_n C_n A_n \qquad (4.74)$$

is the total internal energy. Each diatomic can collide with any one of the host gas atoms, so that

$$[A_N, H] = \sum_{s=1}^{N_g} \sum_{r=1}^{N} A_n [P_{n,r}, V_{r,s}], \qquad (4.75)$$

where s labels the host gas atoms. Thus

$$\mathsf{V}^{-1}\frac{\mathrm{d}\langle A_N\rangle}{\mathrm{d}t} = n_\mathrm{g} N\frac{\mathrm{d}\langle A\rangle}{\mathrm{d}t}, \qquad (4.76)$$

and eqns (4.1) and (4.2) follow, with

$$k_{m,n} = n_\mathrm{g}\,\mathsf{V}^{-1}R_\mathrm{T}(m\leftarrow n) = n_\mathrm{g}\,Z_\mathrm{tr}^{-1}\int \exp(-\beta e)\,\frac{\mathrm{d}\dot{N}(m\leftarrow n)}{\mathrm{d}E}\,\mathrm{d}e \qquad (4.77)$$

and $C_n = p_n N/\mathsf{V}$.

In section 2.8.2 we prove the theorem of microscopic reversibility

$$g_n\frac{\mathrm{d}\dot{N}(m\leftarrow n)}{\mathrm{d}E} = g_m\frac{\mathrm{d}\dot{N}(n\leftarrow m)}{\mathrm{d}E}, \qquad (4.78)$$

where $\dot{N}$ is the degeneracy averaged rate, g_n is the degeneracy of the state n, and the quantum numbers n and m refer to the internal rotational or vibrational energies (in the absence of external fields). Equation (4.78) is valid when both sides are evaluated at the same total energy E. By the change of variables $e = E - E_n$ in eqn (4.77) and $e = E - E_m$ in the similar equation for $k_{n,m}$ we conclude that

$$\left.\begin{matrix}k_{m,n}\\k_{n,m}\end{matrix}\right\} = n_\mathrm{g}\,Z_\mathrm{tr}^{-1}\int \exp(-\beta E)\,\frac{\mathrm{d}\dot{N}(n\leftarrow m)}{\mathrm{d}E}\,\mathrm{d}e \cdot \begin{cases}g_n^{-1}\exp(\beta E_n)\\ g_m^{-1}\exp(\beta E_m)\end{cases}$$

or

$$\frac{k_{m,n}}{k_{n,m}} = \frac{g_m}{g_n}\exp[-\beta(E_m-E_n)]. \qquad (4.79)$$

Equation (4.79) is often referred to as 'detailed balance'. We note that we have defined $k_{m,n}$ in terms of the degeneracy averaged rate $\dot{N}$ so that $k_{m,n}$ is the transition rate from a single state n. The transition rate from all the states n is $g_n k_{m,n}$.

In equilibrium we require that $\langle\dot{P}_m\rangle = 0$ or, from eqn (4.72),

$$\left(\frac{p_n}{p_m}\right)_e = \frac{R_\mathrm{T}(n\leftarrow m)}{R_\mathrm{T}(m\leftarrow n)} = \frac{g_n}{g_m}\exp[-\beta(E_n-E_m)]. \qquad (4.80)$$

The equilibrium condition in a microcanonical ensemble is obtained from eqn (8.63), namely

$$\left(\frac{p_n}{p_m}\right)_e = \frac{g_n\rho(E-E_n)}{g_m\rho(E-E_m)}. \qquad (4.81)$$

Our derivation of the relaxation equation (based on Levine 1966e) rests on several assumptions. We have assumed that the density matrix of the pair before the collision is stationary and does not depend on the previous history of the system (in this connection see section 3.6.2) and that only pairwise collisions take place. Relaxation equations are also referred to as 'master equations' and were first derived by Pauli (1928) using the Golden rule. An extensive discussion of the relaxation of internal energy and of computation of transition rates can be found in Herzfeld and Litovitz (1959), Cottrell and McCoubrey (1961), Herzfeld (1963), Widom (1963), and Osipov and Stupochenko (1963).

2.5. Formal collision theory

2.5.0. It is convenient to cast the formal aspects of collision theory in a form independent of an explicit specification of the initial state. Such a formalism can then be used with an arbitrary initial state, which can be a superposition of states of different total energies.

The Møller wave operator Ω is defined (Møller 1945) to generate a state of H with the (continuous) energy E from an initial state of H_0 of the same energy

$$|E, \mathbf{n}^+\rangle = \Omega^+ |E, \mathbf{n}\rangle \tag{5.1}$$

for any E in the continuous spectrum of H_0. (If H_0 can support bound states we shall require that Ω annihilates these states.) Using our normalization convention

$$\langle E', \mathbf{n}' | E, \mathbf{n}\rangle = \delta(E'-E)\delta_{\mathbf{n}'\mathbf{n}},$$

we can formally write

$$\Omega^+ = \sum_{\mathbf{n}} \int dE |E, \mathbf{n}^+\rangle\langle E, \mathbf{n}|, \tag{5.2}$$

with a similar definition of Ω^-, where

$$|E, \mathbf{n}^-\rangle = \Omega^- |E, \mathbf{n}\rangle. \tag{5.3}$$

The operator $\Omega(\lambda)$, a function of the complex variable λ, is defined by

$$(\lambda - H)\Omega(\lambda) = (\lambda - H_0) \tag{5.4}$$

and satisfies the equations

$$|E, \mathbf{n}^+\rangle = \Omega(E^+) |E, \mathbf{n}\rangle \tag{5.5}$$

$$\text{and}\quad \Omega^+ = \sum_{\mathbf{n}} \int dE\, \Omega(E^+) |E, \mathbf{n}\rangle\langle E, \mathbf{n}| = \int dE\, \Omega(E^+)\delta(E-H_0). \tag{5.6}$$

In stationary collision energy where the given total energy is fixed, we shall deal mainly with $\Omega(E^+)$.

The Hermitian adjoint of the wave operator $\Omega^{+\dagger}$ is, from eqn (5.2),

$$\Omega^{+\dagger} = \sum_{\mathbf{n}} \int dE\, |E, \mathbf{n}\rangle\langle E, \mathbf{n}^+|. \tag{5.7}$$

In Appendix 2.A we prove the normalization

$$\langle E', \mathbf{m}^+ | E, \mathbf{n}^+\rangle = \langle E', \mathbf{m} | E, \mathbf{n}\rangle \tag{5.8}$$

and since

$$\langle E', \mathbf{m} | E, \mathbf{n}\rangle = \delta(E-E')\delta_{\mathbf{mn}}, \tag{5.9}$$

we have that

$$|E, \mathbf{n}\rangle = \Omega^{+\dagger} |E, \mathbf{n}^+\rangle = \Omega^{+\dagger}\Omega^+ |E, \mathbf{n}\rangle. \tag{5.10}$$

The operator $\Omega^\dagger\Omega$ is thus a projection on the continuous spectrum of H_0

$$\Omega^\dagger\Omega = \sum_{\mathbf{n}} \int dE \, |E, \mathbf{n}\rangle\langle E, \mathbf{n}|, \qquad (5.11)$$

but is not necessarily the identity operator since H_0 may have bound states (which are annihilated by Ω, since they are orthogonal to the states $|E, \mathbf{n}\rangle$). An operator Ω such that $\Omega^\dagger\Omega$ is a projection is referred to as a *partial isometry*.§

The possible existence of bound states of H and H_0 implies that Ω is not unitary, but we still can use it to define a transformation function between the continuous spectra of H and H_0, since Ω is isometric between these two subspaces. Let

$$|\Psi^+\rangle = \sum_{\mathbf{n}} \int dE \, |E, \mathbf{n}+\rangle A(E, \mathbf{n}) = \sum_{\mathbf{n}} \int dE \, |E, \mathbf{n}\rangle B(E, \mathbf{n}), \quad (5.12)$$

be two alternative expansions for the state $|\Psi^+\rangle$. Taking the scalar product of the second equation with $\langle E', \mathbf{n}'|$, we have, using the normalization of the initial states

$$B(E', \mathbf{n}') = \sum_{\mathbf{n}} \int dE \, \langle E', \mathbf{n}'|E, \mathbf{n}+\rangle A(E, \mathbf{n})$$

$$= \sum_{\mathbf{n}} \int dE \, \langle E', \mathbf{n}'|\Omega^+|E, \mathbf{n}\rangle A(E, \mathbf{n}). \qquad (5.13)$$

The value for the transformation function is obtained from the L.S. equation as

$$\langle E', \mathbf{n}'|\Omega^+|E, \mathbf{n}\rangle = \delta(E-E')\delta_{\mathbf{n}\mathbf{n}'}+$$
$$+[E^+(\mathbf{n})-E'(\mathbf{n}')]^{-1}\langle E', \mathbf{n}'|V\Omega^+|E, \mathbf{n}\rangle. \quad (5.14)$$

We note that the norm-conserving properties of the wave operator imply that $A(E, \mathbf{n})$ is also the amplitude to observe the state $|E, \mathbf{n}\rangle$ in $|\Phi\rangle$, where $|\Phi\rangle = \Omega^{+\dagger}|\Psi^+\rangle$ is the initial state that corresponds to $|\Psi^+\rangle$. The proof is immediate:

$$\langle E, \mathbf{n}|\Phi\rangle = \langle E, \mathbf{n}|\Omega^\dagger\Omega|\Phi\rangle = \langle E, \mathbf{n}+|\Psi^+\rangle. \qquad (5.15)$$

In the same fashion if ρ_0^i is the density matrix for the system in the H_0 representation, the density matrix in the H representation is

$$\rho_+^i = \Omega^+ \rho_0^i \Omega^{+\dagger}. \qquad (5.16)$$

§ An isometry preserves the norm (metric) of any state, while a partial isometry preserves the norm of any state in a given closed subspace. In our case $\|\Omega\phi\| = \|\phi\|$ for any ϕ in the continuous spectrum since $\|\Omega\phi\|^2 \equiv \langle\Omega\phi|\Omega\phi\rangle = \langle\phi|\Omega^\dagger\Omega|\phi\rangle = \langle\phi|\phi\rangle \equiv \|\phi\|^2$. Therefore $\Omega^\dagger\Omega$ is the identity operator on the range of $\Omega^\dagger$ (i.e. the continuous spectrum of H_0), so that $\Omega\Omega^\dagger$ is also a projection, $\Omega\Omega^\dagger\Omega\Omega^\dagger = \Omega I \Omega^\dagger = \Omega\Omega^\dagger$, on the continuous spectrum of H.

The transition operator T is defined by
$$T = V\Omega. \tag{5.17}$$
In practice we are mostly concerned with the operator $T(E^+)$,
$$T(E^+) = V\Omega(E^+), \tag{5.18}$$
so that $T(E^+)|E, \mathbf{n}\rangle = V|E, \mathbf{n}^+\rangle$. From eqn (5.4) (or eqns (5.5) and (1.8 b))
$$\Omega(E^+) = I + G(E^+)V \tag{5.19a}$$
$$= I + G_0(E^+)V\Omega(E^+) \tag{5.19b}$$
$$= I + G_0(E^+)T, \tag{5.19c}$$
where I is the identity operator on the continuous spectrum of H_0, and so
$$T(E^+) = V + VG(E^+)V \tag{5.20a}$$
$$= V + VG_0(E^+)T(E^+). \tag{5.20b}$$
We also note the identities
$$G(E^+) = \Omega(E^+)G_0(E^+) \tag{5.21a}$$
$$= G_0(E^+)[I + T(E^+)G_0(E^+)], \tag{5.21b}$$
which are derived by inverting eqn (5.4) and by using eqns (5.19c) and (5.21a).

The operation of L-conjugation, $\ddagger$, was introduced as the basic symmetry of the Green's operators (cf. eqn (1.2.18)).
$$G^{\ddagger}(\lambda) = G^{\dagger}(\lambda^*) = G(\lambda), \tag{5.22}$$
so that, from eqn (5.21a),
$$G(E^+) = G_0(E^+)\Omega^{\ddagger}(E^+), \tag{5.21c}$$
where, from eqn (5.19b),
$$\Omega^{\ddagger}(E^+) = I + \Omega^{\ddagger}(E^+)VG_0(E^+). \tag{5.23}$$
Thus from the L.S. equation $|E, \mathbf{n}^-\rangle = |E, \mathbf{n}\rangle + G_0(E^-)V|E, \mathbf{n}^-\rangle$ we have that
$$\langle E, \mathbf{n}^-| = \langle E, \mathbf{n}|\Omega^{\ddagger}(E^+). \tag{5.24}$$
In contrast to the wave operator, $T(E^+)$ is invariant under L-conjugation, as is obvious from eqn (5.20a),
$$T^{\ddagger}(E^+) = \Omega^{\ddagger}(E^+)V = T(E^+). \tag{5.25}$$
Equation (5.25) implies in terms of matrix elements
$$\langle E, \mathbf{m}|V|E, \mathbf{n}^+\rangle = \langle E, \mathbf{m}|T|E, \mathbf{n}\rangle = \langle E, \mathbf{m}^-|V|E, \mathbf{n}\rangle. \tag{5.26}$$
The optical theorem (eqn (4.7)) can also be written in abstract form. From eqn (5.20a) and its Hermitian conjugate we have
$$T(E^+) - T^{\dagger}(E^+) = V[G(E^+) - G(E^-)]V = -2\pi i V\delta(E-H)V, \tag{5.27}$$

while from eqn (5.20b) and its L-conjugate

$$[I+T(E^+)G_0(E^+)]V[I+T(E^+)G_0(E^+)]^\dagger$$
$$= T(E^+)[I+T(E^+)G_0(E^+)]^\dagger$$
$$= [I+T(E^+)G_0(E^+)]T^\dagger(E^+), \quad (5.28)$$

or $\quad T(E^+)-T^\dagger(E^+) = T(E^+)[G_0(E^+)-G_0(E^-)]T^\dagger(E^+)$
$$= -2\pi i T(E^+)\delta(E-H_0)T^\dagger(E^+). \quad (5.29)$$

In both derivations we have used the convention regarding the limit $\epsilon \to 0$, and the symbolic instruction

$$\mathrm{im}\lim_{\epsilon \to +0} (E^+-H)^{-1} = -\pi\delta(E-H).$$

The first Born approximation which consists of replacing T by V will thus violate the optical theorem (the diagonal elements of a real potential are necessarily real).

2.5.1. *The S matrix*

In time-dependent collision theory we shall find it useful to introduce the operator S by
$$S = \Omega^{-\dagger}\Omega^+ \quad (5.30)$$
so that $\quad \langle E, \mathbf{m}|S|E, \mathbf{n}\rangle = \langle E, \mathbf{m}^-|E, \mathbf{n}^+\rangle.$

In Appendix 2.A we show that

$$\langle E', \mathbf{m}^-|E, \mathbf{n}^+\rangle = \delta(E-E')\delta_{\mathbf{nm}}-2\pi i \delta(E-E')\langle E, \mathbf{m}|T|E, \mathbf{n}\rangle, \quad (5.31)$$

so that§ $\quad S = I - 2\pi i \int dE\,\delta(E-H_0)T(E^+)\delta(E-H_0). \quad (5.32)$

We note that the optical theorem is equivalent to the statement
$$S^\dagger S = I = SS^\dagger. \quad (5.33)$$

To prove eqn (5.33) we write, using eqn (5.32),

$$S^\dagger S = I - \int dE\,\delta(E-H_0)\{2\pi i T(E^+)-2\pi i T^\dagger(E^+)+$$
$$+2\pi i T(E^+)\delta(E-H_0)T^\dagger(E^+)2\pi i\}\delta(E-H_0) = SS^\dagger = I.$$

Indeed, the unitarity of the S operator is directly related to the conservation of the norm
$$SS^\dagger = \Omega^{-\dagger}\Omega^+\Omega^{+\dagger}\Omega^- = \Omega^{-\dagger}\Omega^- \quad (5.34\mathrm{a})$$
and $\quad S^\dagger S = \Omega^{+\dagger}\Omega^-\Omega^{-\dagger}\Omega^+ = \Omega^{+\dagger}\Omega^+ \quad (5.34\mathrm{b})$

since $\Omega^\pm\Omega^{\pm\dagger}$ are projection operators on the continuous spectrum of H, and $\Omega^\pm$ generate the states of the continuous spectrum, so that $\Omega^\pm\Omega^{\pm\dagger}\Omega^\pm = \Omega^\pm$. The operators $\Omega^{\pm\dagger}\Omega^\pm$ are projection operators on the

---
§ See also Appendix 2.C, eqn (C.4).

continuous spectrum of H_0, and so are equivalent to the identity operator in their action on the continuous states of H_0.

Matrix elements of the type $\langle E, \mathbf{m}|T(E^+)|E, \mathbf{n}\rangle$ can be considered as elements of the matrix T,

$$(\mathbf{T})_{\mathbf{m},\mathbf{n}} = \langle E, \mathbf{m}|T(E^+)|E, \mathbf{n}\rangle. \qquad (5.35)$$

By definition, the matrix elements are between states of equal energy, which is also equal to the argument of the $T(E^+)$ operator. For this reason it is common to refer to these elements as being 'on the energy shell'. As we have seen, these are the relevant elements for the computations of rates, as the total energy is conserved when we consider the possible final states due to the collision. During the collision, when the system is not under observation, energy need not be conserved or, in formal language, the equations that determine T involve as intermediates, states not on the energy shell. Thus eqn (5.20b) reads

$$\langle E, \mathbf{m}|T(E^+)|E, \mathbf{n}\rangle = \langle E, \mathbf{m}|V|E, \mathbf{n}\rangle +$$
$$+ \sum_{\mathbf{p}} \int dE' \langle E, \mathbf{m}|V|E', \mathbf{p}\rangle \frac{\langle E', \mathbf{p}|T(E^+)|E, \mathbf{n}\rangle}{E^+ - E'}, \quad (5.36)$$

where we have put

$$G_0(E^+) = \sum_{\mathbf{p}} \int dE' \frac{|E', \mathbf{p}\rangle\langle E', \mathbf{p}|}{E^+ - E'}.$$

In the same fashion the transformation function from the H_0 to the H representation (eqn (5.14)) involves the same matrix elements, namely $\langle E', \mathbf{p}|T(E^+)|E, \mathbf{n}\rangle$.

If we label all quantum numbers of the continuous energy states of H_0 by a single letter, $H_0|i\rangle = E_i|i\rangle$, we can define an $\mathbf{S}$ matrix by

$$(\mathbf{S})_{f,i} = \langle f|S|i\rangle = \delta_{fi} - 2\pi i \delta(E_i - E_f)\langle f|T|i\rangle. \qquad (5.37)$$

It is a unitary matrix since $\Omega^{\pm}\Omega^{\pm\dagger}$ are the identity operators in the subspace of continuous energy states of H_0,

$$\langle f|S^\dagger S|i\rangle = \delta_{fi}. \qquad (5.38)$$

For states $|E, \mathbf{n}\rangle$ normalized on the energy scale we can put

$$\langle E', \mathbf{m}|S|E, \mathbf{n}\rangle = \delta(E' - E)(\mathbf{S})_{\mathbf{m},\mathbf{n}}, \qquad (5.39)$$

where $\mathbf{S}$ is a matrix on the energy shell. The relation

$$\langle E', \mathbf{m}|S^\dagger S|E, \mathbf{n}\rangle = \delta(E - E')\delta_{\mathbf{mn}} \qquad (5.40)$$

can be written

$$\langle E', \mathbf{m}|S^\dagger S|E, \mathbf{n}\rangle = \sum_{\mathbf{p}} \int dE'' \, \delta(E' - E'')(\mathbf{S}^\dagger)_{\mathbf{m},\mathbf{p}} \delta(E'' - E)(\mathbf{S})_{\mathbf{p},\mathbf{n}}.$$

Since $\delta(E'-E'')\delta(E''-E) = \delta(E'-E)\delta(E''-E)$ the unitarity of the S matrix can be stated as

$$\sum_p (\mathbf{S}^\dagger)_{m,p}(\mathbf{S})_{p,n} = \delta_{mn}, \tag{5.41}$$

or
$$\mathbf{S}^\dagger \mathbf{S} = \mathbf{I}.$$

Equation (5.37) can thus be written as

$$\mathbf{S} = \mathbf{I} - 2\pi i \mathbf{T}. \tag{5.42}$$

Equation (5.35) can be transformed to other normalization conventions by the introduction of suitable density of states factors. For example, we have found it useful to introduce the states $|\mathbf{k}_n, n\rangle$, normalized on the momentum scale, so that

$$\langle \mathbf{k}_m, m | T | \mathbf{k}_n, n \rangle = [\rho(E-E_m)\rho(E-E_n)]^{-\frac{1}{2}} \langle E, \mathbf{m} | T | E, \mathbf{n} \rangle \tag{5.43}$$

and the scattering amplitude (eqn (3.16)) can be written

$$f_{m,n}(\mathbf{k}_m, \mathbf{k}_n) = -(2\pi)^2 (k_m k_n)^{-\frac{1}{2}} (\mathbf{T})_{m,n}. \tag{5.44}$$

Equations (5.42) and (5.44) were used previously (eqns (3.54)–(3.56)) to introduce the $\mathbf{S}$ matrix in terms of the asymptotic behaviour of the wavefunction.

There are two traditional ways of ensuring that the $\mathbf{S}$ matrix is unitary in a practical computation. If $\mathbf{K}$ is an arbitrary Hermitian matrix, $\mathbf{K} = \mathbf{K}^\dagger$, then a matrix of the form

$$\mathbf{S} = (\mathbf{I} - i\pi \mathbf{K})(\mathbf{I} + i\pi \mathbf{K})^{-1} \tag{5.45}$$

is clearly a unitary matrix. Comparing with eqn (5.42) we see that $\mathbf{K}$ is related to $\mathbf{T}$ by

$$\mathbf{T} = \mathbf{K} - i\pi \mathbf{K}\mathbf{T}. \tag{5.46}$$

Equation (5.46) is known as Heitler's (1954) equation. If we define the $K(E)$ operator by

$$K(E) = V + VPv(E-H_0)^{-1}K(E) \tag{5.47}$$

and use the resolution of $G_0(E^+)$ into the principal value part and the singularity we can see that§

$$T(E^+) = K(E) - i\pi K^\dagger(E)\delta(E-H_0)T(E^+). \tag{5.48}$$

§ To prove eqn (5.48) we put, without indicating explicitly the argument E^+,
$$T = V + VPv(E-H_0)^{-1}T - i\pi V\delta(E-H_0)T$$
so that
$$T - K = VPv(E-H_0)^{-1}(T-K) - i\pi V\delta(E-H_0)T = -i\pi K^\dagger \delta(E-H_0)T$$
where in the last line we noted that $K^\dagger = [I - VPv(E-H_0)^{-1}]^{-1}V$, a relation obtained from the Hermitian adjoint of eqn (5.47).

The matrix **K**, with elements

$$(\mathbf{K})_{\mathbf{m},\mathbf{n}} = \langle E, \mathbf{m}|K(E)|E, \mathbf{n}\rangle$$
$$= \langle E, \mathbf{m}|V|E, \mathbf{n}\rangle + \sum_{\mathbf{p}} Pv \int dE' \langle E, \mathbf{m}|V|E', \mathbf{p}\rangle \frac{\langle E', \mathbf{p}|K(E)|E, \mathbf{n}\rangle}{E-E'}, \tag{5.49}$$

is Hermitian, and eqn (5.46) is the matrix representation of eqn (5.48). It is important to note that it is an equation on the energy shell.

The operator $K(E)$ is directly related to the so-called 'standing wave' solution,
$$|E, \mathbf{n}^0\rangle = |E, \mathbf{n}\rangle + Pv(E-H_0)^{-1}V|E, \mathbf{n}^0\rangle. \tag{5.50}$$

The label of this solution refers to the asymptotic form of the principal value in the coordinate representation (eqn (1.2.35 a)), namely, a standing wave. In terms of this solution

$$\langle E, \mathbf{m}|K(E)|E, \mathbf{n}\rangle = \langle E, \mathbf{m}|V|E, \mathbf{n}^0\rangle. \tag{5.51}$$

If we repeat the derivation of the last footnote, omitting, however, the factor V on the left, we have that

$$|E, \mathbf{n}^+\rangle = |E, \mathbf{n}^0\rangle - i\pi \sum_{\mathbf{m}} |E, \mathbf{m}^0\rangle(\mathbf{T})_{\mathbf{m},\mathbf{n}}, \tag{5.52}$$

where we put $T(E^+) = V\Omega(E^+)$ and $\delta(E-H_0) = \sum_{\mathbf{m}} |E, \mathbf{m}\rangle\langle E, \mathbf{m}|$. Regarding eqn (5.52) as a matrix equation (in the quantum numbers) we can invert it, using Heitler's equation, to give

$$|E, \mathbf{n}^0\rangle = \sum_{\mathbf{m}} |E, \mathbf{m}^+\rangle(\mathbf{I}-i\pi\mathbf{K})_{\mathbf{m},\mathbf{n}}. \tag{5.53}$$

Equation (5.53) determines the normalization of the standing wave solutions from the known normalization of the outgoing solution. It is important to note that solutions with the same energy but different values of all other quantum numbers are not orthogonal.

An alternative, but closely related, method to guarantee a unitary **S** matrix in an actual computation is based on the observation that the **S** matrix can be diagonalized by a unitary transformation, and as a unitary matrix **S** will have eigenvalues that can be written as $\exp(2i\delta)$ where δ is real. The number δ, which is a function of the energy (since S is) is the *eigenphase shift*. From eqns (5.42) and (5.46) the same transformation will bring **K** and **T** to diagonal forms with the eigenvalues $-\pi^{-1}\tan\delta$ and $-\pi^{-1}\sin\delta\exp(i\delta)$ respectively. These last two results can be used to establish a connection between the number δ and the potential V and thus obtain numerical values for the phase shift. In an actual computation these numbers may be evaluated by an approximate procedure (say, Born), but the resulting **S** matrix is bound to be unitary and the optical theorem will be satisfied (see eqn (5.58) below).

Let $|\kappa_j\rangle$ be the jth eigenvector of the K and T operators

$$T|\kappa_j\rangle = -\pi^{-1}\sin\delta_j \exp(i\delta_j)|\kappa_j\rangle = \tau_j|\kappa_j\rangle. \tag{5.54}$$

The vector $\mathbf{\varkappa}_j$ with components $\langle E, \mathbf{n}|\kappa_j\rangle$ is thus an eigenvector of the **T** matrix with eigenvalue τ_j. These vectors are orthonormal,

$$\sum_{\mathbf{n}} \langle \kappa_l|E, \mathbf{n}\rangle\langle E, \mathbf{n}|\kappa_j\rangle = \delta_{lj}. \tag{5.55a}$$

In terms of these vectors and the diagonal matrix $\mathbf{\tau}$, $(\mathbf{\tau})_{lj} = \tau_j\delta_{lj}$,

$$\mathbf{T} = \mathbf{\varkappa}\mathbf{\tau}\mathbf{\varkappa}^\dagger, \tag{5.56}$$

so that, from eqn (4.9),

$$R_{f,i} = \frac{2\pi}{\hbar}\left|\sum_j \langle f|\kappa_j\rangle\langle\kappa_j|i\rangle\tau_j\right|^2 \delta(E_f - E_i). \tag{5.57}$$

The orthonormality relation (5.55a) can also be written

$$\langle\kappa_l|\delta(E-H_0)|\kappa_j\rangle = \sum_f \langle\kappa_l|f\rangle\delta(E-E_f)\langle f|\kappa_j\rangle = \delta_{lj}, \tag{5.55b}$$

so that

$$\sum_f R_{f,i} = \frac{2}{\pi\hbar}\sum_j |\langle i|\kappa_j\rangle|^2 \sin^2\delta_j. \tag{5.58}$$

Equation (5.58) is equivalent to the optical theorem, eqn (4.7), which states that

$$\sum_f R_{f,i} = -\frac{2}{\hbar}\text{im}\langle i|T|i\rangle = \frac{2}{\pi\hbar}\sum_j |\langle i|\kappa_j\rangle|^2 \sin^2\delta_j. \tag{5.59}$$

Crude bounds on transition rates can be obtained by taking $\sin^2\delta_j = 1$. A more reasonable estimate is to take the average value of $\sin^2\delta_j$ as $\frac{1}{2}$. (In this connection see Massey and Mohr (1934) and the general discussion in section 3.5.0.) The total rate of transitions from all states of a given energy E is obtained using eqn (5.55a) as

$$-\frac{2}{\hbar}\sum_{\mathbf{n}} \text{im}\langle E, \mathbf{n}|T|E, \mathbf{n}\rangle = \frac{4}{2\pi\hbar}\sum_j \sin^2\delta_j. \tag{5.60}$$

For scattering by a central potential, eqn (5.60) is indeed equivalent to combining eqns (4.59) and (2.77), as the angular momentum for relative motion commutes with H, and hence with T, so that the partial waves provide a suitable basis to diagonalize T. (Another example was discussed in section 2.3.1.) We can thus try to improve on the Born approximation, by using it to compute the phase shifts only (eqn (5.99)) and then using these to construct a **K** matrix. Thus, using eqns (4.59) and (4.60) we obtain an improved (probability conserving) Born approximation

$$\sigma = \frac{4\pi}{k^2} \sum_l (2l+1)\sin^2(\delta_l)_B$$

as compared to using eqn (2.99) directly in eqn (2.77).

2.5.2. Scattering by two potentials

It is often convenient to resolve the interaction V into two parts, $V = V_1 + V_2$, which are not equivalent in terms of their physical significance. For example, consider the vibrational excitation of a diatomic molecule by an atom, and take V_2 as that part of the interaction that can cause vibrational excitation, while V_1 can lead to elastic collisions only. One way to define V_1 is to consider it as the part of V that depends on the relative separation, R, only

$$V(\mathbf{R}, \mathbf{r}) = V_1(R) + V_2(\mathbf{R}, \mathbf{r}), \tag{5.61}$$

where $\mathbf{r}$ is the internuclear distance in the diatomic. An explicit form for $V_1(R)$ can be obtained by expanding $V(\mathbf{R}, \mathbf{r})$ in a Taylor series about the equilibrium separation r_e. If $|n\rangle$ is a vibrational state of the diatomic, $\langle \mathbf{k}', m|V_1|\mathbf{k}, n\rangle = 0$ if $n \neq m$. Note, however, that if we write the matrix representation of V_2 in the basis of vibrational states,

$$V_2 = \sum_{m,n} |m\rangle\langle m|V_2|n\rangle\langle n|,$$

then the diagonal elements are also ineffective in causing vibrational transitions. Thus an alternative definition of V_1, which is applicable to other types of internal excitations, is§

$$V_1' = \sum_n |n\rangle\langle n|V|n\rangle\langle n|, \tag{5.62}$$

so that $\qquad \langle \mathbf{k}', m|V_1'|\mathbf{k}, n\rangle = 0, \quad m \neq n.$

In systems where the off-diagonal elements of V_2 in the h representation are large, one can introduce a third definition of V_1, namely the 'adiabatic' interaction. (See below, section 3.2.1. In the most general sense, one can regard V_1 as an 'equivalent potential' for elastic scattering, as defined in section 3.2.0.)

Let $H_1 = H_0 + V_1$, and let∥ G_1 be the Green's operator and Ω_1 the wave operator for H_1,

$$G_1 = (E^+ - H_1)^{-1} = \Omega_1 G_0. \tag{5.63}$$

§ V_1' includes V_1 as a special case. However, if we keep only the first term in the Taylor expansion of V and regard the vibrational states as harmonic oscillator states, the two definitions are equivalent. The two approximations above are often made in practice following the work of Landau and Teller (1936).

∥ In the interests of compact notation we shall omit, in the following derivation, the argument E^+ of the relevant operators. Thus $G_1 = G_1(E^+)$, etc.

Using the identities
$$G = G_1 + G_1 V_2 G = G_1 + G V_2 G_1 \tag{5.64}$$
we can write the transition operator T as
$$\begin{aligned} T &= V + V G V \\ &= V_1 + V_1 G_1 V_1 + (I + V_1 G_1) V_2 (I + G V) \\ &= T_1 + \Omega_1^\dagger T_2 \Omega_1, \end{aligned} \tag{5.65}$$
where $T_1 = V_1 \Omega_1 = V_1 + V_1 G_1 V_1$ is the transition operator due to V_1 alone, and
$$T_2 = V_2 + V_2 G V_2 \tag{5.66}$$
is the transition operator between the states of H_1 due to the perturbation V_2. If Ω is the wave operator for V, $\Omega = I + GV$, we have from eqn (5.64),
$$V_2 \Omega = V_2 (I + G V_2)(I + G_1 V_1) = T_2 \Omega_1, \tag{5.67}$$
thus providing two alternative forms for the second term of eqn (5.65), which was introduced by Watson (1952). Watson has also shown that if we write the second term as $\Omega_1^\dagger V_2 \Omega$ the limits $\epsilon \to 0$, in the two wave operators, can be taken independently of one another.

Let $|\mathbf{k}, n_1^+\rangle$ be the solution of the L.S. equation for the interaction V_1,
$$|\mathbf{k}, n_1^+\rangle = |\mathbf{k}, n\rangle + G_1 V_1 |\mathbf{k}, n\rangle = \Omega_1 |\mathbf{k}, n\rangle. \tag{5.68}$$
In the example of vibrational excitation we know that by our selection of V_1,
$$\chi_{\mathbf{k},n}^+(\mathbf{R}, \mathbf{r}) = \langle \mathbf{R}, \mathbf{r} | \mathbf{k}, n_1^+\rangle = L_{\mathbf{k}}^n(\mathbf{R}) \varphi_n(\mathbf{r}), \tag{5.69}$$
where $\varphi_n(\mathbf{r})$ is the vibrational state, that was occupied in the initial state
$$\langle \mathbf{R}, \mathbf{r} | \mathbf{k}, n\rangle = \exp(i\mathbf{k} \cdot \mathbf{R}) \varphi_n(\mathbf{r}) \tag{5.70}$$
and, from eqn (5.68),
$$L_{\mathbf{k}}^n(\mathbf{R}) = \exp(i\mathbf{k} \cdot \mathbf{R}) + \int G_0^+(\mathbf{R}, \mathbf{R}') V_1(\mathbf{R}') L_{\mathbf{k}}^n(\mathbf{R}') \, d\mathbf{R}'. \tag{5.71}$$

The mutual interaction V_1, which cannot cause internal excitation, can cause a distortion of the relative motion of the atom and the diatomic and thus modify the probability density of finding them at a given relative separation. In other words, $|L_{\mathbf{k}}^n(\mathbf{R})|^2 d\mathbf{R}$ will depend on R and we expect, on semi-classical grounds, that the value of this density will be enhanced in those regions where $V_1(\mathbf{R})$ is attractive and diminished where $V_1(\mathbf{R})$ is repulsive. $L_{\mathbf{k}}^n(\mathbf{R})$ depends also on the initial wave vector $\mathbf{k}$, and by the same reasoning we expect that with increasing k the effects of $V_1(R)$ on the shape of $L_{\mathbf{k}}^n(\mathbf{R})$ will diminish (since the semiclassical momentum is proportional to $[E-V]^{\frac{1}{2}}$).

Let $|\mathbf{k}, n^+\rangle$ be the solution of the L.S. equation with the full interaction V and the initial state $|\mathbf{k}, n\rangle$. Using the first equation of (5.67) (without the V_2 term on the left),

$$\begin{aligned}|\mathbf{k}, n^+\rangle &= \Omega|\mathbf{k}, n\rangle \\ &= (I+GV_2)(I+G_1V_1)|\mathbf{k}, n\rangle \\ &= (I+GV_2)|\mathbf{k}, n_1^+\rangle \\ &= |\mathbf{k}, n_1^+\rangle + G_1 V_2|\mathbf{k}, n^+\rangle.\end{aligned} \quad (5.72)$$

The last two lines are the two alternative forms of the L.S. equation with $|\mathbf{k}, n_1^+\rangle$ as the initial state and V_2 as the perturbation. It is clear that only the second term in these equations can lead asymptotically to internally excited states. To obtain the asymptotic form of the state $|\mathbf{k}, n^+\rangle$ we have to determine the asymptotic forms of the Green's operators G_1 or G. This requires some care since the relevant Hamiltonians contain not only the kinetic energy operator for relative motion but also mutual interaction terms (whether V_1 or V). From the L-conjugate of eqn (5.63)

$$G_1 = G_0 \Omega_1^\ddagger$$

and, premultiplying by $\langle \mathbf{R}, n|$,

$$\langle \mathbf{R}, n|G_1 = \int \langle \mathbf{R}, n|G_0|\mathbf{k}, n\rangle (2\pi)^{-3}\, \mathrm{d}\mathbf{k}\, \langle \mathbf{k}, n|\Omega_1^\ddagger.$$

Introducing a complete set in coordinate space,

$$\langle \mathbf{R}, n|G_1 = \int \langle \mathbf{R}, n|G_0|\mathbf{R}', n\rangle\, \mathrm{d}\mathbf{R}' \langle \mathbf{R}', n|\mathbf{k}, n\rangle (2\pi)^{-3}\, \mathrm{d}\mathbf{k}\, \langle \mathbf{k}, n|\Omega_1^\ddagger,$$

we take the limit $\epsilon \to 0$ and then the asymptotic limit, $R \to \infty$, using eqns (2.24) and (3.7):

$$\lim_{R\to\infty} \langle \mathbf{R}, n|G_1$$
$$= -\frac{\mu \exp(i k_n R)}{2\pi \hbar^2 R} \int \exp(-i\mathbf{k}'.\mathbf{R}')\exp(i\mathbf{k}.\mathbf{R}')(2\pi)^{-3}\, \mathrm{d}\mathbf{k}\mathrm{d}\mathbf{R}'\langle \mathbf{k}, n|\Omega_1^\ddagger,$$

where $\mathbf{k}' = k_n \hat{\mathbf{R}}$ and $E - E_n = \hbar^2 k_n^2/2\mu$. The integration over $\mathbf{R}'$ leads to $\delta(\mathbf{k}' - \mathbf{k})$, so that

$$\lim_{R\to\infty} \langle \mathbf{R}, n|G_1 = -(\mu/2\pi\hbar^2)\frac{\exp(i k_n R)}{R} \langle \mathbf{k}', n|\Omega_1^\ddagger$$
$$= -(\mu/2\pi\hbar^2)\frac{\exp(i k_n R)}{R} \langle \mathbf{k}', n_1^-|, \quad (5.73)$$

where we have used eqns (5.68) and (5.24). Equation (5.73) is of considerable use as it enables us to establish the asymptotic form of a Green's function for interacting systems. Thus, by an identical argument,

$$\lim_{R\to\infty} \langle \mathbf{R}, n|G = -(\mu/2\pi\hbar^2)\frac{\exp(i k_n R)}{R} \langle \mathbf{k}', n^-|. \quad (5.74)$$

Two aspects of this limit deserve further consideration. First, the vector $\mathbf{k}'$ is determined both in direction (that of $\hat{\mathbf{R}}$) and in magnitude, $E = E_n + \hbar^2 k'^2/2\mu$, and energy is conserved. Second, the state vector on the left is the ingoing solution of the L.S. equation. In the operator derivation, eqn (5.65), this last point occurs since the wave operator is not L-invariant. In the coordinate representation it followed from the proper sequence of operations: take matrix elements, let $\epsilon \to 0$, take the asymptotic limit. It is possible, by interchanging the order of operations to arrive at other, incorrect, forms of eqn (5.73).

From eqns (5.72) and (5.73) with $m \neq n$

$$\lim_{R \to \infty} \langle \mathbf{R}, m | \mathbf{k}, n^+ \rangle = \lim_{R \to \infty} \int \varphi_m^*(\mathbf{r}) \psi_{\mathbf{k},n}^+(\mathbf{R}, \mathbf{r}) \, d\mathbf{r}$$

$$= -(\mu/2\pi\hbar^2) \frac{\exp(ik_m R)}{R} \langle \mathbf{k}', m_1^- | V_2 | \mathbf{k}, n^+ \rangle$$

$$= -(\mu/2\pi\hbar^2) \frac{\exp(ik_m R)}{R} \langle \mathbf{k}', m^- | V_2 | \mathbf{k}, n_1^+ \rangle. \quad (5.75)$$

The scattering amplitude for $m \neq n$ is thus

$$f_{m,n}(\mathbf{k}', \mathbf{k}) = -(\mu/2\pi\hbar^2) \langle \mathbf{k}', m_1^- | V_2 | \mathbf{k}, n^+ \rangle$$

$$= -(\mu/2\pi\hbar^2) \langle \mathbf{k}', m | \Omega_1^\ddagger V_2 \Omega | \mathbf{k}, n \rangle. \quad (5.76)$$

The second line in eqn (5.76) confirms our abstract operator computation. For $m = n$ there is, of course, an added contribution from the asymptotic form of $\langle \mathbf{R}, n | \mathbf{k}, n_1^+ \rangle$. Our formal result, eqn (5.65), is independent of any physical interpretation that we may wish to assign to V_1 and so, in general, with $E = E_n + \hbar^2 k^2/2\mu = E_m + \hbar^2 k'^2/2\mu$,

$$\langle \mathbf{k}', m | T | \mathbf{k}, n \rangle = \langle \mathbf{k}', m | T_1 | \mathbf{k}, n \rangle + \langle \mathbf{k}', m_1^- | T_2 | \mathbf{k}, n_1^+ \rangle. \quad (5.77)$$

Alternative formal forms are clear, for example,

$$\langle \mathbf{k}', m | T_1 | \mathbf{k}, n \rangle = \langle \mathbf{k}', m | V_1 | \mathbf{k}, n_1^+ \rangle = \langle \mathbf{k}', m_1^- | V_1 | \mathbf{k}, n \rangle \quad (5.78)$$

and

$$\langle \mathbf{k}', m_1^- | T_2 | \mathbf{k}, n_1^+ \rangle = \langle \mathbf{k}', m_1^- | V_2 | \mathbf{k}, n^+ \rangle = \langle \mathbf{k}', m^- | V_2 | \mathbf{k}, n_1^+ \rangle. \quad (5.79)$$

Eqns (5.78) and (5.79) are the matrix elements of the statements that T_1 and T_2 are L-conjugation invariant (cf. eqn (5.26)).

We can expand the total wavefunction as

$$\psi_{\mathbf{k},n}^+(\mathbf{R}, \mathbf{r}) = \sum_m F_{\mathbf{k},n}^m(\mathbf{R}) \varphi_m(\mathbf{r}), \quad (5.80)$$

so that, using eqns (5.75) and (5.69) with $m \neq n$,

$$\langle \mathbf{k}', m | T | \mathbf{k}, n \rangle = \sum_p{}' \int L_{-\mathbf{k}'}^{m*}(\mathbf{R}) V_{m,p}(\mathbf{R}) F_{\mathbf{k},n}^p(\mathbf{R}) \, d\mathbf{R}, \quad (5.81)$$

where $V_{m,p} = \langle m|V_2|p\rangle$ and
$$L^{m*}_{-\mathbf{k}'}(\mathbf{R}) = \langle \mathbf{k}', m_1^-|\mathbf{R}, m\rangle.$$
It is also interesting to write eqn (5.81) in the momentum representation
$$\langle \mathbf{k}', m|T|\mathbf{k}, n\rangle = \sum_{p \neq m} \iint \langle \mathbf{k}', m_1^-|\mathbf{l}', m\rangle V_{mp}(\mathbf{l}'-\mathbf{l})\langle \mathbf{l}, p|\mathbf{k}, n^+\rangle (2\pi)^{-6}\, \mathrm{d}\mathbf{l}\mathrm{d}\mathbf{l}', \tag{5.82}$$
where $\qquad V_{mp}(\mathbf{l}'-\mathbf{l}) = \langle \mathbf{l}', m|V_2|\mathbf{l}, p\rangle$
and in both equations we have assumed that V_2 is a local potential in R.

The numerical magnitude of these matrix elements is clearly dependent on the potential V_1, which distorts the relative motion of the two molecules in the initial and final states. More often than not V_2 is primarily a short-range potential, so that the distortion effects of V_1 are particularly important. In a similar fashion the distortion due to V_1 affects the relative momentum distribution. Another feature is the implicit role of V_2. The transition operator is not linear in V_2 in that the amplitudes $F^p_{\mathbf{k},n}(\mathbf{R})$ have been obtained under the combined influence of V_1 and V_2. We have previously seen (section 2.3.0) that even though these amplitudes may be, for $p \neq n$, asymptotically small (or identically zero), they need not be small for finite R. Thus the transition $n \to m$ can go via the sequence $n \to p \to m$, or $n \to p \to p' \to m$ and so on. In the lowest order in V_2 we can replace $|\mathbf{k}, n^+\rangle$ by $|\mathbf{k}, n_1^+\rangle$ and eqns (5.81) reduce to a single term,
$$\langle \mathbf{k}', m|T|\mathbf{k}, n\rangle \approx \int L^{m*}_{-\mathbf{k}'}(\mathbf{R}) V_{m,n}(\mathbf{R}) L^n_{\mathbf{k}}(\mathbf{R})\, \mathrm{d}\mathbf{R}. \tag{5.83}$$
This approximation neglects completely the possibility of multiple transitions due to V_2. Model computations (Sharp and Rapp 1965, Lester and Bernstein 1967) show that such transitions are indeed important and cannot be neglected without an extensive justification. We shall show later that some justification can be provided by replacing $V_{m,n}(\mathbf{R})$ by an effective potential which takes partial account of the possibility of such transitions. In other words, if we are willing to reinterpret the matrix element $V_{m,n}(\mathbf{R})$, we can justify the functional form (5.83) as a reasonable approximation. A rigorous form is, of course, to replace $\langle m, \mathbf{R}|V|n, \mathbf{R}\rangle$ by $\langle m, \mathbf{R}'|T_2|n, \mathbf{R}\rangle$. This latter 'potential' is, however, in principle non-local and non-real.

The approximation in eqn (5.83) is often referred to as 'the distorted-wave Born approximation' (DWB). It is a Born approximation with respect to V_2 (see also eqn (5.87) below) and corresponds to replacing the second term of eqn (5.65) by
$$T_{\mathrm{DWB}} = T_1 + \Omega_1^\ddagger V_2 \Omega_1. \tag{5.84}$$

As is clear, T_{DWB} is L-conjugation invariant, but the second term will fail to satisfy the optical theorem.§ We shall later derive T_{DWB} from a variational principle.

In contrast to the usual Born approximation the matrix element of V_2 is evaluated between states which are 'distorted' by the potential V_1. In favourable situations (for example, Devonshire 1937) the matrix elements can be evaluated analytically, otherwise one has to replace the set of exact coupled equations for the relative motion by their DWB approximation,
$$\Omega_{\text{DWB}} = \Omega_1 + G_1 V_2 \Omega_1,$$
and integrate numerically. By the definition of V_2 the second term in Ω_{DWB} leads to inelastic processes only, so that the presence of V_2 does not affect the elastic scattering. The total wavefunction is thus
$$\psi^+_{\mathbf{k},n\text{DWB}} = L^n_{\mathbf{k}}(\mathbf{R})\varphi_n(\mathbf{r}) + \sum_{m \neq n} F^m_{\mathbf{k},n}(\mathbf{R})_{\text{DWB}}\,\varphi_m(\mathbf{r}),$$
where $F^m_{\mathbf{k},n}(\mathbf{R})_{\text{DWB}} = \langle m, \mathbf{R} | G_1 V_2 | \mathbf{k}, n_1^+ \rangle$, or
$$[E - E_m - K - V_{m,m}(\mathbf{R})]F^m_{\mathbf{k},n}(\mathbf{R})_{\text{DWB}} = V_{m,n}(\mathbf{R})L^n_{\mathbf{k}}(\mathbf{R}).$$

In practice it is often the case that the potential V is expressed as a function of a small number of parameters (Morse, Lennard-Jones, etc.). The initial and final states do not depend on these parameters, so that if z denotes one such parameter
$$\frac{d}{dz}\langle \mathbf{k}', m | T | \mathbf{k}, n \rangle = \left\langle \mathbf{k}', m \left| \frac{dT}{dz} \right| \mathbf{k}, n \right\rangle.$$

An operator expression for dT/dz can be obtained from $T = V + VGV$ and $G = G_0 + G_0 VG$ since, by assumption, H_0 does not depend on z. We then find
$$\frac{dG}{dz} = G\frac{dV}{dz}G, \tag{5.85}$$
and
$$\frac{dT}{dz} = (I+VG)\frac{dV}{dz}(I+GV) = \Omega^{\ddagger}\frac{dV}{dz}\Omega. \tag{5.86}$$

(Note that in eqn (5.86) V, G, and Ω are functions of z.) In particular, if $V = V_1 + zV_2$ and V_1 and V_2 do not depend on z, we have that
$$\left(\frac{dT}{dz}\right)_{z=0} = \Omega_1^{\ddagger} V_2 \Omega_1, \tag{5.87}$$
where Ω_1 is the wave operator for V_1. Our previous approximation of replacing T_2 in eqn (5.65) by V_2 (eqn (5.83)) is thus equivalent to the lowest

§ From the definition of V_2, and from eqn (5.69), $\langle \mathbf{k}, n|\Omega_1^{\ddagger} V_2 \Omega_1|\mathbf{k}, n\rangle = 0$.

FORMAL COLLISION THEORY

term in the expansion of T in powers of V_2. The relation

$$\frac{d}{dz}\langle \mathbf{k}', m | T | \mathbf{k}, n \rangle = \left\langle \mathbf{k}', m^- \left| \frac{dV}{dz} \right| \mathbf{k}, m^+ \right\rangle \tag{5.88}$$

can be considered as a 'Hellmann–Feynman' type theorem (Levine 1966a). A similar equation can be established for the K operator (eqn (5.47)). An interesting consequence of these equations is the fact that the eigenphases δ_j (the phase shifts) increase monotonically when we increase the magnitude of the potential.

For a central potential we can perform a partial wave analysis of eqn (5.86). Each wave is independent, and one can show that

$$\frac{d\delta_l}{dz} = -(2\mu/\hbar^2) \int_0^\infty k|U_{kl}(R)|^2 \frac{dV}{dz} \, dR, \tag{5.89}$$

where U_{kl} is defined by eqn (2.83).§ Thus as the potential becomes more attractive ($dV/dz < 0$) the phase shift increases. Numerical results for a potential of the form $zf(R)$ are presented in Fig. 2.8.

To cast eqn (5.89) in an operator form we use eqns (5.86), (5.30), and (5.11) to conclude that

$$S^\dagger \frac{dT}{dz} = \Omega^\dagger \frac{dV}{dz} \Omega. \tag{5.90}$$

2.5.3. *The density of states*

The density of states at a particular energy can be defined as

$$\rho(E) = \mathrm{Tr}[\delta(E-H)], \tag{5.91}$$

when we recall that $\rho(E)\,dE$ is the number of states of total energy between E and $E+dE$, and each state of energy E contributes unity to the trace in eqn (5.91). Using eqn (1.12) we can write

$$\rho(E) = -\pi^{-1}\mathrm{im}\,\mathrm{Tr}[(E^+-H)^{-1}]$$

$$= -\pi^{-1}\mathrm{im}\sum_n (E^+-E_n)^{-1}$$

$$= -\pi^{-1}\mathrm{im}\,\frac{d}{dE}\ln \prod_n (E^+-E_n)$$

$$= -\pi^{-1}\mathrm{im}\,\frac{d}{dE}\ln \det(E^+-H), \tag{5.92}$$

where in the second line we took the trace in a basis that diagonalizes H, and in the last line we wrote the determinant as a product of its eigenvalues.

§ We recall again that δ_l is a function of z and eqn (5.89) gives $d\delta_l/dz$ for the same value of z used to evaluate $U_{kl}(R)$.

The change in the density of states when the Hamiltonian is modified from H_0 to H, $H = H_0+V$, is thus given by

$$\Delta(E) = \rho(E)-\rho_0(E)$$
$$= -\pi^{-1}\mathrm{im}\frac{\mathrm{d}}{\mathrm{d}E}[\ln\det(E^+-H)-\ln\det(E^+-H_0)]$$
$$= -\pi^{-1}\mathrm{im}\frac{\mathrm{d}}{\mathrm{d}E}\ln\det[I-G_0(E^+)V], \qquad (5.93)$$

where we have written

$$\det(E^+-H) = \det\{(E^+-H_0)[I-G_0(E^+)V]\}$$
$$= \det(E^+-H_0)\det[I-G_0(E^+)V].$$

The final determinant in eqn (5.93) is the Fredholm determinant $D(E^+)$,

$$D(E^+) = \det[I-G_0(E^+)V] \qquad (5.94)$$

(see DeWitt 1956, Baker 1958). Equation (5.92) can also be written in terms of $\mathrm{Tr}(E^--H)^{-1}$, by changing the sign in front. Proceeding in the same steps that led to eqn (5.93) and noting that the determinant of a product of two matrices is the product of the determinants we obtain

$$\Delta(E) = (2\pi)^{-1}\mathrm{im}\frac{\mathrm{d}}{\mathrm{d}E}\ln\det\left[\frac{I-G_0(E^-)V}{I-G_0(E^+)V}\right]. \qquad (5.95)$$

The determinant in eqn (5.95) can be rewritten following DeWitt as

$$\frac{D(E^-)}{D(E^+)} = \det\left[\frac{I-G_0(E^-)V}{I-G_0(E^+)V}\right]$$
$$= \det\{I+[G_0(E^+)-G_0(E^-)]V[I-G_0(E^+)V]^{-1}\}. \qquad (5.96\,\mathrm{a})$$

Using eqns (5.20) and (5.32) this can be written as§

$$\det[I-2\pi i\delta(E-H_0)T(E^+)] = \det S(E^+) = \prod_j \exp(2i\delta_j), \qquad (5.96\,\mathrm{b})$$

where in the last line we have again written the determinant in terms of its eigenvalues. We can conclude that∥

$$\Delta(E) = (2\pi)^{-1}\mathrm{im}\frac{\mathrm{d}}{\mathrm{d}E}\ln\det S(E^+) = \pi^{-1}\sum_j\frac{\mathrm{d}\delta_j}{\mathrm{d}E}. \qquad (5.97)$$

The mutual interaction V may be sufficiently attractive to support

---

§ Equation (5.69 b) corresponds to $D(E^+) = \prod_j D_j(E^+)$, where $D_j(E^+)$ is the Fredholm determinant in the subspace spanned by the eigenfunctions $|k_j\rangle$ that diagonalize the T matrix.

∥ See also Smith (1960).

bound states.§ The number n of these states can be determined as

$$n = \int_{-\infty}^{0} \Delta(E)\, dE, \qquad (5.98)$$

where the upper limit is determined by our convention that the continuous spectrum of H begins at zero energy. From eqn (5.94) we observe that
$$\lim_{E \to -\infty} D(E^+) = \lim_{V \to 0} D(E^+) = 1$$
or, using eqn (5.96),
$$\lim_{E \to -\infty} \delta_j(E) = \lim_{V \to 0} \delta_j(E) = 0$$
and so, using eqn (5.97), we can perform the integration of $\Delta(E)$ to get

$$n = \pi^{-1} \sum_j \delta_j(0). \qquad (5.99)$$

Equation (5.99) is an example of Levinson's (1949) theorem.∥ Not all the bound states predicted by eqn (5.99) are observable. For example, the Pauli principle excludes the possibility of seven electrons in the $2p$ orbitals of the neon atom, a configuration which, energetically, is bound. Thus in electron–neon scattering $\delta_1(E \to 0) \to 2\pi$ even though there is no real bound state (cf. Swan 1955).

At the energy E_0, determined by

$$\delta_j(E_0) = \tfrac{1}{2}\pi,$$

the matrix element $|\tau_j(E_0)|^2$ (eqn (5.54)) contributes its maximum possible value to the transition rate (eqn (5.58)) and the cross-section. As a function of E, $\delta_j(E)$ can either increase or decrease around $E = E_0$. Only in the former case, we have from eqn (5.97)

$$\Delta_j(E_0) = \pi^{-1}\left(\frac{d\delta_j(E)}{dE}\right)_{E=E_0} > 0 \qquad (5.100)$$

or, the presence of the potential causes an increase in the density of states, i.e. the maximum in the cross-section is a resonance.

§ In fact, even the weak van der Waals's attraction between rare-gas atoms has an attractive well which can support a few bound states.

∥ The present proof is not rigorous. We have assumed that all the relevant determinants exist, while in fact $D(E^+)$ is divergent for a three-dimensional problem with a continuous spectrum. We can, however, perform a partial wave resolution, and consider only a particular l value when $D_l(E^+)$ exists provided $V(R)$ is smaller than R^{-2} at the origin and at infinity. (The hard core of intermolecular potentials may violate the required behaviour near the origin, cf. eqn (2.88).) Moreover, the s-wave ($l = 0$) phase shift need not be continuous at $E = 0$. In fact if V can support an s-wave bound state of zero energy, $\delta_0(+0) - \delta_0(-0) = \tfrac{1}{2}\pi$. Since only $\delta(E > 0)$ is measurable one sometimes replaces n in eqn (5.99) by $n + \tfrac{1}{2}$. We have also assumed that the continuous spectrum of H has just a single threshold, so that only elastic collisions are considered.

As an example we consider various manifestations of the Ramsauer effect (1921); see also Peterson (1962). As a model we take an attractive square well of radius a and depth V_0. To evaluate the s-wave phase shift we put

$$U_{k,0}(R) = \begin{cases} k^{-1}\sin(KR) & (R < a), \\ k^{-1}\sin(kR+\delta_0) & (R > a), \end{cases}$$

where $(\hbar^2 K^2)/2\mu = E+V_0$, and require that $U_{k,0}(R)$ and its derivative be continuous at $R = a$, so that

$$\delta_0 = \tan^{-1}\left(\frac{k}{K}\tan Ka\right) - ka.$$

At low energies (cf. eqns (2.86) and (2.87)) only the s-wave contributes to the cross-section, and at such energies that δ_0 passes through π, $\sigma = (4\pi/k^2)\sin^2\delta_0$ will have a minimum. Such an effect will be observed when $\delta_0 \to n\pi$ at zero energy, starts increasing slightly as E increases, and then starts to decrease as E increases further. Such behaviour is experimentally observed in the low-energy scattering of electrons by rare-gas atoms A, Kr, and Xe. For He and Ne, δ_0 approaches $n\pi$ ($n = 1, 2$ respectively) from below and the effect is absent. The phase shift δ_0 does indeed show such behaviour for different values of V_0. An example is plotted in Fig. 2.8.

At high energies $k \simeq K$ and

$$\delta_0 \simeq (K-k)a, \tag{5.101}$$

a decreasing function of the energy. At such energies it is no longer valid to neglect the higher l-waves, but we can estimate their phase shift by the following geometrical argument. When E is above the centrifugal barrier, $E > \hbar^2 l(l+1)/2\mu R^2$, $l \ll kR$ and $b \simeq l/k \ll R$ and the lth partial wave sees a well of about the same length§ as the s-wave. Thus eqn (5.101) is approximately correct for all δ_l, when $l \ll kR$ and predicts a phase shift of $\frac{1}{2}\pi$ at‖

$$(K-k)a = (n-\tfrac{1}{2})\pi \quad (n \text{ integer}). \tag{5.102}$$

Expanding $(K-k)$ in powers of k, we obtain for the phase shift

$$\delta \approx \frac{V_0 a}{\hbar v} \tag{5.103}$$

---

§ Cf. Fig. 2.1. The well-length is equal to the length of the chord intercepted by the sphere of radius a, at the impact parameter b.

‖ Many partial wave cross-sections will thus be simultaneously large, and the total cross-section will show an extremum which is not a resonance. A precise analysis determines δ_l as the maximum ('glory') phase as a function of l.

and for the velocity corresponding to the nth extremum

$$\frac{V_0 a}{\hbar v_n} \approx (n-\tfrac{1}{2})\pi, \tag{5.104}$$

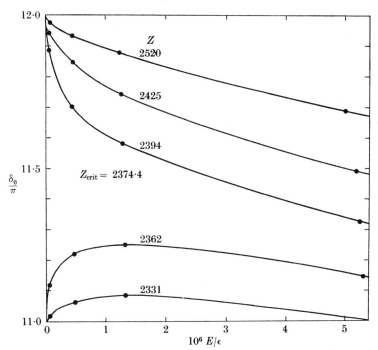

Fig. 2.8. Low-energy behaviour of an s-wave phase shift for a L.-J. (12, 6) potential of depth ϵ and equilibrium position r_m (Harrison and Bernstein 1963). $Z = 2\mu\epsilon r_m^2/\hbar^2$ is the reduced parameter of the system, and Z_{crit} is the value of Z above which the potential can support an additional bound state (from 11 to 12 states). The numerical method used ensures that the radial wavefunction vanishes at and in the near vicinity of the origin, and computes the phase shift as the (asymptotic) difference in the position of the zeros of the radial wavefunction and the spherical Bessel function. Thus in the limit $k \to 0$ we obtain directly the number of nodes of the radial wavefunction and Levinson's theorem is satisfied despite the steep repulsive part of the potential. Since the L.-J. potential depends monotonically on ϵ, the phase shift is monotonic in Z (cf. eqn (5.89)). The scattering length a (eqn (2.89)) changes sign on the introduction of another bound state.

where $v = \hbar k/\mu$. Bernstein (1961) predicted maxima in the total elastic-collision cross-section, and on the basis of a semiclassical analysis (1963) obtained for the Lennard-Jones (12, 6) potential§

$$0.95\frac{\epsilon\sigma}{\hbar v_n} \simeq (n-\tfrac{3}{8})\pi.$$

§ Equation (5.103) can be derived as an estimate to the integral in eqn (2.99), showing that the detailed shape of the potential well is unimportant for the present argument.

These extrema have now been experimentally observed (see Bernstein 1965) and are indeed the most significant feature of the currently available (poor resolution) total cross-sections for atom–atom collisions. The range of values of n in eqns (5.102) and (5.103) is partly determined by Levinson's theorem, which implies that δ_0 can decrease through $\tfrac{1}{2}\pi$ at least§ n_0 times where n_0 is the number of s-wave bound states. All observations thus far conform to $\max(n) \leqslant n_0$, where $\max(n)$ is the highest value of n observed. There are no experimental results on the very low-energy region for atom–atom collisions. (This region is determined by $ka < 1$ and the high atomic mass (as compared to electrons), makes the attainment of low k values very difficult.)

2.5.4. *Multiple-scattering theory*

Our discussion of collision events thus far has been restricted to binary collisions only. In this section we consider explicitly a system of N molecules. Our aim is to write down an expansion for the wave operator and derived operators which is appropriate to the low-density limit, namely, to situations where uncorrelated binary collisions are the most important physical process. Such an expansion will not only provide justification for some of the steps we have taken on intuitive grounds in section 2.4.3 but will also provide the correction terms to our description.

The total Hamiltonian H for the system of N molecules is resolved as $H = H_0 + V$, where H_0 is the sum of the Hamiltonians of the isolated molecules

$$H_0 = \sum_{r=1}^{N} H_r \tag{5.105a}$$

and V is the sum of pair interactions between the molecules

$$V = \tfrac{1}{2} \sum_{r \neq s} \sum V_{r,s}. \tag{5.105b}$$

A pair of indices (r, s) defines a particular pair of molecules, and if we denote a pair index by a small Greek letter,

$$V = \sum_{\alpha} V_{\alpha}, \qquad H_{\alpha} = H_r + H_s + V_{r,s} \tag{5.106}$$

and α runs over the $\tfrac{1}{2}N(N-1)$ possible pairs.

An eigenstate of H_r will be denoted by i_r, $|i_r\rangle = |\mathbf{k}_r, n_r\rangle$, where $\hbar \mathbf{k}_r$ is the linear momentum of the centre of mass and n_r the internal state of

---

§ At a resonance $\delta_l(E)$ increases with energy by π (cf. eqns (2.59) and (5.100) and Fig. 2.3). Since Levinson's theorem should hold, $\delta_l(E)$ can decrease by π $(n_l + m_l)$ times, where m_l is the number of resonances. For an s-wave there can be no resonances due to the centrifugal barrier, and in the adiabatic approximation there can be no resonances due to internal excitations of the colliding atoms (section 3.2.0).

molecule r. Considering the molecules distinguishable, we can write an eigenstate of H_0 as $|i^N\rangle$,

$$|i^N\rangle = \prod_{r=1}^{N} |i_r\rangle, \qquad (E-H_0)|i^N\rangle = 0 \qquad (5.107\,\text{a})$$

and, from eqn (5.105 a),

$$E = \sum_r (E_{n_r}+\hbar^2 k_r^2/2\mu_r) = \sum_r E_r. \qquad (5.107\,\text{b})$$

We have specified the initial state not only by its total energy, but also by the internal and translational motion of each molecule.

Let Ω_α and T_α be the wave and transition operators for the pair α,

$$\Omega_\alpha = I+(E^+-H_0)^{-1}V_\alpha\Omega_\alpha = I+(E^+-H_0)^{-1}T_\alpha, \qquad (5.108)$$

where we have not indicated explicitly the argument (E^+) of the operators. We use the notation $r \neq \alpha$ to imply that r differs from both indices of the pair α so that, with $\alpha = (s,t)$,

$$\Omega_\alpha |i^N\rangle = \prod_{r \neq \alpha} |i_r\rangle |i_s, i_t^+\rangle, \qquad (5.109)$$

where, using eqn (5.107), we can write

$$|i_s, i_t^+\rangle = |i_s, i_t\rangle + \left(E^+ - \sum_{r \neq \alpha} E_r - H_s - H_t\right)^{-1} V_\alpha |i_s, i_t^+\rangle,$$

so that $|i_s, i_t^+\rangle$ is the conventional L.S. equation for a binary collision with the energy E_s+E_t.

We seek to determine the higher terms in the expansion

$$\Omega = \sum_\alpha \Omega_\alpha + \ldots = I+(E^+-H_0)^{-1}\sum_\alpha T_\alpha + \ldots,$$

where Ω is the total wave operator,

$$\Omega = I+(E^+-H_0)^{-1}V\Omega. \qquad (5.110)$$

Following Watson (1953, 1956) we try the closed form

$$\Omega = I+(E^+-H_0)^{-1}\sum_\alpha T_\alpha \Omega^\alpha, \qquad (5.111)$$

where the operators Ω^α correct the isolated binary-collisions picture. To determine the operators Ω^α we require that (5.111) be the solution of (5.110) or

$$\sum_\alpha T_\alpha \Omega^\alpha = V\Omega = \sum_\alpha V_\alpha \Omega = \sum_\alpha V_\alpha\left[I+(E^+-H_0)^{-1}\sum_\beta T_\beta \Omega^\beta\right].$$

Using the definition $T_\alpha = V_\alpha + V_\alpha G_0^+ T_\alpha$ in the left-hand side, we find

$$\Omega^\alpha = I+(E^+-H_0)^{-1}\sum_{\beta \neq \alpha} T_\beta \Omega^\beta. \qquad (5.112)$$

The set of eqns (5.112) can be iterated, starting with $\Omega^\beta = I$ on the right-hand side, leading to§

$$\Omega = I+(E^+-H_0)^{-1}\sum_\alpha T_\alpha+(E^+-H_0)^{-1}\sum_\alpha T_\alpha(E^+-H_0)^{-1}\sum_{\beta\neq\alpha}T_\beta+\dots. \quad (5.113)$$

Recalling the identity $G^+ = \Omega G_0^+$ we can multiply eqn (5.113) from the right by $(E^+-H_0)^{-1}$ to obtain the 'binary collision expansion' of Siegert and Teramoto (1958). This expansion can be used in eqn (5.92) to generate expansions for other functions of H, in particular the partition function (see Watson 1956, Glassgold, Heckrotte, and Watson 1959). In a similar manner, defining the transition operator by $\Omega = 1+G_0^+ T$ we have

$$T = \sum_\alpha T_\alpha + \sum_\alpha T_\alpha G_0^+ \sum_{\beta\neq\alpha}T_\beta+\dots. \quad (5.114)$$

There is a considerable body of literature concerned with the derivation of closed forms of integral equations that can be iterated to the form (5.114). (See, for example, Faddeev 1961, 1963, Weinberg 1964, Lovelace 1964, and Rosenberg 1965. For $N = 3$, eqns (5.112) are essentially Faddeev's equations. In specific applications it is often more convenient to use the method of channel projections as outlined in the following section. See also Malfliet and Ruijgrok 1967 and van Nieuwenhuizen and Ruijgrok 1967.)

To indicate the volume dependence of the terms in the iteration of T, we recall from section 2.4.1 that a matrix element of T that changes the state (whether translational or internal) of p molecules has a volume dependence of $\mathsf{V}^{-(p+1)}$. (Recall that the factor V is due to the conservation of the total centre of mass momentum.) The iteration (5.114) corresponds therefore to a density expansion. The first term represents independent pair collisions, while higher-order terms represent collisions involving more than two molecules (see also section 2.7.1). Explicitly

$$\langle f^N|T_\alpha|i^N\rangle = \prod_{r\neq\alpha} \langle f_r|i_r\rangle \delta(\mathbf{k}'_s+\mathbf{k}'_t-\mathbf{k}_s-\mathbf{k}_t)\langle \mathbf{k}', n'_s, n'_t|T_\alpha|\mathbf{k}, n_s, n_t\rangle,$$

where $\mathbf{k}$ is the initial relative momentum, the delta function represents the conservation of the momentum of the centre of mass of the pair α, and the primed quantities refer to the final state.

It is of interest to examine the same problem from a density matrix point of view. Our discussion follows Snider (1960). We define $\rho^{(N)}$ as the N particle density matrix normalized to $N!$ and define the m particle

§ If by the symbol $\sum_{\beta\neq\alpha}$ we mean summation over both β and α, excluding $\beta = \alpha$, then the third term in eqn (5.113) should be written $G_0^+ \sum_{\beta\neq\alpha} T_\alpha G_0^+ T_\beta$.

reduced density matrix $\rho^{(m)}$ by

$$\rho^{(m)} = [(N-m)!]^{-1}\mathrm{Tr}_{N-m}\rho^{(N)}, \qquad (5.115)$$

where the trace is over $N-m$ molecules.

In particular the equation of motion

$$i\hbar\frac{d\rho^{(N)}}{dt} = [H, \rho^{(N)}]$$

leads to $\quad i\hbar\dfrac{d\rho^{(2)}}{dt} = [H_{1,2}, \rho^{(2)}] + \mathrm{Tr}_3[(V_{1,3}+V_{2,3}), \rho^{(3)}], \qquad (5.116)$

where $\rho^{(2)}$ is the density matrix for the pair of molecules.

In the lowest binary approximation the evolution of a pair of molecules is unaffected by the rest of the system and eqn (5.116) can be approximated by

$$i\hbar\frac{d\rho^{(2)}}{dt} = [H_{1,2}, \rho^{(2)}], \qquad (5.117)$$

$$H_{1,2} = H_1 + H_2 + V_{1,2}.$$

Equation (5.117) has the solution

$$\rho^{(2)}(t) = \exp[-iH_{1,2}(t-t')/\hbar]\rho^{(2)}(t')\exp[iH_{1,2}(t-t')/\hbar]$$

and, using the techniques of section 2.7.2, it can be shown that

$$\rho^{(2)}(t) = \Omega_{1,2}\rho_0^{(2)}(t)\Omega_{1,2}^\dagger, \qquad (5.118)$$

where $\rho_0^{(2)}(t)$ is the density matrix for the non-interacting pair. We have previously used eqn (5.118) to discuss Ehrenfest's theorem for ensembles, in the low-density limit.

2.6. Reactive collisions

2.6.0. The resolution of the total Hamiltonian as H_0+V was achieved by requiring that $V(R)$, the mutual interaction, decreases as the relative separation between the molecules increases. When we deal with molecules which possess internal degrees of freedom the above specification need not be unique, in that there is more than one way of separating the whole system into two parts. For example, consider a three-atom system, say KHBr, then we can think of K+HBr or KBr+H or several other possibilities when we recognize that the atoms are themselves complex entities. There are thus several possibilities of writing

$$H = H_{0,i}+V_i = H_{0,f}+V_f$$

depending on the relative coordinate that we increase beyond any finite value. Experimentally it is, of course, well known that processes of the type

$$\text{K+HBr} \to \text{H+KBr}, \tag{A}$$

can take place, but equally well the process

$$\text{K+HBr}(v=n) \to \text{K+HBr}(v=m), \tag{B}$$

can take place, where v, say, is a vibrational quantum number (assuming no vibration–rotation interaction) which can either remain the same ($m=n$) or change due to the collision. The process (A) is a reactive collision, while (B) is a non-reactive collision (which need not, however, be elastic since vibrational or rotational or, under suitable conditions, internal electronic excitation may occur).

In a certain formal sense the non-uniqueness in the resolution of the total Hamiltonian which we noted in connection with reactive collisions is already present when non-reactive but inelastic collisions can take place. Consider, for example, a collision of a diatomic molecule with an atom. Let $|n\rangle$ be the lowest internal state of the diatomic, of energy E_n,§ and let the atom be structureless. An initial state for such a system is denoted $|\mathbf{k}_n, n\rangle$, where $E = E_n + \hbar^2 k_n^2/2$. This state can be described by the Hamiltonian $H_{0,n}$,

$$H_{0,n} = K_i + E_n, \tag{6.1 a}$$

where K_i is the kinetic energy operator for the relative motion. As we increase k_n we reach a value where $E > E_m$ so that the internal state $|m\rangle$ can be observed as a final internal state of the collision. $|\mathbf{k}_m, m\rangle$ is

---

§ Our previous convention for the zero of energy is $E_n = 0$.

described by the Hamiltonian $H_{0,m}$,
$$H_{0,m} = K_i + E_m. \tag{6.1b}$$
Further increase of the total energy will lead to other possible final states and, in general,
$$H_{0,l} = K_i + E_l \quad (l = n, m, p ...) \tag{6.1c}$$
is a Hamiltonian for a possible final state where l is an index of the bound states of the diatomic.

Above some energy threshold we can consider the possibility of an atom exchange between the diatomic and the atom,
$$A + BC \to AB + C.$$
A possible final state of such a reactive collision can be described by the Hamiltonian
$$H_{0,r} = K_f + E_r, \tag{6.2}$$
where K_f is the kinetic energy operator for the relative motion in the new coordinate and E_r is a possible internal energy ($E > E_r$). For the exchange reaction above, let R_i be the distance from A to the centre of mass of BC and R_f the distance from C to the centre of mass of AB, then in the coordinate representation
$$K_i = -(\hbar^2/2\mu_i)\nabla^2_{R_i}, \quad K_f = -(\hbar^2/2\mu_f)\nabla^2_{R_f},$$
where μ is the appropriate reduced mass
$$\mu_i = M_A(M_B + M_C)/(M_A + M_B + M_C)$$
and similarly for μ_f.

To summarize, with each detailed possible state of the reactants or products we can associate a Hamiltonian which is a sum of a kinetic energy operator for relative motion and an internal energy. In the conventional terminology each one of those Hamiltonians defines a channel. In this sense a channel corresponds to specifying a complete set of quantum numbers for the state. Collisions, where the final state may be found in more than one channel, are referred to as 'multi-channel' collisions. In this sense non-reactive but inelastic collisions are multi-channel collisions. In a collision of an atom with a rigid rotor the pair of quantum numbers j, l can be considered as defining a channel.

We can, however, consider together all the possible initial states. These correspond to the well-separated atom A and bound molecule B–C and describe them by the Hamiltonian H_0^i,
$$H_0^i = K_i + \sum_l |l\rangle E_l \langle l| \quad (l = m, n, p ...), \tag{6.3}$$

where l enumerates the bound internal states of BC (by assumption A is structureless). In general, the internal energy operator in H_0^i is not equal to h_i, the Hamiltonian operator for the internal motion. The operator h_i has also a continuous spectrum which corresponds to the dissociated diatomic. (In reality atoms possess internal structure and the continuous spectrum of h_i is much more complicated in that the atom or the diatomic or its fragments can ionize, and various charged fragments can thus appear.) If C_i is the projection operator on the bound states of the diatomic§ we have that

$$H_0^i = H_{0,i} C_i = K_i + h_i C_i, \qquad (6.4\,\text{a})$$

is the appropriate Hamiltonian for the initial states of the type $A+BC$ where BC is bound. We must therefore conclude that if C_f is the projection on the bound states of AB (or in general on the bound states of h_f), then $H_{0,i} C_f$ does not necessarily vanish. From a formal point of view the states of $H_{0,i}$ form a complete set, and so they do provide, in principle, a basis for the expansion of states of the type $AB+C$, namely the final states. The point is that in practice such an expansion may not be very convenient (see Castillejo, Percival, and Seaton 1960) and it is conceptually simpler to describe the possible final states of reactive collisions by the Hamiltonian H_0^f,

$$H_0^f = H_{0,f} C_f = K_f + h_f C_f. \qquad (6.4\,\text{b})$$

We can associate with each Hamiltonian H_0^j a reaction channel and refer to the states of H_0^j as states of the jth channel. The channel label j can be explicitly introduced in the labelling of the possible initial and final states. Thus

$$\begin{aligned}
H_0^i |\mathbf{k}, n, i\rangle &= (\hbar^2 k^2 / 2\mu_i + E_n) |\mathbf{k}, n, i\rangle \\
&= H_{0,i} |\mathbf{k}, n, i\rangle \\
&= H_{0,n} |\mathbf{k}, n, i\rangle,
\end{aligned} \qquad (6.5)$$

where n is a label‖ of the internal state and $H_{0,n} = K_i + E_n$.

From an experimental point of view the channel label j is certainly meaningful in that, after the collision is over and the molecules are well separated and non-interacting, we can determine whether a final state belongs to the channel j. Such an experiment provides a 'yes' or 'no'

---

§ In general C_i is the projection on the bound internal states of h_i.

‖ From the last line in eqn (6.5) n is also a channel label, and in this sense the set of quantum numbers (n, i) is the proper channel labelling, and if no reactive collisions are possible, n by itself is a channel label. We shall reserve, however, the term channel label to the index of the reactive channels only, even though our formal considerations regarding this index are equally applicable to the index n as well.

answer to the question 'was the collision reactive?'. It can be performed when the molecules have separated to a considerable distance apart and so cannot affect the evolution of the system during the collision. We therefore expect (following von Neumann 1955 and Golden 1957) that the classification of final channels can be formally introduced by means of projection operators. Let C_f be the operator that rejects asymptotic states that are not in the channel f,

$$C_f|E,\mathbf{n},j\rangle = \delta_{jf}|E,\mathbf{n},j\rangle. \qquad (6.6)$$

The eigenvalues of C_f provide the answer to the question 'Was the collision reactive?' when we associate 'yes' or 'no' with one or zero respectively. Since a final state must belong to some reaction channel,

$$\sum_f C_f = I. \qquad (6.7)$$

We therefore conclude that the different possible final channels provide mutually exclusive alternatives for the question 'What is the final state of the collision?' A mathematically elegant discussion of these questions has been given by Jauch (1958) and Jauch and Marchand (1966), and a simplified summary is given in section 2.7.4. One can also discuss these questions explicitly in the coordinate representation (section 2.6.4). In principle, however, the answer is provided by the experimental ability to distinguish the different reactive channels.

2.6.1. Reaction rates

Let $|E,\mathbf{n},i\rangle$ be an initial state in the channel i and let $G_{0,i}(E^+)$ be the Green's operator for the Hamiltonian $H_{0,i}$, $G_{0,i}(E^+) = (E^+ - H_{0,i})^{-1}$, so that the L.S. equation is

$$|E,\mathbf{n},i^+\rangle = |E,\mathbf{n},i\rangle + G_{0,i}(E^+)V_i|E,\mathbf{n},i^+\rangle. \qquad (6.8)$$

In this form it is not explicitly clear that the state vector $|E,\mathbf{n},i^+\rangle$ describes the whole range of events including the reactive collisions into the channel f. We recall that eqn (6.8) was obtained from

$$(E^+ - H)|E,\mathbf{n},i^+\rangle = (E^+ - H_{0,i})|E,\mathbf{n},i\rangle \qquad (6.9)$$

by premultiplying by $G_{0,i}(E^+)$. We can, however, premultiply eqn (6.9) by $G_{0,f}(E^+)$, the Green's operator for $H_{0,f}$, to obtain

$$|E,\mathbf{n},i^+\rangle = G_{0,f}(E^+)(E^+ - H_{0,i})|E,\mathbf{n},i\rangle + G_{0,f}(E^+)V_f|E,\mathbf{n},i^+\rangle, \qquad (6.10)$$

where we put $H = H_{0,f} + V_f$. Equation (6.10) was first written by Lippmann (1956) in order to indicate explicitly the presence of an outgoing

Green's function for $H_{0,f}$. We now have that

$$\langle f, \mathbf{m}, E'|E, \mathbf{n}, i^+\rangle = \frac{i\epsilon}{E^+ - E'}\langle f, \mathbf{m}, E'|E, \mathbf{n}, i\rangle +$$
$$+ (E^+ - E')^{-1}\langle f, \mathbf{m}, E'|V_f|E, \mathbf{n}, i^+\rangle, \quad (6.11)$$

where we recall that $\quad (E - H_{0,i})|E, \mathbf{n}, i\rangle = 0$
and that $\quad (E' - H_{0,f})|E', \mathbf{m}, f\rangle = 0$
and have put

$$\langle f, \mathbf{m}, E'|(E^+ - H_{0,f})^{-1} = (E^+ - E')^{-1}\langle f, \mathbf{m}, E'|.$$

In the limit $\epsilon \to 0$ the first term does not contribute. Note that this conclusion does not depend on the vanishing of the overlap integral. Since the channels i and f are different reaction channels we expect that at least one particle has been exchanged in the collision. We thus expect that
$$\langle f, \mathbf{m}, E'|E, \mathbf{n}, i\rangle = A\delta_{f,i} + B(f, i), \quad (6.12)$$

where B is finite (or often zero) also in the limit $E' \to E$. Even though the first term in eqn (6.11) does not contribute in the channel f in the limit $\epsilon \to 0$ it is inconsistent to drop the first term in eqn (6.10) during any formal manipulations, prior to taking matrix elements and the limit $\epsilon \to 0$. This point has been repeatedly emphasized (Foldy and Tobocman 1957, Epstein 1957) as dropping this term leads to a homogeneous equation for $|E, \mathbf{n}, i^+\rangle$ which is unacceptable in that we can add any multiple of the solution of this homogeneous equation to the solution of the inhomogeneous equation

$$|E, \mathbf{m}, f^+\rangle = |E, \mathbf{m}, f\rangle + G_{0,f}(E^+)V_f|E, \mathbf{m}, f^+\rangle,$$

thus removing the uniqueness of the solution of the L.S. equation.

The rate of reactive transitions out of the initial state $|E, \mathbf{n}, i\rangle$ to the final state $|E', \mathbf{m}, f\rangle$ is given by Ehrenfest's theorem with A the projection on that final state

$$\frac{d}{dt}\langle E, \mathbf{n}, i^+|A|E, \mathbf{n}, i^+\rangle = (2/\hbar)\text{im}\langle E, \mathbf{n}, i^+|AV_f|E, \mathbf{n}, i^+\rangle$$
$$= (2/\hbar)\text{im}\langle E, \mathbf{n}, i^+|E', \mathbf{m}, f\rangle\langle E', \mathbf{m}, f|V_f|E, \mathbf{n}, i^+\rangle, \quad (6.13)$$

where we have noted that the projection on the final state commutes with $H_{0,f}$. Using eqn (6.11) this rate can be written as

$$(2\pi/\hbar)|\langle E', \mathbf{m}, f|V_f|E, \mathbf{n}, i^+\rangle|^2\delta(E' - E), \quad (6.14)$$

which is identical in form to the results for non-reactive collisions.§

§ Note that according to our convention the set of quantum numbers **n** includes the direction of the relative momentum, whereas in the notation $|\mathbf{k}_n, n, i\rangle$, n labels only internal quantum numbers.

We can now proceed to write down expressions for rates of different processes by complete analogy to the treatment of non-reactive collisions (section 2.4.1). The rate per unit volume of the transition

$$(\mathbf{k}_n, n, i) \to (\mathbf{k}_m, m, f)$$

per unit density of reactants is thus given by

$$\mathsf{V}^{-1} R(\mathbf{m} \leftarrow \mathbf{n}) = \frac{2\pi}{\hbar} |\langle \mathbf{k}_m, m, f | V_f | \mathbf{k}_n, n, i^+\rangle|^2 \rho_f(E-E_m), \qquad (6.15)$$

where $E = E_n + \hbar^2 k_n^2/2\mu_i = E_m + \hbar^2 k_m^2/2\mu_f$ and $\rho_f(E)$ is the density of translational states in the channel f at the translational energy E where

$$\rho_f(E - E_m) = \mu_f \hbar k_m/(2\pi\hbar)^3.$$

The differential cross-section associated with the above transition can thus be written as

$$\frac{d\sigma_{m,n}}{d\omega_m} = R(\mathbf{m} \leftarrow \mathbf{n})/v_n \mathsf{V}, \quad v_n = \hbar k_n/\mu_i.$$

A scattering amplitude can be introduced by

$$f_{m,n}(\mathbf{k}_m, \mathbf{k}_n) = -(\mu_f/2\pi\hbar^2)\langle \mathbf{k}_m, m, f | V_f | \mathbf{k}_n, n, i^+\rangle, \qquad (6.16)$$

so that
$$\frac{d\sigma_{m,n}}{d\omega_m} = \frac{v_m}{v_n} |f_{m,n}(\mathbf{k}_m, \mathbf{k}_n)|^2. \qquad (6.17)$$

Equation (6.16) introduces the scattering amplitude using Ehrenfest's theorem. We shall show below (eqn (6.81)) that an identical definition follows from the asymptotic behaviour of the solution of the L.S. equation in the coordinate representation.

The rate per unit volume of reactive transitions from the initial state $|\mathbf{k}_n, n, i\rangle$ to the internal state $|m\rangle$ in the final channel, irrespective of the direction of the final momentum, is obtained by integrating eqn (6.15) over $\hat{\mathbf{k}}_m$ and, using eqn (6.16),

$$\mathsf{V}^{-1} R(m \leftarrow \mathbf{n}) = (\hbar k_m/\mu_f) \int |f_{m,n}(\mathbf{k}_m, \mathbf{k}_n)|^2 \, d\hat{\mathbf{k}}_m, \qquad (6.18)$$

and the associated cross-section $\sigma_{m,n}(\mathbf{k}_m)$ is obtained by dividing by v_n. We can also sum over the internal-states quantum numbers m to obtain the total rate of reactive transitions out of the state $|\mathbf{k}_n, n, i\rangle$. The associated cross-section is the total reactive cross-section. The corresponding observable is clearly $A = C_f$. In deriving the rates we have used three different measurable features of the final states $A = |\mathbf{k}_m, m, f\rangle\langle\mathbf{k}_m, m, f|$, $A = |m, f\rangle\langle m, f|$, and $A = C_f$, respectively. In general, we can associate a rate with any measurable feature of the final states.

Finally, by analogy to eqn (4.59) we can introduce the rate of reactive transitions in the microcanonical ensemble

$$\frac{\mathrm{d}\dot{N}(m \leftarrow n)}{\mathrm{d}E} = (2\pi\hbar)^{-1}\pi^{-1}k_n^2\,\sigma_{m,n}, \qquad (6.19)$$

where we have put $\mathsf{V}^{-1}R(m \leftarrow \mathbf{n}) = v_n\,\sigma_{m,n}$ and E is the total energy in the centre of mass system.

In thermal equilibrium the probability (per unit volume) $P(E)\,\mathrm{d}E$ of observing the internal state n with total energy E to $E+\mathrm{d}E$ is given by the Boltzmann distribution (cf. eqn (4.80)),

$$P(E)\,\mathrm{d}E = (Z_{\text{int}}\,Z_{\text{tr}})^{-1}g_n\exp(-\beta E)\,\mathrm{d}E, \qquad (6.20)$$

where Z_{int} is the internal partition function for the reactants, Z_{tr} is the translational partition function (per unit volume) for the relative motion of the non-interacting reactants and g_n is the degeneracy of the state n. From the discussion of section 2.4.3 the rate per unit volume and unit density of reactants of the reactive transition $n \to m$ is

$$\mathsf{V}^{-1}R_{\text{c}}(m \leftarrow n) = (Z_{\text{int}}\,Z_{\text{tr}})^{-1}\int \exp(-\beta E)\frac{\mathrm{d}\dot{N}(m \leftarrow n)}{\mathrm{d}E}\,\mathrm{d}E, \qquad (6.21)$$

where c stands for the canonical ensemble.

Changing the variable of integration to $e = E - E_n$ and noting that $\dot{N}(m \leftarrow n)$ vanishes for $E < E_n$ we can write

$$\mathsf{V}^{-1}R_{\text{c}}(m \leftarrow n) = Z_{\text{int}}^{-1}(\pi\mu_i)^{-\frac{1}{2}}(2/\mathbf{k}T)^{\frac{3}{2}}\times$$
$$\times g_n\exp(-\beta E_n)\int_0^\infty e\sigma_{m,n}(e)\exp(-\beta e)\,\mathrm{d}e. \qquad (6.22)$$

The only difference in the formal form of the results between reactive and non-reactive rates is in the significance of the labels n and m. In the present case n and m are internal labels for states in different reactive channels. We also note that, as expected,

$$\mathsf{V}^{-1}R_{\text{c}}(m \leftarrow n) = (p_n)_{\text{c}}\mathsf{V}^{-1}R_{\text{T}}(m \leftarrow n), \qquad (6.23)$$

where $(p_n)_{\text{c}} = Z_{\text{int}}^{-1}g_n\exp(-\beta E_n)$ is the occupation probability of the internal state n in the canonical ensemble, and $R_{\text{T}}(m \leftarrow n)$ is defined by analogy to eqn (4.63). To obtain the rate of reactive collisions from channel i to channel f in a canonical ensemble we sum eqn (6.22) over m and n:

$$\mathsf{V}^{-1}R_{\text{c}}(f \leftarrow i) = \sum_{n,m}\mathsf{V}^{-1}R_{\text{c}}(m \leftarrow n)$$
$$= \sum_{n,m}(p_n)_{\text{c}}\mathsf{V}^{-1}R_{\text{T}}(m \leftarrow n) \qquad (6.24)$$
$$= \sum_{\mathbf{n}}(p_{\mathbf{n}})_{\text{c}}\sum_{\mathbf{m}}\mathsf{V}^{-1}R(\mathbf{m} \leftarrow \mathbf{n}),$$

where

$$(p_{\mathbf{n}})_{\text{c}}\,\mathrm{d}\mathbf{k}_n = P(E)(2\pi)^{-3}\,\mathrm{d}\mathbf{k}_n. \qquad (6.25)$$

REACTIVE COLLISIONS

It is clearly desirable to introduce a compact derivation of the necessary summations and averaging involved in eqn (6.24). Towards this aim we introduce, in section 2.6.3, a compact formulation of our discussion.

2.6.2. *Operator formulation of the theory of reactive collisions*

When reactive collisions are possible the Møller wave operator should carry a channel label, so that Ω_i is the operator that generates a solution of the L.S. equation from an initial state in the channel i,

$$|E, \mathbf{n}, i^+\rangle = \Omega_i |E, \mathbf{n}, i\rangle. \tag{6.26}$$

The transition matrix element§ for reactive collisions can be written

$$\langle E', \mathbf{m}, f | T_{f,i} | E, \mathbf{n}, i \rangle,$$

with $T_{f,i} = V_f \Omega_i$. As usual, we shall be mainly concerned with $T_{f,i}(E^+)$, where

$$T_{f,i}(E^+) = V_f \Omega_i(E^+) = V_f + V_f G(E^+) V_i, \tag{6.27a}$$

where we have used eqn (6.9) to write

$$\Omega_i(E^+) = I + G(E^+) V_i. \tag{6.28}$$

The operator $T_{f,i}(E^+)$ is not invariant under L-conjugation since V_f is different from V_i. However, we can write

$$[T_{i,f}(E^+) + H_{0,i}]^{\ddagger} = [H_{0,i} + V_i + V_i G(E^+) V_f]^{\ddagger}$$
$$= H + V_f G(E^+) V_i \tag{6.29}$$
$$= H_{0,f} + T_{f,i}(E^+).$$

If we take the matrix elements of eqn (6.29) on the energy shell we obtain

$$\langle E, \mathbf{m}, f | T_{f,i}(E^+) | E, \mathbf{n}, i \rangle = \langle E, \mathbf{m}, f | \Omega_f^{\ddagger} V_i | E, \mathbf{n}, i \rangle$$
$$= \langle E, \mathbf{m}, f^- | V_i | E, \mathbf{n}, i \rangle \tag{6.30}$$
$$= \langle E, \mathbf{m}, f | V_f | E, \mathbf{n}, i^+ \rangle,$$

where, from eqn (6.27),

$$[T_{i,f}(E^+)]^{\ddagger} = [V_i \Omega_f(E^+)]^{\ddagger} = \Omega_f^{\ddagger}(E^+) V_i \tag{6.31}$$

and

$$\langle E, \mathbf{m}, f | \Omega_f^{\ddagger}(E^+) = \langle E, \mathbf{m}, f^- |. \tag{6.32}$$

The symmetry property, eqn (6.30), holds only between states of equal energy. In a similar fashion the identity

$$H = H_{0,i} + V_i = H_{0,f} + V_f$$

yields, on the energy shell only,

$$\langle E, \mathbf{m}, f | V_i | E, \mathbf{n}, i \rangle = \langle E, \mathbf{m}, f | V_f | E, \mathbf{n}, i \rangle. \tag{6.33}$$

§ Note that this is a 'matrix element' between eigenstates of different Hamiltonians.

The Born approximation is thus uniquely defined. In practice, due to the inherent inaccuracies of available wavefunctions, eqn (6.33) is not exactly satisfied. Another useful identity connects the total Green's operator to the Green's operators for $H_{0,i}$ and $H_{0,f}$, namely,§

$$G(E^+) = \Omega_i(E^+)G_{0,i}(E^+) \qquad (6.34\,\text{a})$$
$$= \Omega_f(E^+)G_{0,f}(E^+). \qquad (6.34\,\text{b})$$

Multiplying eqn (6.34) from the left by V_i we have that

$$T_{i,i}(E^+)G_{0,i}(E^+) = T_{i,f}(E^+)G_{0,f}(E^+). \qquad (6.35)$$

Armed with this identity we can proceed to prove the optical theorem for rearrangement collisions. Our proof follows Lippmann (unpublished), and for simplicity we omit the argument E^+. Using the definition

$$T_{i,i} = V_i \Omega_i = V_i + V_i G_{0,i} T_{i,i} \qquad (6.36)$$

and following the steps from eqns (5.27) to (5.29),

$$T_{i,i} - T^\dagger_{i,i} = -2\pi i T_{i,i} \delta(E - H_{0,i}) T^\dagger_{i,i}, \qquad (6.37)$$

where $\delta(E - H_{0,i})$ is the density of the asymptotic states. We can resolve these states into their respective channels using the resolution, eqn (6.7),

$$\sum_f C_f = I. \qquad [(6.7)]$$

Then, from eqns (6.37) and (1.12),

$$T_{i,i} - T^\dagger_{i,i} = \sum_f T_{i,i}[G_{0,i}(E^+) - G_{0,i}(E^-)]C_f T^\dagger_{i,i}$$
$$= \sum_f T_{i,i} G_{0,i}(E^+)(-2i\epsilon)C_f G^\dagger_{0,i}(E^+) T^\dagger_{i,i}, \qquad (6.38)$$

where we recall that $G_{0,i}(E^-) = G^\dagger_{0,i}(E^+)$ (L-conjugation invariance). From eqn (6.35) and its Hermitian conjugate

$$T_{i,i} - T^\dagger_{i,i} = \sum_f T_{i,f} G_{0,f}(E^+)(-2i\epsilon)C_f G^\dagger_{0,f}(E^+) T^\dagger_{i,f}$$
$$= -2\pi i \sum_f T_{i,f} \delta(E - H_{0,f}) C_f T^\dagger_{i,f}, \qquad (6.39)$$

where we have put

$$-2\pi i \delta(E - H_{0,f}) C_f = \lim_{\epsilon \to +0} \frac{-2i\epsilon C_f}{(E^+ - H_{0,f})(E^- - H_{0,f})}$$
$$= \lim_{\epsilon \to +0} G_{0,f}(E^+)(-2i\epsilon) C_f G^\dagger_{0,f}(E^+).$$

The optical theorem is the diagonal, on the energy shell, matrix element of eqn (6.39), namely,

$$-(2/\hbar)\text{im}\langle E, \mathbf{n}, i | T_{i,i} | E, \mathbf{n}, i \rangle$$
$$= (2\pi/\hbar) \sum_f \left[\sum_\mathbf{m} \int dE' |\langle E, \mathbf{n}, i | T_{i,f} | E', \mathbf{m}, f \rangle|^2 \delta(E - E') \right]. \qquad (6.40)$$

§ Both forms are derived from our basic identity $(A-B)^{-1} = A^{-1} + (A-B)^{-1}BA^{-1}$, where to get (6.34 a) we have taken $A = (E^+ - H_{0,i})$ and $B = V_i$.

Each square bracket above is the total rate of transitions into a particular channel. Indeed in section 2.4.0 we obtained the optical theorem from the rate of change of the identity operator. A similar derivation holds in the present case when we employ eqn (6.7). The total rate of transitions into the channel f is thus given by

$$\frac{d}{dt}\langle \mathbf{k}, n, i^+|C_f|\mathbf{k}, n, i^+\rangle = \sum_m (\hbar k_m/\mu_f) \int |f_{m,n}(\mathbf{k}_m, \mathbf{k})|^2 \, d\hat{\mathbf{k}}_m$$
$$= \mathsf{V}^{-1} \sum_m R(m \leftarrow \mathbf{n}), \tag{6.41}$$

where we have used eqns (6.15), (6.16), and (6.18).

As another application of eqn (6.35) we note that $T_{i,i}$ is invariant under L-conjugation, so that, using eqns (6.35), (6.34), and (6.27a), we can write

$$T_{f,i} = V_f + V_f G_{0,i} T_{i,i} = V_f + T_{f,f} G_{0,f} V_i. \tag{6.27b}$$

Using the second of these equations and eqn (6.7) we obtain a set of coupled equations,

$$T_{f,i} = V_f + \sum_j T_{f,j} G_{0,j} C_j V_i \tag{6.27c}$$

for the transition operators.

To consider the behaviour of the transition amplitude in the vicinity of a new channel threshold or over a small energy interval it is convenient to introduce the effective range theory (Ross and Shaw 1961). Over a small energy interval the variation of the transition amplitude with energy is primarily due to the momentum dependence of the initial and final states. For interactions of finite range the significant contributions will come from the low R values where (provided $kR < [2(2l+3)]^{\frac{1}{2}}$), $j_l(kR)$ is proportional to $(kR)^l$. We thus perform a partial wave analysis and define a matrix $\mathbf{M}$ by

$$\mathbf{K} = \mathbf{k}^{l+\frac{1}{2}} \mathbf{M}^{-1} \mathbf{k}^{l+\frac{1}{2}},$$

where $(\mathbf{k}^{l+\frac{1}{2}})_{i,j} = \delta_{ij} k_i^{l_i+\frac{1}{2}}$, k_i is the relative momentum, and l_i is the orbital angular momentum in channel i. Ross and Shaw (1961) have shown that $\mathbf{M}$ is an even function of all the momenta k_j, and so remains real even when some of the channels are closed (k_i pure imaginary).

According to the effective range theory $\mathbf{M}$ has the following energy dependence near any energy E_0:

$$\mathbf{M}(E) = \mathbf{M}(E_0) + \tfrac{1}{2}\mathbf{R}[\mathbf{k}^2 - \mathbf{k}^2(E_0)] + ...,$$

where $\mathbf{R}$ is real, energy independent, and diagonal matrix. If there is only one channel we have (taking $\mathbf{K} = \tan \delta_l$)

$$M = k^{2l+1} \cot \delta_l$$

and at $E_0 = 0$ we can write
$$k^{2l+1}\cot\delta_l = -a_l^{-1} + \tfrac{1}{2}r_l k^2,$$
where a_l is known as the (lth) 'scattering length' (compare with eqn (2.89)), r_l is the 'effective range'. Using eqn (5.46) we can write
$$\mathbf{T} = \mathbf{k}^{l+\frac{1}{2}}(\mathbf{M} - i\mathbf{k}^{2l+1})^{-1}\mathbf{k}^{l+\frac{1}{2}}.$$

At a threshold of a new channel this equation predicts that the dominant energy dependence of the transition matrix element is given by $(k_f)^{l_f+\frac{1}{2}}$, where k_f is the relative momentum in the new channel. On the other hand, for exothermic reactions the energy dependence is determined primarily by the momentum in the initial channel.

When the interaction is not of a finite range the factorization is no longer useful, and one has to consider explicitly the momentum dependence of the radial wave functions for the particular problem. (See, for example, Gailitis 1964.)

2.6.3. *The yield function—absolute rate theory*

In general, the initial state in a collision does not have a sharp energy. If, however, it is stationary it can be characterized by a density matrix ρ_0^i which is diagonal in energy (so that $[\rho_0^i, H_0^i] = 0$),
$$\rho_0^i = \int \mathrm{d}E\, \mathrm{f}(E)\delta(E - H_{0,i})C_i, \tag{6.42}$$
where $\mathrm{f}(E)\,\mathrm{d}E$ is the density function. More explicit forms for ρ_0^i are obtained by writing down the spectral resolutions,
$$\delta(E - H_{0,i})C_i = \sum_n |E, \mathbf{n}, i\rangle\langle E, \mathbf{n}, i|$$
$$= \sum_n \int |\mathbf{k}_n, n\rangle g_n \rho_i(E - E_n)\,\mathrm{d}\hat{\mathbf{k}}_n \langle \mathbf{k}_n, n|, \tag{6.43}$$
where $|n\rangle$ is the internal state of the reactants, g_n is the statistical weight of this state, and $\rho_i(E - E_n)$ is the density of translational states.

The density matrix for the interacting molecules which corresponds to ρ_0^i as the initial state is
$$\rho_+^i = \Omega_i \rho_0^i \Omega_i^\dagger.$$
The rate of change of the expectation value of an observable A is given by the equation of motion
$$\frac{\mathrm{d}}{\mathrm{d}t}\mathrm{Tr}(\rho_+^i A) = (-i/\hbar)\mathrm{Tr}\{[H, \rho_i^+]A\}$$
$$= (-i/\hbar)\mathrm{Tr}\{\rho_i^+[A, H]\}$$
$$= (-i/\hbar)\mathrm{Tr}\{\Omega_i \rho_0^i \Omega_i^\dagger[A, H]\}$$
$$= \mathrm{Tr}(\rho_0^i Y_A^i), \tag{6.44}$$
where
$$Y_A^i = (-i/\hbar)\Omega_i^\dagger[A, H]\Omega_i$$

and we have used the cyclic invariance of the trace
$$\mathrm{Tr}(ABC) = \mathrm{Tr}(BCA).$$
If A is measured in the channel f, $[A, H] = [A, V_f]$.

It is convenient to define a yield function (Coulson and Levine 1967) by
$$Y_A^i(E) = 2\pi\hbar\,\mathrm{Tr}[\delta(E-H_{0,i})C_i Y_A^i] \tag{6.45}$$
and express the rate of change of A, due to the collisions with an initial state in the channel i, as

$$\begin{aligned}\langle \dot{A}\rangle_i &= \frac{\mathrm{d}}{\mathrm{d}t}\mathrm{Tr}(\rho_+^i A)/\mathrm{Tr}(\rho_+^i) \\ &= \int \mathrm{f}(E)\mathrm{Tr}[\delta(E-H_{0,i})C_i Y_A^i]\,\mathrm{d}E/\mathrm{Tr}(\rho_0^i) \\ &= (2\pi\hbar)^{-1}\int \mathrm{f}(E)Y_A^i(E)\,\mathrm{d}E/\mathrm{Tr}(\rho_0^i).\end{aligned} \tag{6.46}$$

As an example consider $A = C_f$ ($f \neq i$). If we denote ther elevant yield function by $Y_f^i(E)$ then

$$\begin{aligned}Y_f^i(E) &= -2\pi i \sum_\mathbf{n} g_n \langle E, \mathbf{n}, i|\Omega_i^+[C_f, H]\Omega_i|E, \mathbf{n}, i\rangle \\ &= 2\pi\hbar \sum_\mathbf{n} g_n \frac{\mathrm{d}}{\mathrm{d}t}\langle E, \mathbf{n}, i^+|C_f|E, \mathbf{n}, i^+\rangle \\ &= 2\pi\hbar\mathsf{V}^{-1}\sum_\mathbf{n}\sum_\mathbf{m} g_n R(\mathbf{m}\leftarrow\mathbf{n})\rho_i(E-E_n),\end{aligned} \tag{6.47}$$

where we have used eqns (6.41) and (6.15). The yield function contains an implicit summation over all possible final states and an averaging over all initial states of the same energy. From eqns (6.47) and (6.19) we can also write
$$Y_f^i(E) = 2\pi\hbar\sum_n\sum_m g_n \frac{\mathrm{d}\dot{N}(m\leftarrow n)}{\mathrm{d}E} \tag{6.48a}$$
$$= \sum_n \pi^{-1}g_n k_n^2 \sigma_n, \tag{6.48b}$$
where σ_n is the total reactive cross-section, with the internal state n initially occupied.

The yield function $Y_f^i(E)$ has the dimension 'number' and is the probability of a reactive collision (from channels i to f, $f \neq i$) when the initial state is a microcanonical ensemble.§ We shall prove this below. Ross and Mazur (1961), Smith (1962a) and Marcus (1966a, b) have obtained alternative forms for the reaction rate, eqn (6.49), from which the same significance of the yield function is obtained. There have been several recent attempts to obtain a closed form for the yield function,

§ In a microcanonical ensemble the statistical operator is $\delta(E-H)$ where H is the relevant Hamiltonian, in our case H_0^i. Other ensembles are easily derived from this by the method of spectral resolution (cf. eqn (6.46)).

since in principle it is all we need to compute rate constants in stationary systems. In particular, if the initial distribution is canonical ($f(E) = \exp(-\beta E)$) the reaction rate is proportional to the Laplace transform of the yield function (eqn (6.46) with $A = C_f$)

$$V^{-1}R_c(f \leftarrow i) = (2\pi\hbar)^{-1} \int_0^\infty \exp(-\beta E) Y_f^i(E) \, dE / Z_i, \quad (6.49)$$

where $Z_i = \text{Tr}(\rho_0^i)$ is the product§ of the internal partition functions of the reactants and the partition function for their relative motion. Using eqn (6.48) in eqn (6.49) we see the equivalence of the two results (6.48) and (6.24). Equation (6.49) with $Y_f^i(E)$, defined by eqn (6.48 b), was introduced by Eliason and Hirschfelder (1959).

In general,||
$$\lim_{E \to +0} Y_f^i(E) \to 0 \quad (6.50)$$

and, by integration by parts (recall that $\beta = (kT)^{-1}$),

$$(2\pi\hbar)^{-1} \int_0^\infty \exp(-\beta E) Y_f^i(E) \, dE$$

$$= (2\pi\hbar\beta)^{-1} \int_0^\infty \exp(-\beta E)[dY_f^i(E)/dE] \, dE$$

$$= \frac{kT}{h} \exp(-A_\ddagger / kT). \quad (6.51)$$

The quantity $A_\ddagger$ is defined by the last line and can be interpreted as a 'free energy' by analogy to the conventional definition

$$Z = \exp(-\beta A) = \int_0^\infty \exp(-\beta E) \rho(E) \, dE, \quad (6.52)$$

where $\rho(E)$ is the density of states per unit energy. Thus, from eqn (6.49),

$$V^{-1}R_c(f \leftarrow i) = \frac{kT}{h} \exp[-(A_\ddagger - A_i)/kT], \quad (6.53)$$

where A_i is the free energy of the initial state. Equation (6.53) forms the basis of the statistical-thermodynamics theory of reaction rates (Glasstone, Laidler, and Eyring 1941). We see that it can be derived

---

§ $Z_i = \text{Tr}[\exp(-\beta H_{0,i})C_i] = \text{Tr}\{\exp[-\beta(K_i + h_1 + h_2)]C_i\}$
$= \text{Tr}[\exp(-\beta K_i)]\text{Tr}[\exp(-\beta h_1)C_{i,1}]\text{Tr}[\exp(-\beta h_2)C_{i,2}]$,

where h_1 and h_2 are the internal Hamiltonians for the two molecules, which commute with one another and with K_i.

|| If the reaction is endothermic the yield function is necessarily zero below some finite threshold. If the reaction is exothermic then, from eqn (6.17), $k_n^2 \sigma_n$ is proportional to k_n. Physically, the vanishing of the yield function at $E = 0$ is thus due to the vanishing of the initial density. (Recall that $\hbar k_n \sigma_n / \mu_i$ is the rate and the second factor of k_n is from the density of states at the energy E.)

without any assumptions regarding the dynamics of the reaction. It does rely, however, on the assumption that the initial state is drawn from a canonical ensemble. Moreover, the interpretation of $\mathrm{d}Y_j^i(E)/\mathrm{d}E$ as a 'density of states' is not required by the present derivation, although it is dimensionally consistent and is useful as an approximation scheme (sections 3.4.0 and 3.5.0).

The experimental rate-constant is defined, for an elementary reaction, as the observed rate of transitions per unit volume divided by the product of densities of the reactants. The identification of the experimental rate-constant with $\mathsf{V}^{-1}R_\mathrm{c}(f \leftarrow i)$ can only be made when the distribution of reactants is nearly canonical.

2.6.4. *The theory of reactive collisions in the coordinate representation*

In this section we consider an alternative approach to collision theory which is based explicitly on the asymptotic behaviour in coordinate space of the solution of the L.S. equation. This approach is based on the work of Glauber and Schomaker (1953), Gerjuoy (1958a), and McElroy and Hirschfelder (1963) on the use of Green's theorem in collision theory. This is a multi-dimensional extension of the conventional three-dimensional Green's theorem, and can be proved by integration by parts. The theorem states that given two functions f and g (which are twice differentiable with respect to the coordinate R_i) then, with $K_i = -(\hbar^2/2\mu_i)\nabla^2_{R_i}$,

$$\int \mathrm{d}\mathsf{V}\,\{fK_i g - g K_i f\} = -(\hbar^2/2\mu_i) \int \mathrm{d}\mathsf{V}_i' \int \mathrm{d}\mathbf{S}_i \cdot \mathbf{W}_i(f,g), \quad (6.54)$$

where the volume integration $\mathrm{d}\mathsf{V}_i'$ is over all coordinates excluding R_i and $\mathbf{W}_i(f,g)$ is the vector Wronskian of f and g with respect to the coordinate R_i,

$$\mathbf{W}_i(f,g) = (f\nabla_{R_i} g - g\nabla_{R_i} f). \quad (6.55)$$

We also note that

$$(\hbar/2i\mu_i)\mathbf{W}_i(f,g) = \langle f^* | \mathbf{j}_i | g \rangle, \quad (6.56)$$

where $\mathbf{j}_i$ is the quantum-mechanical current operator with respect to R_i.

The surface integral in eqn (6.54) is over a surface that encloses the volume integration over R_i. The element of surface area is defined by $\mathrm{d}\mathbf{S}_i = \mathbf{n}_i\,\mathrm{d}S_i$, where $\mathbf{n}_i$ is the unit outward normal to the surface. For a sphere $\mathbf{n}_i = \hat{\mathbf{R}}_i$ and $\mathrm{d}S_i = R_i^2\,\mathrm{d}\omega_i$ so that we only need W_i, the radial component of $\mathbf{W}_i$,

$$W_i(f,g) = \hat{\mathbf{R}}_i \cdot \mathbf{W}_i(f,g) = f\frac{\partial g}{\partial R_i} - g\frac{\partial f}{\partial R_i}. \quad (6.57)$$

We shall use the notation $\{f,g\}_i$ to denote the integral of $W_i(f,g)$ over a surface of a sphere (of radius ρ_i) with respect to ω_i and over the volume

V'_i with respect to all other variables:

$$\{f,g\}_i = \int dV'_i \int W_i(f,g)|_{R_i=\rho_i} \rho_i^2 \, d\omega_i$$
$$= \int dV'_i \int W_i(R_i f, R_i g)|_{R_i=\rho_i} \, d\omega_i. \tag{6.58}$$

If the volume of integration over R_i is enclosed by a sphere of radius ρ_i, we can write Green's theorem as

$$\int dV \{fK_i g - gK_i f\} = -(\hbar^2/2\mu_i)\{f,g\}_i. \tag{6.59}$$

In particular, if we integrate over all space we should take the limit $\rho_i \to \infty$ of the right-hand side. We can also use the notation

$$F_i(f,g) = -(i\hbar/2\mu_i)\{f,g\}_i, \tag{6.60}$$

where F_i is the surface integral of the quantum-mechanical probability current.

If we introduce a Hamiltonian H by $H = \sum_i K_i + V$, where V is real, then eqn (6.54) is generalized to

$$\int dV \{fHg - gHf\} = -\sum_i (\hbar^2/2\mu_i) \lim_{\rho_i \to \infty} \{f,g\}_i$$
$$= -\sum_i i\hbar \lim_{\rho_i \to \infty} F_i(f,g), \tag{6.61}$$

where on the left the volume integration is over all space. There are clearly no contributions to the right-hand side unless both f and g are non-vanishing for large R_i. We shall see below that a much stronger statement holds, namely, in order that $\lim_{\rho_i \to \infty} F_i(f^*,g)$ be finite, f and g should have, for large R_i, a common component. We also indicate below (eqn (6.64) with $A = I$) that this statement is closely related to the conservation of probability.

To see explicitly the significance of the probability current F_i we consider eqn (6.61) with $f = A\psi_1$ and $g = \psi_2$, where ψ_1 and ψ_2 are two degenerate solutions of the Schrödinger equation. Then

$$\langle \psi_2|[A,H]|\psi_1 \rangle = \int dV \{(A\psi_1)(H\psi_2^*) - \psi_2^* H A \psi_1\}$$
$$= -i\hbar \sum_i \lim_{\rho_i \to \infty} F_i(A\psi_1, \psi_2^*). \tag{6.62}$$

In writing down the second line above we have used the identity

$$\int (H-E)\psi_2^* A\psi_1 \, dV = \int \psi_2^* A(H-E)\psi_1 \, dV,$$

where $(H-E)\psi_1 = (H-E)\psi_2 = 0$ and H is real, so that

$$\int (H\psi_2^*)(A\psi_1) \, dV = \int \psi_2^* A H \psi_1 \, dV. \tag{6.63}$$

In particular,
$$\frac{d}{dt}\langle\psi|A|\psi\rangle = \sum_i \lim_{\rho_i\to\infty} F_i(\psi^*, A\psi). \tag{6.64}$$

In section 2.4.0 we obtained the optical theorem by the substitution $A = I$ in the equation of motion. The same procedure can be used in the coordinate representation. Let the coordinate representation of $|i^+\rangle$, the solution of the L.S. equation, be denoted ψ_i^+ and let ϕ_i be the coordinate representation of the initial state. The L.S. equation can be summarized by
$$\psi_i^+ = \phi_i + \psi_{sc}, \tag{6.65}$$
where ψ_{sc} is the scattered wave. Putting $A = I$ we have
$$\sum_i \lim_{\rho_i\to\infty} F_i(\psi_i^{+*}, \psi_i^+) = 0. \tag{6.66}$$
Moreover,
$$\sum_i \lim_{\rho_i\to\infty} F_i(\phi_i^*, \phi_i) = 0. \tag{6.67}$$
Subtracting the two we have that
$$\sum_i \lim_{\rho_i\to\infty} F_i(\psi_{sc}^*, \psi_{sc}) = -2\,\text{im}\sum_i \lim_{\rho_i\to\infty} F_i(\phi_i^*, \psi_{sc}). \tag{6.68}$$

To see the significance of eqn (6.68) as the optical theorem we consider the collision of structureless particles, with $\phi_i(\mathbf{R}) = \exp(i\mathbf{k}\cdot\mathbf{R})$, so that from eqn (2.28),
$$\psi_{sc}(\mathbf{R}) \underset{R\to\infty}{\to} f(\mathbf{k}', \mathbf{k})\exp(ikR)/R,$$
where $\hat{\mathbf{k}}' = \hat{\mathbf{R}}$. Then§
$$\lim_{R\to\infty} F_i(\psi_{sc}^*, \psi_{sc}) = (\hbar k/\mu)\int |f(\mathbf{k}', \mathbf{k})|^2\,d\hat{\mathbf{k}}'. \tag{6.69}$$

To evaluate the interference current we note that ϕ_i propagates in the direction $\hat{\mathbf{k}}$ (the incident direction) only. We therefore expect that, for large values of R, $F_i(\phi^*, \psi_{sc})$ will vanish unless $\mathbf{k}' = \mathbf{k}$ ($k' = k$ by energy conservation). Indeed we find‖ •
$$\lim_{R\to\infty} F_i(\phi_i^*, \psi_{sc}) = \frac{2\pi\hbar f(\mathbf{k}, \mathbf{k})}{i\mu} + \text{c.c.} \tag{6.70}$$

§ The radial component of the probability current for the scattered wave $j_{sc}(R)$ is given from eqn (6.56) for large R by
$$j_{sc}(R) = -(i\hbar/2\mu)\{R^{-1}f^*\exp(-ikR)(-R^{-2}f + ikR^{-1}f)\exp(ikR) - \text{c.c.}\} = (\hbar k/\mu)f^*f/R^2.$$
Since $\hat{\mathbf{k}}' = \hat{\mathbf{R}}$, the element of solid angle on the sphere can be written as $d\hat{\mathbf{k}}'$, and c.c. denotes the complex conjugate term.

‖ We find by differentiation that
$$(\hbar/2i\mu)W_i(\phi_i^*, \psi_{sc}) = (\hbar k/2\mu R)(1+\cos\theta)\{f\exp[ikR(1-\cos\theta)] + \text{c.c.}\},$$
where $\cos\theta = \cos(\hat{\mathbf{k}}.\hat{\mathbf{R}}) = \cos(\hat{\mathbf{k}}.\hat{\mathbf{k}}')$. For large values of R the surface integral of this

or, from eqns (6.68)–(6.70),

$$\operatorname{im} f(\mathbf{k}, \mathbf{k}) = (k/4\pi) \int |f(\mathbf{k}', \mathbf{k})|^2 \, d\hat{\mathbf{k}}' = (k/4\pi)\sigma(\mathbf{k}), \qquad (6.71)$$

which is the optical theorem. From a physical point of view our computation implies that the scattered wave interferes with the incidence wave. The depletion of the incident wave in the forward direction is equal, according to the optical theorem, to the gain in any other direction.

The asymptotic form of the solution of the L.S. equation, when reactive collisions can take place, depends on the particular relative coordinate that we let increase beyond any finite limit. Each reactive channel j is associated with a different R_j such that $V_j(R_j) \to 0$ as $R_j \to \infty$, where R_j is the relative separation of the two molecules in the channel j.

Gerjuoy (1958b) and Thorson (1962) have shown that the coordinate representation of the solution of the L.S. equation can be uniquely determined by the condition that the scattered wave has the same asymptotic behaviour as the outgoing (or alternatively the incoming) Green's function for the Hamiltonian H. The outline of the argument is as follows. Let $\mathbf{r}$ be the set of all coordinates of the problem and put

$$\psi_i^+(\mathbf{r}) = \phi_i(\mathbf{r}) + \psi_{\text{sc}}(\mathbf{r}), \qquad (6.72)$$

where $\qquad [E - H(\mathbf{r})]\psi_i^+(\mathbf{r}) = 0, \qquad [E - H_{0,i}(\mathbf{r})]\phi_i(\mathbf{r}) = 0, \qquad (6.73)$

and $H = H_{0,i} + V_i$, so that

$$[E - H(\mathbf{r})]\psi_{\text{sc}}(\mathbf{r}) = V_i(\mathbf{r})\phi_i(\mathbf{r}). \qquad (6.74)$$

The outgoing Green's function for H satisfies the differential equation

$$[E - H(\mathbf{r})]G^+(\mathbf{r}, \mathbf{r}') = \delta(\mathbf{r} - \mathbf{r}') \qquad (6.75)$$

and the boundary condition that asymptotically it behaves as outgoing radial waves in all channels. Multiplying eqn (6.74) on the left by G^+, eqn (6.75) on the left by ψ_{sc}, subtracting, and integrating over all $\mathbf{r}$ we obtain from Green's theorem

$$\psi_{\text{sc}}(\mathbf{r}) = \int d\mathbf{r}' \, G^+(\mathbf{r}, \mathbf{r}')V_i(\mathbf{r}')\phi_i(\mathbf{r}') + \sum_i \lim_{\rho_i \to \infty} F_i(\psi_{\text{sc}}, G^+). \qquad (6.76)$$

current vanishes everywhere apart from the forward ($\theta = 0$) direction. If we refer all directions to an axis determined by $\mathbf{k}$ then $d\hat{\mathbf{k}}' = d\cos\theta \, d\phi$ and for large R

$$F_i(\phi_i^*, \psi_{\text{sc}}) = (R\hbar k/2\mu) \int_0^{2\pi} d\phi \int_{-1}^{1} d\cos\theta \, \{(1 + \cos\theta)f \exp[ikR(1 - \cos\theta)] + \text{c.c.}\}.$$

Integrating by parts and neglecting terms of order R^{-1},

$$F_i(\phi_i^*, \psi_{\text{sc}}) = (\hbar/\mu) \int_0^{2\pi} d\phi \, \{i(1 + \cos\theta)f \exp[ikR(1 - \cos\theta)]\}_{-1}^{1} + \text{c.c.}$$
$$= (2\pi\hbar/i\mu)f(\cos\theta = 0) + \text{c.c.}$$

REACTIVE COLLISIONS

The uniqueness of ψ_{sc} is obtained by the condition that it behaves asymptotically as G, so that the surface term above vanishes (since the Wronskian of two functions with identical asymptotic behaviour vanishes asymptotically). Thus

$$\psi_i^+(\mathbf{r}) = \phi_i(\mathbf{r}) + \int d\mathbf{r}'\, G^+(\mathbf{r},\mathbf{r}')V_i(\mathbf{r}')\phi_i(\mathbf{r}'), \qquad (6.77)$$

which we recognize as the coordinate representation of the L.S. equation.

To obtain the asymptotic behaviour of G in any particular channel we recall that (cf. eqn (6.34))

$$G = \Omega_j G_{0,j} = G_{0,j}\Omega_j^\ddagger. \qquad (6.78)$$

We can now repeat the discussion in section 2.5.2. If $|m\rangle$ is the internal state in the channel j then we obtain

$$\lim_{R_j\to\infty} \langle \mathbf{R}_j, m, j | G^+ = -\frac{\mu_j \exp(ik_m R_j)}{2\pi\hbar^2 R_j}\langle \mathbf{k}'_m, m, j^-|, \qquad (6.79)$$

where $E = E_m + \hbar^2 k_m^2/2\mu_j$ and $\mathbf{k}'_m = k_m \hat{\mathbf{R}}_j$. $\langle \mathbf{k}'_m, m, j^-|$ is the incoming solution of the L.S. equation with an incident wave in the channel j,

$$\langle \mathbf{k}_m, m, j^-| = \langle \mathbf{k}_m, m, j|\Omega_j^\ddagger. \qquad (6.80)$$

We can now introduce the scattering amplitude from channel i to channel j ($j \neq i$), by the definition

$$\psi_i^+(\mathbf{r}) \underset{R_j\to\infty}{\to} \sum_m f_{m,n}(\mathbf{k}_m,\mathbf{k}_n)\varphi_{m,j}\exp(ik_m R_j)/R_j, \qquad (6.81)$$

where $$\phi_i(\mathbf{r}) = \exp(i\mathbf{k}_n\cdot\mathbf{R}_i)\varphi_{n,i}$$

and $\varphi_{n,i}$ is an internal state in the channel i. Using eqns (6.79) and (6.77), this definition is in agreement with eqn (6.16) (recall eqn (6.30)).

Using the asymptotic behaviour of ψ_{sc} one can now write down the optical theorem for reactive collisions. Realizing that $\lim F_i(f^*,g)$ vanishes unless both f and g are outgoing in the channel i, we find from eqns (6.68) and (6.81),

$$(4\pi\hbar/\mu_i)\mathrm{im}\, f_{n,n}(\mathbf{k}_n,\mathbf{k}_n) = \sum_f (\hbar/\mu_f)\sum_m k_m \int |f_{m,n}(\mathbf{k}_m,\mathbf{k}_n)|^2\, d\hat{\mathbf{k}}_m$$
$$= -(2/\hbar)\mathrm{im}\langle \mathbf{k}_n, n, i | T_{i,i} | \mathbf{k}_n, n, i\rangle, \qquad (6.82)$$

where in the last line we have used eqn (6.16). Using the same equation we can clearly transform the second line to get eqn (6.40). We also note, from eqn (6.41), that each term in the summation over the channels can be equated to $\langle \mathbf{k}_n, n, i^+|\hat{C}_f|\mathbf{k}_n, n, i^+\rangle$ and interpreted as the total rate of reactive collisions from the initial state.

To conclude, in the coordinate space representation the distinction between the different reactive channels is based on the asymptotic vanishing of $F_i(f^*, g)$ whenever either f or g or both are not asymptotically outgoing in the channel i. Moreover, this procedure distinguishes the internal states of the same reaction channel, in that these states are characterized by different values of the wave vector k. In this sense, therefore, the internal state label can also be considered a channel label.

2.7. Time-dependent collision theory

2.7.0. IN this section we follow the collision event in time. The collision begins at some time t' in the past with two non-interacting molecules and terminates at some time t'' in the future when two (the same or different) molecules are again well separated. We shall not discuss many new results but only present a new point of view which is complementary to that of stationary-collision theory. It should be mentioned, however, that a very large number of the basic papers on collision theory were written using a time-dependent point of view. The present discussion is in the spirit of Eckstein (1956) and Jauch (1958).

We begin our discussion for the case where the resolution of H, $H = H_0 + V$ is unique. At the time t' the wavefunction for the system is $\Phi_i(t')$. In general, we can put in the coordinate representation

$$\Phi_i(\mathbf{R},\mathbf{r};t') = \sum_n \int dE\, A(\mathbf{n}, E, t') \varphi_n(\mathbf{r}) \langle \mathbf{R} | E, \hat{\mathbf{k}}_n \rangle, \qquad (7.1)$$

where $\mathbf{R}$ is the relative coordinate, $\mathbf{r}$ is the set of internal coordinates, n is an internal quantum number, and $A(\mathbf{n}, E, t')$ is the amplitude to find the molecules with relative momentum $\hbar \mathbf{k}_n$ and in the internal state n at the time t'. In the absence of interaction n and $\mathbf{k}_n$ remain good quantum numbers and, from the Schrödinger time-dependent equation,

$$\Phi_i(\mathbf{R},\mathbf{r};t) = \exp[-iH_0(t-t')/\hbar]\Phi_i(\mathbf{R},\mathbf{r};t') \qquad (7.2)$$

or
$$A(\mathbf{n}, E, t) = \exp[-iE(t-t')/\hbar] A(\mathbf{n}, E, t'), \qquad (7.3)$$

where $E = E_n + \hbar^2 k_n^2 / 2\mu$. The function $A(\mathbf{n}, E)$ is determined by the experimentalist when he prepares the initial state.§

Due to the mutual interaction the evolution of the physical system is governed by the full Hamiltonian H, so that the wavefunction at time t, $(t > t', t' \to -\infty)$, is given by

$$\Psi_i^+(t) = \exp[-iH(t-t')/\hbar]\Phi_i(t'), \quad t > t'$$
$$= U^+(t-t')\Phi_i(t'), \qquad (7.4)$$

where $U^+(t-t')$ is the forward evolution operator

$$U^+(t-t') = \theta(t-t')\exp[-iH(t-t')/\hbar], \qquad (7.5)$$

and $\theta(t-t')$ is the unit step function that vanishes for negative values

---

§ At the risk of stating the obvious, we remark that Φ_i is a pure state. The different components have definite phase relations determined by $A(\mathbf{n}, E)$.

of its arguments (eqn (1.2.37)). The U^+ evolution operator was shown to satisfy the differential equation (1.2.39), which, using the U_0^+ evolution operator (for the Hamiltonian H_0), can be written as the integral equation

$$U^+(t-t') = U_0^+(t-t') - (\mathrm{i}/\hbar) \int_{-\infty}^{\infty} \mathrm{d}t''\, U_0^+(t-t'')V U^+(t''-t'), \quad (7.6)$$

where, due to the inherent time ordering introduced by the step functions, the actual range of t'' is $t > t'' > t'$. We also recall that (cf. eqn (1.2.36))

$$G_0(E^+) = (-\mathrm{i}/\hbar) \int_{-\infty}^{\infty} \mathrm{d}t\, U_0^+(t)\exp(\mathrm{i}Et/\hbar) \quad (7.7)$$

and that a similar relation holds between $G(E^+)$ and $U^+(t)$ so that eqn (7.6) can be considered as the transform§ of

$$G(E^+) = G_0(E^+) + G_0(E^+)V G(E^+). \quad (7.8)$$

Operating with both sides of eqn (7.6) on $\Phi_i(t')$ we obtain in the limit $t' \to -\infty$

$$\Psi_i^+(t) = \Phi_i(t) - (\mathrm{i}/\hbar) \int_{-\infty}^{\infty} \mathrm{d}t''\, U_0^+(t-t'')V\Psi_i^+(t''), \quad (7.9)$$

where we have used the definition of $\Psi_i^+(t)$, eqn (7.4), and

$$\Phi_i(t) = U_0^+(t-t')\Phi_i(t'). \quad (7.10)$$

To establish a connection with the stationary point of view we need to resolve eqn (7.9) into its energy components. We can do this by Fourier transforming, namely, by defining

$$\Psi_{i,E}^+ = (2\pi\hbar)^{-1} \int_{-\infty}^{\infty} \mathrm{d}t\, \exp(\mathrm{i}Et/\hbar)\Psi_i^+(t) \quad (7.11)$$

when we note the identity

$$\delta(E-E') = (2\pi\hbar)^{-1} \int_{-\infty}^{\infty} \mathrm{d}t\, \exp[\mathrm{i}(E-E')t/\hbar]. \quad (7.12)$$

Thus, from eqns (7.12) and (7.1),

$$\Phi_{i,E}(\mathbf{R},\mathbf{r}) = \sum_{\mathbf{n}} A(\mathbf{n}, E)\varphi_\mathbf{n}(\mathbf{r})\langle \mathbf{R}|E, \hat{\mathbf{k}}_\mathbf{n}\rangle, \quad (7.13)$$

and, using the convolution theorem (eqn (B.2)) and eqns (7.9) and (7.13) we have that

$$\Psi_{i,E}^+ = \Phi_{i,E} + G_0(E^+)V\Psi_{i,E}^+ = \sum_{\mathbf{n}} A(\mathbf{n}, E)|E, \mathbf{n}^+\rangle. \quad (7.14)$$

Equation (7.14) is the conventional L.S. equation, and the component

§ Note that the second term in eqn (7.6) is the convolution of U_0^+ and U^+. On taking the transform of this term we shall obtain a product of the transforms (cf. Appendix B, eqn (B. 2)).

analysis, eqn (7.11), shows that we can write

$$\Psi_i^+(t) = \sum_{\mathbf{n}} \int dE\, A(\mathbf{n}, E, t)|E, \mathbf{n}^+\rangle. \tag{7.15}$$

With $t = 0$, eqn (7.15) is identical to eqn (5.12). This suggests introducing an amplitude $B(\mathbf{n}, E, t)$ by

$$\Psi_i^+(t) = \sum_{\mathbf{n}} \int dE\, B(\mathbf{n}, E, t)|E, \mathbf{n}\rangle \tag{7.16}$$

so that $\langle E, \mathbf{m}; t|\Psi_i^+(t)\rangle = \exp(iEt/\hbar)B(\mathbf{m}, E, t)$. The physical information of interest is clearly the limit of this amplitude as $t \to \infty$, namely, the amplitude to find the state $|E, \mathbf{m}\rangle$ occupied after the collision is over. As a consistency check we note from eqn (7.9) that as $t \to -\infty$ the second term does not contribute (recall that $t'' < t$) so that

$$\lim_{t \to -\infty} \exp(iEt/\hbar)B(\mathbf{n}, E, t) \to A(\mathbf{n}, E), \tag{7.17}$$

as expected.

From its definition

$$\exp(iEt/\hbar)B(\mathbf{n}, E, t)$$
$$= \sum_{\mathbf{m}} \int dE'\, \exp[-i(E'-E)t/\hbar]A(\mathbf{m}, E')\langle \mathbf{n}, E|E', \mathbf{m}^+\rangle, \tag{7.18}$$

where the matrix element $\langle \mathbf{n}, E|E', \mathbf{m}^+\rangle$ was previously evaluated as

$$\langle \mathbf{n}, E|E', \mathbf{m}^+\rangle = \delta(E-E')\delta_{\mathbf{n},\mathbf{m}} + (E'^+ - E)^{-1}\langle \mathbf{n}, E|T|E', \mathbf{m}\rangle. \tag{7.19}$$

Using the result (eqn (B.8))

$$\lim_{t \to \mp\infty} \lim_{\epsilon \to +0} \int g(\omega)\frac{\exp(i\omega t)}{\omega - i\epsilon}\, d\omega \to \begin{cases} 0 \\ 2\pi i g(\omega = 0), \end{cases} \tag{7.20}$$

we have, from eqn (7.18) (with $\omega = E - E'$),

$$\lim_{t \to \mp\infty} \exp(iEt/\hbar)B(\mathbf{n}, E, t) \to \begin{cases} A(\mathbf{n}, E) \\ \sum_{\mathbf{m}} (\delta_{\mathbf{n},\mathbf{m}} - 2\pi i\langle E, \mathbf{n}|T|E, \mathbf{m}\rangle)A(\mathbf{m}, E). \end{cases} \tag{7.21}$$

If we define a column vector $\mathbf{B}$ with components $B(\mathbf{n}, E)$,

$$B(\mathbf{n}, E) = \lim_{t \to \infty} \exp(iEt/\hbar)B(\mathbf{n}, E, t),$$

we can write eqn (7.21) as $\quad \mathbf{B} = \mathbf{S}\mathbf{A} \tag{7.22}$

where $\mathbf{S}$ is the S matrix (eqn (5.42)) and $\mathbf{A}$ is the column vector with components $A(\mathbf{n}, E)$. The S matrix determines, therefore, the probability to observe the state $|E, \mathbf{m}\rangle$ after the collision.

2.7.1. *The time evolution*

The wave function $\Psi_i^+(t)$ describes the evolution of the system during the collision, from the remote past $t \to -\infty$, when it is identical to the initial state, to the remote future, when the molecules are again no longer interacting. In particular, we note that the choice

$$A(\mathbf{m}, E') = \delta_{\mathbf{n},\mathbf{m}} \cdot \delta(E-E')$$

corresponds to the initial state in stationary-collision theory. In this sense the plus solution of the L.S. equation describes the forward evolution in time of the collision. In this section we consider the description of the system during the collision.

For any finite time t there is a contribution from the second term in the equation,

$$\Psi_i^+(t) = \Phi_i(t) - (\mathrm{i}/\hbar) \int_{-\infty}^{\infty} \mathrm{d}t''\, U_0^+(t-t'') V \Psi_i^+(t''), \tag{7.9}$$

which corresponds to the scattered wave. From t'' onwards the scattered wave evolves under the Hamiltonian H_0, so that t'' is the last instant of time at which the molecules were under the influence of their mutual interaction. The scattered wave is obtained by summing over all the possible values of $t'' < t$. We can iterate the integral equation using the Born–Neumann§ series for the integral equation (7.9),

$$\Psi_i^+(t) = \Phi_i(t) - (\mathrm{i}/\hbar) \int_{-\infty}^{\infty} \mathrm{d}t''\, U_0^+(t-t'') V \Phi_i(t'') +$$

$$+ (\mathrm{i}/\hbar)^2 \iint_{-\infty}^{\infty} \mathrm{d}t'\, \mathrm{d}t''\, U_0^+(t-t'') V U_0^+(t''-t') V \Phi_i(t') + \ldots. \tag{7.23}$$

Each term in this series is inherently time-ordered. For example, in the third term $-\infty < t' < t''$, $t' < t'' < t$. The first term corresponds to the initial state; in the second term the system evolved unperturbed

---

§ We recall that truncating this series at the nth term is equivalent to iterating the integral equation (7.6) $n-1$ times, starting with the approximation of replacing U^+ by U_0^+. The results are equivalent to the Born iteration in stationary-collision theory. In particular first-order time-dependent perturbation theory is equivalent to the (first) Born approximation. In this order, we replace T in eqn (7.19) by V, and so eqn (7.21) is unmodified, save for the same replacement. We can obtain the same conclusion by retaining only the lowest order in eqn (7.23) so that

$\exp(\mathrm{i}Et/\hbar) B(\mathbf{n}, E, t)_\mathrm{B}$

$$= \sum_{\mathbf{m}} \int \mathrm{d}E'\, A(\mathbf{m}, E')(-\mathrm{i}/\hbar) \int_{-\infty}^{\infty} \mathrm{d}t'\, \langle E, \mathbf{n}|\exp[\mathrm{i}(E-E')t'/\hbar] V | E', \mathbf{m}\rangle \theta(t-t').$$

Equation (7.21) with T replaced by V is obtained on using eqn (7.12) to perform the limit $t \to \infty$.

TIME-DEPENDENT COLLISION THEORY 133

up to the time t'' when§ the mutual interaction operated once, and then continued to evolve under the Hamiltonian H_0. The net contribution is obtained by summing over all the possible values of t''. In the third term the mutual interaction operated twice, at t' and at t'', where $t'' > t'$. In general, in the $(n+1)$th term the interaction operates n times at successive instants. In the way we have set the integral equation, time increases from right to left or, in other words, in any term in the iteration (7.23) an interaction on the left occurs later than an interaction to the right. The first mutual interaction is the first one on the right. The last mutual interaction is the last one on the left.

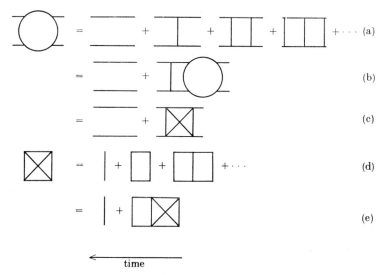

FIG. 2.9. Iterations and summations of integral equations using diagrammatic methods.

It is often very convenient to summarize such discussions by a diagram. If we denote the unperturbed evolution of a single molecule by a single horizontal line and the mutual interaction by a single vertical line connecting the evolution lines, we can summarize the Born–Neumann series by the first diagram in Fig. 2.9, where the total evolution operator is represented by the symbol on the left. We can also sum these diagrams into the form of Fig. 2.9(b) which corresponds to the integral equation (7.6). In Fig. 2.9(c) we have introduced another operator, denoted by the crossed box which, as is clear from the diagram, provides the complete account of all the mutual interactions during the collision. (As the

§ We have assumed that neither H_0 nor V are explicitly time-dependent. Thus the mutual interaction is instantaneous.

reader may suspect, this operator should be the transform of the transition operator.) Comparing diagrams (a) and (c) we obtain an iterated form for this operator, which demonstrates explicitly that its action is not instantaneous. Diagram (d) can now be summed to give diagram (e).

2.7.2. *The wave operator*

It is convenient to relate $\Psi_i^+(t)$ to the wavefunction $\Phi_i(t)$, which corresponds to the physical situation that the system evolved up to the time t in the absence of any mutual interaction. Putting

$$\Phi_i(t') = \exp[-iH_0(t'-t)/\hbar]\Phi_i(t), \quad t > t'$$
$$= U_0^-(t'-t)\Phi_i(t), \qquad (7.24)$$

where $\quad U_0^-(t'-t) = \theta(t-t')\exp[-iH_0(t'-t)/\hbar]$

is the backward evolution operator, we can rewrite eqn (7.4) as

$$\Psi_i^+(t) = \lim_{t' \to -\infty} U^+(t-t')U_0^-(t'-t)\Phi_i(t) = \Omega^+\Phi_i(t), \qquad (7.25)$$

where the operator Ω^+ is defined by this equation. From its definition

$$\Omega^+ = \lim_{\tau \to -\infty} U^+(-\tau)U_0^-(\tau) = \lim_{\tau \to -\infty} \exp(iH\tau/\hbar)\exp(-iH_0\tau/\hbar). \qquad (7.26)$$

From the basic symmetry of the evolution operators (note that this is exactly equivalent to L-conjugation for the Green's operators)

$$U^-(-\tau) = U^+(\tau)^\dagger, \qquad (7.27)$$

we can easily show that

$$\Phi_i(t) = \Omega^{+\dagger}\Psi_i^+(t). \qquad (7.28)$$

The proof is based on the observation that for $t' \to -\infty$ $\Psi_i^+(t')$ should tend to $\Phi_i(t')$. Thus, from eqn (7.10),

$$\Phi_i(t) = \lim_{t' \to -\infty} U_0^+(t-t')\Psi_i^+(t')$$
$$= \lim_{t' \to -\infty} U_0^+(t-t')U^-(t'-t)\Psi_i^+(t). \qquad (7.29)$$

The identification of Ω as the Møller wave operator is based on comparing eqns (7.25), (7.15), and (7.1),

$$\Psi_i^+(t) = \sum_\mathbf{n} \int dE \, A(\mathbf{n}, E, t) |E, \mathbf{n}^+\rangle$$
$$= \Omega^+\Phi_i(t)$$
$$= \sum_\mathbf{n} \int dE \, A(\mathbf{n}, E, t)\Omega^+|E, \mathbf{n}\rangle. \qquad (7.30)$$

To examine the long-time behaviour of the collision, we let the collision run its course without interruption and obtain $\Psi_i^+(t')$ for $t' \to \infty$. We now let the molecules evolve 'backward' from t' to some finite time t under the action of the Hamiltonian H_0. This evolution cannot lead to any transitions since the mutual interaction does not operate. The probability of finding the state $|E, \mathbf{m}\rangle$ at this finite time t is thus the same as the probability to find it after the collision is over. We represent the final state of the collision, which evolved under H_0 to the time t, by the limit

$$\lim_{t'\to+\infty} U_0^-(t-t')\Psi_i^+(t') = \lim_{t'\to\infty} U_0^-(t-t')U^+(t'-t)\Psi_i^+(t)$$
$$= \Omega^{-\dagger}\Omega^+\Phi_i(t), \qquad (7.31)$$

where
$$\Omega^{-\dagger} = \lim_{t'\to\infty} U_0^-(t-t')U^+(t'-t)$$
$$= \lim_{\tau\to\infty} U_0^-(-\tau)U^+(\tau)$$
$$= \lim_{\tau\to\infty} \exp(iH_0\tau/\hbar)\exp(-iH\tau/\hbar). \qquad (7.32)$$

The S operator can therefore be defined by

$$S = \Omega^{-\dagger}\Omega^+. \qquad (7.33)$$

The present operator derivation is thus in agreement with the explicit evaluation of the long-time behaviour, eqn (7.22).

In our discussion thus far we have specified the initial state from which the system evolved. We can also ask the complementary question, namely, 'What is the wavefunction $\Psi_f^-(t)$ at some finite time t that will evolve into a particular final state in the future?'. Using the 'backward' evolution operator we can determine this function by letting the specified final state evolve from the future to the present under the action of the full Hamiltonian H,

$$\Psi_f^-(t) = \lim_{t'\to+\infty} U^-(t-t')\Phi_f(t')$$
$$= \lim_{t'\to+\infty} U^-(t-t')U_0^+(t'-t)\Phi_f(t)$$
$$= \Omega^-\Phi_f(t), \qquad (7.34)$$

where Ω^- is defined by the last line. Equation (7.34) provides the time-dependent significance of the minus solutions of the L.S. equation. From a time-dependent point of view $\exp(-iEt/\hbar)|E, \mathbf{m}^-\rangle$ is the state that as $t \to +\infty$ will evolve to $|E, \mathbf{m}\rangle$. We can also obtain this conclusion explicitly from the L.S. equation. In a formal fashion we have

$$\lim_{t\to\mp\infty} \lim_{\epsilon\to+0} \exp[-i(E-E')t/\hbar]\langle E', \mathbf{n}|E, \mathbf{m}^-\rangle$$
$$= \lim_{t\to\mp\infty} \lim_{\epsilon\to+0} \exp(i\omega t/\hbar)\left[\delta(\omega)\delta_{\mathbf{nm}} - \frac{\langle E', \mathbf{n}|T|E, \mathbf{m}\rangle}{\omega+i\epsilon}\right]$$
$$\to \begin{cases} \delta(E-E')[\delta_{\mathbf{nm}} - 2\pi i\langle E, \mathbf{n}|T|E, \mathbf{m}\rangle] \\ \delta(E-E')\delta_{\mathbf{nm}} \end{cases} \qquad (7.35)$$

with $\omega = E' - E$. We see that the minus solutions describe the course of the collision backwards, starting with a specified initial state in the future and leading to the range of possible products in the past.

Our result for the elements of the S matrix

$$(S)_{m,n} = \langle E, \mathbf{m}^- | E, \mathbf{n}^+ \rangle \tag{7.36}$$

can now be interpreted from a new point of view. $|E, \mathbf{n}^+\rangle$ is the state that evolved from $|E, \mathbf{n}\rangle$. $|E, \mathbf{m}^-\rangle$ is the state that will evolve to $|E, \mathbf{m}\rangle$, so that $(S)_{m,n}$ is the amplitude that if the system evolved from $|E, \mathbf{n}\rangle$ it will evolve into $|E, \mathbf{m}\rangle$, in agreement with eqn (7.21). In this sense the transformation (7.22) can be thought of as the transformation from the H_0 representation of the total wavefunction before the collision to the H_0 representation after the collision.

2.7.3. *The change in observables*

When we consider the **S** matrix as a transformation function the question naturally comes to mind whether we can relate the expectation values of an operator A before and after a collision. Following Lippmann (1965b) we put

$$(i\hbar)^{-1} \int_{-\infty}^{\infty} \langle \Psi_i^+(t) | [A, H] | \Psi_i^+(t) \rangle \, \mathrm{d}t = \int_{-\infty}^{\infty} \frac{\mathrm{d}}{\mathrm{d}t} \langle \Psi_i^+(t) | A | \Psi_i^+(t) \rangle \, \mathrm{d}t$$

$$= \langle \Psi_i^+(+\infty) | A | \Psi_i^+(+\infty) \rangle - \langle \Psi_i^+(-\infty) | A | \Psi_i^+(-\infty) \rangle$$

$$= \langle \Phi_i | S^\dagger A S - A | \Phi_i \rangle = \Delta \langle A \rangle. \tag{7.37}$$

These equations clearly correspond to integrating Heisenberg's equation. In particular, the expectation value of A after the collision is obtained by evaluating the expectation value of $S^\dagger A S$ with respect to the initial state (as is consistent with the Heisenberg 'picture', it is the operator that changes with time). We also note that taking A to be the identity we obtain the unitarity of the S matrix (which is equivalent to the optical theorem).

To obtain the yield of transitions into some specified final states we can take A to be the projection operator on these states. If the initial state has a sharp energy the rate of production of the final states is time-independent, so that their final yield increases linearly with time. For the initial state $|i\rangle$ with $A = |f\rangle\langle f|$ we obtain (for $f \neq i$),

$$\Delta \langle A \rangle = |\langle i | S | f \rangle|^2 = |2\pi \delta(E_i - E_f) \langle i | T | f \rangle|^2, \tag{7.38}$$

where we have used eqn (5.37). We can rewrite our result as

$$\Delta \langle A \rangle = 2\pi \hbar \delta(E_i - E_f)(2\pi/\hbar)|\langle i | T | f \rangle|^2 \delta(E_i - E_f)$$

$$= 2\pi \hbar \delta(E_i - E_f) R_{f,i}. \tag{7.39}$$

The long-time yield is seen to be divergent as expected. Note that $2(\pi\hbar\delta E_i - E_f)$ is indeed the duration of the experiment

$$2\pi\hbar\delta(E_i - E_f) = \int_{-\infty}^{\infty} \exp[i(E_i - E_f)t/\hbar] \, dt = \int_{-\infty}^{\infty} dt. \qquad (7.40)$$

The fault is clearly with our 'experimental arrangement', which was designed to give constant rates. To measure a yield we start with an initial state with some spread in its energy so that the collision will be over in a finite time interval.§ In this case the offending δ-function will be integrated over during the integration over the initial energy, thus ensuring that transitions occur only between states of the same energy. To summarize, the yield of a transition from a state of well-defined energy is given by the rate of the transition times the density of states of the initial state. (See also eqns (7.43) and (7.61).)

If the operator A is measurable, we can write the increment $\Delta\langle A\rangle$ as

$$\Delta\langle A\rangle = 2\pi\hbar \sum_f \delta(E_i - E_f)(A_f - A_i)R_{f,i}, \qquad (7.41)$$

where we have used the optical theorem, eqn (4.7), and the spectral resolution of A, $A = \sum_f |f\rangle A_f \langle f|$.

As an example let $|i\rangle$ be a state of total energy E with the internal state $|n\rangle$ occupied,

$$|i\rangle = \int d\hat{\mathbf{k}}_n \rho^{\frac{1}{2}}(E - E_n)|\mathbf{k}_n, n\rangle. \qquad (7.42)$$

Then with
$$A = \sum_m |m\rangle A_m \langle m|$$

$$\Delta\langle A\rangle = 2\pi\hbar \int d\hat{\mathbf{k}}_n \rho(E - E_n) \sum_m (A_m - A_n) V^{-1} R(m \leftarrow \mathbf{n}).$$

In particular, if $A_{m'} = \delta_{m, m'}$

$$\Delta\langle A\rangle = \pi^{-1} k_n^2 \sigma_{m,n} = 2\pi\hbar \frac{d\dot{N}(m \leftarrow n)}{dE}, \qquad (7.43)$$

where we have used eqn (4.58). With our selection of A, $\Delta\langle A\rangle$ is the yield of the inelastic transitions $n \to m$ in the microcanonical ensemble, and eqn (7.43) confirms our previous remarks regarding the connection between the rate and the yield function.

2.7.4. *Operator theory of reactive collisions—the Jauch resolution*

In our previous discussions of multichannel collisions we have been able to resolve the solution of the L.S. equation into components that belong to different channels in an asymptotic sense only. Using the

§ The culprit here is the time-energy uncertainty relation which prevents us from measuring finite time intervals when the energy has a sharp value.

evolution operator we can try to trace the evolution of these components and so obtain a resolution of the total wavefunction into components, such that each component will evolve into (or has evolved from) a particular channel.

We begin our discussion with an initial state $\Phi_i(t')$ in a particular channel i, at some time t' in the past. The wavefunction that evolved from this state, $\Psi_i^+(t')$, can be defined by

$$\Psi_i^+(t) = \lim_{t' \to -\infty} U^+(t-t')\Phi_i(t'). \qquad (7.44)$$

In the absence of interaction $\Phi_i(t')$ will evolve under the Hamiltonian $H_{0,i}$,

$$\Phi_i(t') = \exp[-iH_{0,i}(t'-t)/\hbar]\Phi_i(t) \quad (t' < t)$$
$$= U_{0,i}^-(t'-t)\Phi_i(t), \qquad (7.45)$$

where $U_{0,i}^-(\tau) = \theta(-\tau)\exp[-iH_{0,i}\tau/\hbar]$, and so eqn (7.44) can be rewritten as

$$\Psi_i^+(t) = \lim_{t' \to -\infty} U^+(t-t')U_{0,i}^-(t'-t)\Phi_i(t) = \Omega_i^+ \Phi_i(t), \qquad (7.46)$$

where Ω_i^+ is defined by the above. Using the methods of the previous section, Ω_i^+ can be identified as the Møller wave-operator for an initial state in the channel i. Along similar lines we can also introduce $\Psi_i^-(t)$ for the state that will evolve into Φ_i in the remote future:

$$\Psi_i^-(t) = \lim_{t' \to +\infty} U^-(t-t')U_{0,i}^+(t'-t)\Phi_i(t) = \Omega_i^- \Phi_i(t). \qquad (7.47)$$

The adjoint $\Omega_i^{+\dagger}$ is introduced by analogy to eqn (7.28), namely,

$$\Phi_i(t) = \lim_{t' \to -\infty} U_{0,i}^+(t-t')U^-(t'-t)\Psi_i^+(t) = \Omega_i^{+\dagger}\Psi_i^+(t), \qquad (7.48)$$

or, in words, as $t' \to -\infty$, $\Psi_i^+(t')$ tends to the initial state, in the channel i. If all the information at our disposal was that $\Psi_i^+(t' \to -\infty)$ tends to an initial state, we could try to determine the relevant channel by letting this state evolve forward from t' to t in the absence of interaction. We would then examine a series of limits, each of the form

$$\lim_{t' \to -\infty} U_{0,j}^+(t-t')\Psi_i^+(t') = \lim_{t' \to -\infty} U_{0,j}^+(t-t')U^-(t'-t)\Psi_i^+(t)$$
$$= \Omega_j^{+\dagger}\Omega_i^+ \Phi_i(t), \qquad (7.49)$$

and expect to obtain a contribution from the term with $j = i$ only. If we put $\Psi_j^+(t) = \Omega_j^+ \Phi_j(t)$ then the above expectation is equivalent to requiring that

$$\langle \Psi_j^+(t)|\Psi_i^+(t)\rangle = \delta_{ji}\langle \Phi_i(t)|\Phi_i(t)\rangle, \qquad (7.50)$$

or states that evolved from initial states in different channels are orthogonal, and in the same way we expect that states that will evolve into final states in different channels are orthogonal,

$$\langle \Psi_j^-(t)|\Psi_i^-(t)\rangle = \delta_{ji}\langle \Phi_i(t)|\Phi_i(t)\rangle. \qquad (7.51)$$

For an extensive discussion and proof of these expectations we refer the reader to the original work of Jauch (1958), and Jauch and Marchand (1966). A proof can be constructed, however, in the coordinate representation, since the functions Ψ_i^+ and $(\Psi_j^+)^*$ for $i \neq j$ have different asymptotic behaviour in all channels. Thus from eqn (6.62) (with $A = 1$) the above expectations hold.

Equations (7.51) and (7.46) provide us with a formal way of examining the 'past' of the state $\Psi_j^+(t)$ (or the 'future' of $\Psi_j^-(t)$) in that

$$\Omega_i^{+\,\dagger}\Psi_j^+(t) = \Omega_i^{+\,\dagger}\Omega_j^+\Phi_j(t) = \delta_{ij}\Phi_i(t). \tag{7.52}$$

We can introduce the operators $R_i^\pm$ that establish the channel label directly by§

$$R_i^\pm = \Omega_i^\pm \Omega_i^{\pm\,\dagger} \tag{7.53}$$

in that $\quad R_i^\pm \Psi_j^\pm(t) = \delta_{ij}\Psi_j^\pm(t).$

These operators are projectors since, using eqn (7.52),

$$R_i^\pm R_j^\pm = \Omega_i^\pm \Omega_i^{\pm\,\dagger}\Omega_j^\pm \Omega_j^{\pm\,\dagger} = R_i^\pm \delta_{ij}. \tag{7.54}$$

The conservation of norm implies that the projectors R commute with the total Hamiltonian H.

We can thus perform a channel assignment during the collision process, in the following sense. Let $\Psi(t)$ be a function that evolves under the full Hamiltonian

$$i\hbar\frac{\partial \Psi(t)}{\partial t} = H\Psi(t),$$

which is arbitrary, apart from the absence of any bound state (of H) components. Then we can put

$$\Psi(t) = \sum_i R_i^+ \Psi(t), \tag{7.55}$$

where $R_i^+ \Psi(t)$ is that component of $\Psi(t)$ that evolved from the channel i. In the same fashion we can write

$$\Psi(t) = \sum_i R_i^- \Psi(t), \tag{7.56}$$

where $R_i^- \Psi(t)$ is the component that will evolve into the channel i. This assignment does not depend on the time t since R and H commute. Thus the resolutions (7.55) and (7.56) are into components that are not merely orthogonal but also non-interacting. It is perhaps worth while to stress that $R_i^+ \Psi(t)$ has components that *will evolve* into all possible channels and, in the same way, $R_i^- \Psi(t)$ has components that have evolved from all possible channels. R_{ji}^- is by no means necessarily orthogonal to R_i^+ (and vice versa). In fact, since $R_j^- R_i^+ \Psi(t)$ is the

§ In these equations either the upper or the lower sign is to be taken throughout the equation.

component of $\Psi(t)$ that has evolved from channel i and will evolve to channel j, $R_j^- R_i^+$ has to be non-null if transitions between the two channels can take place. (They may be forbidden by selection rules, for example.)

By analogy to eqn (7.31) we can consider the behaviour of $\Psi_i^+(t \to \infty)$ in the channel f from

$$\lim_{t' \to \infty} U_{0,f}^-(t-t')\Psi_i^+(t') = \lim_{t' \to \infty} U_{0,f}^-(t-t')U^+(t'-t)\Psi_i^+(t)$$
$$= \Omega_f^{-\dagger}\Psi_i^+(t) = \Omega_f^{-\dagger}\Omega_i^+\Phi_i(t). \quad (7.57)$$

We can thus regard $\Omega_f^{-\dagger}\Omega_i^+$ as the S operator between the channels i and f. From our remarks above we can also consider an operator S', $S' = R_f^- R_i^+$, such that

$$\langle \Psi_f^- | S' | \Psi_i^+ \rangle = \langle \Phi_f | S | \Phi_i \rangle. \quad (7.58)$$

In Appendix 2.C we show that for $f \neq i$

$$\Omega_f^{-\dagger}\Omega_i^+ = -2\pi i \int dE\, C_f \delta(E-H_{0,f}) T_{fi}(E^+) \delta(E-H_{0,i}) C_i. \quad (7.59)$$

This result is an obvious extension of the results for non-reactive collisions. In particular, the operator

$$(\Omega_f^{-\dagger}\Omega_i^+)^\dagger C_f \Omega_f^{-\dagger}\Omega_i^+ \quad (7.60)$$

determines the yield of reactive collisions from the channel i to the channel f (cf. eqn (7.37)). For the initial state $|E, \mathbf{n}, i\rangle$ we find

$$\Delta \langle C_f \rangle = (2\pi)^2 \langle E, \mathbf{n}, i | T_{fi}^\dagger(E^+) \delta(E-H_{0,f}) C_f T_{fi}(E^+) | E, \mathbf{n}, i \rangle$$
$$= 2\pi\hbar \langle E, \mathbf{n}, i^+ | \dot{C}_f | E, \mathbf{n}, i^+ \rangle$$
$$= 2\pi\hbar \rho(E-E_n) \langle \mathbf{k}_n, n, i^+ | \dot{C}_f | \mathbf{k}_n, n, i^+ \rangle$$
$$= 2\pi\hbar \rho(E-E_n) V^{-1} \sum_m R(m \leftarrow \mathbf{n}), \quad (7.61)$$

thus establishing for reactive collisions the connection between rates and yields. Comparing eqns (7.61) and (6.47) we obtain a proof of our previous statement that $Y_f^i(E)$ is the yield of reactive collisions in the microcanonical ensemble,

$$Y_f^i(E) = 2\pi\hbar \sum_{m,n} g_n \frac{d\dot{N}(m \leftarrow n)}{dE}. \quad (7.62)$$

It is convenient to define an operator $\mathbf{T}_{fi}$ for $f \neq i$ by

$$\mathbf{T}_{fi} = \Omega_f^{-\dagger}\Omega_i^+, \quad (7.63)$$

so that an S operator can be written as

$$S = I + \sum_{f,i} \mathbf{T}_{fi}. \quad (7.64)$$

TIME-DEPENDENT COLLISION THEORY

The optical theorem for reactive collisions (eqn (6.40)) guarantees that S is unitary.

Very often (sections 2.3.1, 2.8.3) we encounter a resolution of the $\mathbf{T}_{fi}$ operator in the form

$$\mathbf{T}_{fi} = \sum_{\gamma,\gamma'} |\gamma\rangle \mathbf{T}_{fi}(\gamma,\gamma')\langle\gamma'|, \qquad (7.65)$$

so that we can write

$$\frac{\mathrm{d}\dot{N}(\gamma' \leftarrow \gamma)}{\mathrm{d}E} = (2\pi\hbar)^{-1}|\mathbf{T}_{fi}(\gamma',\gamma)|^2. \qquad (7.66)$$

As an example we consider a case of non-reactive rotational excitation. Equation (7.64) has only one term, so that for $\gamma \neq \gamma'$ we have

$$\frac{\mathrm{d}\dot{N}^J(\gamma' \leftarrow \gamma)}{\mathrm{d}E} = (2\pi\hbar)^{-1}(2J+1)|S^J(\gamma',\gamma)|^2, \qquad (7.67)$$

where we have summed over all values of M, and the superscript J is a reminder that the total angular momentum is conserved, and γ is the pair of indices j, l. We can now sum§ over all l and l' values consistent with a given J and a given j and j' respectively to get

$$\frac{\mathrm{d}\dot{N}^J(j' \leftarrow j)}{\mathrm{d}E} = (2\pi\hbar)^{-1} \sum_{l=|J-j|}^{J+j} \sum_{l'=|J-j'|}^{J+j'} (2J+1)|S^J(\gamma',\gamma)|^2. \qquad (7.68)$$

Summing over all values of J and averaging with respect to the initial state to obtain a degeneracy averaged rate

$$2\pi\hbar \frac{\mathrm{d}\dot{N}(j' \leftarrow j)}{\mathrm{d}E} = (2j+1)^{-1} \sum_{J=0}^{\infty} (2J+1) \frac{\mathrm{d}\dot{N}^J(j' \leftarrow j)}{\mathrm{d}E} = \pi^{-1}k_j^2 \sigma_{j',j}. \qquad (7.69)$$

Using the transformation (3.57) one can perform other chains of summations towards the same final result.|| Equation (7.69) can also be obtained by evaluating the cross-section via the scattering amplitude.

§ There are no cross-terms since the operator $\mathbf{l}$ is diagonalized by our basis, cf. eqn (3.33).
|| Smith's (1962a) results for the rate of reactive collisions were written in terms of the T matrix defined by analogy to eqn (3.57).

2.8. Symmetry

2.8.0. THE role of symmetry in collision theory can be conveniently divided into three aspects. Two of these, time reversal and permutation symmetry of identical particles, have a general validity in the field of atomic and molecular collisions. They differ, however, in that the time-reversal operator is non-linear. The third type involves symmetry elements that are typical of the system under discussion.

2.8.1. *Time reversal*

The operation of time reversal§ is formally introduced in terms of the operator
$$\theta = UK, \tag{8.1}$$
where U is unitary and K is the (non-linear) operator of complex conjugation,
$$K(C_i|i\rangle + C_j|j\rangle) = C_i^*|i\rangle + C_j^*|j\rangle, \tag{8.2}$$
where the Cs are constant. In terms of a complete orthonormal set $\{|n\rangle\}$
$$\theta|i\rangle = \sum_n U|n\rangle\langle n|i\rangle^*. \tag{8.3}$$
We introduce the notation $|\bar{i}\rangle = \theta|i\rangle$,
$$|\bar{i}\rangle = U|i^*\rangle, \tag{8.4}$$
where
$$|i^*\rangle = \sum_n |n\rangle\langle n|i\rangle^* = K|i\rangle \tag{8.5}$$
and $K^2 = 1$.

Let $|i\rangle$ be the state of the system at $t = 0$. We let the system evolve undisturbed to the time t, when we apply the operator θ and let the system evolve for another period of length t (to the time $2t$), when we find the state
$$|j\rangle = \exp[-iH(2t-t)/\hbar]\theta \exp[-iHt/\hbar]|i\rangle. \tag{8.6}$$
Provided $[\theta, H] = 0$ we have that $|j\rangle = |\bar{i}\rangle$, since
$$\theta \exp[-iHt/\hbar] = \exp[iHt/\hbar]\theta. \tag{8.7}$$
In other words, starting with the state $|\bar{i}\rangle$ at time zero, letting it evolve to the time $-t$, and operating with θ^{-1} we obtain the same state as starting with $|i\rangle$ and letting it evolve from the time zero to the time t:
$$\theta^{-1}\exp[iHt/\hbar]|\bar{i}\rangle = \exp[-iHt/\hbar]|i\rangle. \tag{8.8}$$
Thus if $\psi(t)$ is a solution of the time-dependent Schrödinger equation so is $\theta\psi(-t) = \theta\exp[iHt/\hbar]\psi(0) = \exp[-iHt/\hbar]\theta\psi(0)$.

§ See Wigner (1959) and Messiah (1961).

From eqn (8.5) and the fact that U is unitary we obtain that
$$\langle \bar{i}|\bar{f}\rangle = \sum_{n,m} \langle i|n\rangle^* \langle n|m\rangle \langle m|f\rangle^*$$
$$= \langle i^*|f^*\rangle$$
$$= \langle f|i\rangle. \qquad (8.9)$$

Equation (8.9) suggests that the time reversed operator $\bar{A}$ that corresponds to the operator A be defined by
$$\langle \bar{i}|\bar{A}|\bar{f}\rangle = \langle f|A|i\rangle$$
$$= \langle A^\dagger f|i\rangle = \langle i|A^\dagger|f\rangle^*$$
$$= \langle i^*|A^\mathrm{T}|f^*\rangle$$
$$= \langle i^*|U^{-1}UA^\mathrm{T}U^{-1}U|f^*\rangle$$
$$= \langle \bar{i}|UA^\mathrm{T}U^{-1}|\bar{f}\rangle, \qquad (8.10)$$

so that
$$\bar{A} = UA^\mathrm{T}U^{-1} = UK^2 A^\mathrm{T} U^{-1}$$
$$= \theta K A^\mathrm{T} U^{-1}$$
$$= \theta A^\dagger \theta^{-1}, \qquad (8.11)$$

where $\theta^{-1} = KU^{-1}$ and A^T is the transpose of A. (Recall that the formal definition of $A^\dagger$ is $KA^\mathrm{T} = A^\dagger K$.)

An operator is invariant under time reversal if $\bar{A} = A$. In particular, we shall always assume that the total Hamiltonian H and the Hamiltonian for the initial state H_0 are invariant under time reversal. Since for any complex number $\theta \lambda^* \theta^{-1} = \lambda$ it follows that
$$\bar{G}(\lambda) = \theta G^\dagger(\lambda)\theta^{-1} = \theta G(\lambda^*)\theta^{-1} = G(\lambda) \qquad (8.12)$$
and similarly
$$\bar{T}(\lambda) = T(\lambda). \qquad (8.13)$$

Equation (8.13) is the formal statement of the reciprocity theorem.

To appreciate the physical significance of our discussion we consider the operation θ on particular states. From eqn (8.3),
$$\theta\psi(\mathbf{r}) = \langle \mathbf{r}|\theta\psi\rangle = \int d\mathbf{r}' \langle \mathbf{r}|U|\mathbf{r}'\rangle \langle \mathbf{r}'|\psi\rangle^*. \qquad (8.14)$$

From eqn (8.8), if $\psi(\mathbf{r}; t)$ satisfies the Schrödinger equation with a real Hamiltonian then so does $\psi^*(\mathbf{r}; -t)$, or
$$\theta\psi(\mathbf{r}) = \psi^*(\mathbf{r}) = \langle \mathbf{r}|\psi\rangle^* \qquad (8.15)$$

and so we can conclude on comparing eqns (8.15) and (8.14) that
$$\theta|\mathbf{r}\rangle = |\mathbf{r}\rangle. \qquad (8.16)$$

We now see that for the momentum representation

$$\theta|\mathbf{p}\rangle = \int d\mathbf{r}\ |\mathbf{r}\rangle\langle\mathbf{r}|\mathbf{p}\rangle^*$$
$$= \int d\mathbf{r}\ |\mathbf{r}\rangle\langle\mathbf{r}|-\mathbf{p}\rangle$$
$$= |-\mathbf{p}\rangle, \qquad (8.17)$$

or $\bar{\psi}(\mathbf{p};t) = \psi^*(-\mathbf{p};-t)$ and $\bar{\mathbf{p}} = -\mathbf{p}$.

Equations (8.16) and (8.17) have classical analogues that are best seen in terms of eqn (8.6). If we let an isolated system evolve for a period t and then reverse all the momenta and let the system evolve for another period t, we expect classically to find the system back in its original position but with all the momenta reversed. Since Newton's equations of motion for $\mathbf{r}(t)$ and $\mathbf{p}(t)$ are of the second and first order in time respectively, we indeed see that if $\mathbf{r}(t)$ and $\mathbf{p}(t)$ are possible solutions so are $\mathbf{r}(-t)$ and $-\mathbf{p}(-t)$. Similar remarks apply to Hamilton's equation in classical mechanics, while their quantum analogue is the Heisenberg equation

$$i\hbar\frac{dA_H(t)}{dt} = [A_H(t), H]. \qquad (8.18)$$

Operating with θ on the left and θ^{-1} on the right and changing t to $-t$ we have

$$i\hbar\frac{d\bar{A}_H(-t)}{dt} = [\bar{A}_H(-t), H], \qquad (8.19)$$

in agreement with eqn (8.8).

In the Schrödinger picture we have

$$i\hbar\frac{d}{dt}\langle \bar{i}^+|\bar{A}|\bar{i}^+\rangle = \langle \bar{i}^+|[\bar{A},H]|\bar{i}^+\rangle. \qquad (8.20)$$

For the angular momentum $\mathbf{J}$ we put

$$\bar{\mathbf{J}} = \theta\mathbf{J}\theta^{-1} = -\mathbf{J}. \qquad (8.21)$$

If $\mathbf{J}$ is an orbital angular momentum, eqn (8.21) follows from our previous discussion, since

$$\hbar\bar{\mathbf{l}} = \theta(\mathbf{r}\wedge\mathbf{p})\theta^{-1} = \mathbf{r}\wedge\theta\mathbf{p}\theta^{-1} = -\mathbf{r}\wedge\mathbf{p} = -\hbar\mathbf{l}. \qquad (8.22)$$

For a spin angular momentum $\mathbf{S}$ eqn (8.21) is imposed by analogy to the orbital case.

Let $|l, m_l\rangle$ be a state of a rigid rotor where m_l is the projection of $\mathbf{l}$ on a fixed z axis. Since

$$\langle\hat{\mathbf{r}}|l, m_l\rangle = Y_l^{m_l}(\hat{\mathbf{r}}), \qquad (8.23)$$

and by our phase convention

$$Y_l^{m_l}(\hat{\mathbf{r}}) = (-1)^{m_l}Y_l^{-m_l}(\hat{\mathbf{r}}) \quad (m_l > 0), \qquad (8.24)$$

we have that§
$$\theta|l, m_l\rangle = (-1)^{m_l}|l, -m_l\rangle. \tag{8.25}$$

In general, if A is self-adjoint and A and $\bar{A}$ differ at most by a phase factor, one can conclude that eigenvectors of A are converted into other eigenvectors by the action of θ. Thus by operating with θ on
$$A|a\rangle = a|a\rangle, \tag{8.26}$$
we get
$$\theta A \theta^{-1} \theta |a\rangle = \bar{A}|\bar{a}\rangle = a|\bar{a}\rangle,$$
and if A and $\bar{A}$ differ in phase only we see that $|\bar{a}\rangle$ is an eigenvector of A (to within a phase factor) so that
$$\theta|a\rangle = C_a|a'\rangle, \tag{8.27}$$
where C_a is a phase factor, $|C_a| = 1$.

2.8.2. *Reciprocity and microscopic reversibility*

Consider the L.S. equation (in the centre of mass system) for the collision of structureless particles
$$|\mathbf{k}^+\rangle = |\mathbf{k}\rangle + G(E^+)V|\mathbf{k}\rangle,$$
for the case when H_0 and V commute with θ, so that
$$\theta|\mathbf{k}^+\rangle = |-\mathbf{k}\rangle + G(E^-)V\theta|\mathbf{k}\rangle, \tag{8.28}$$
where we have used eqns (8.12) and (8.17). Thus
$$\theta|\mathbf{k}^+\rangle = |-\mathbf{k}^-\rangle. \tag{8.29}$$

We recall that $|\mathbf{k}^+\rangle$ is the state that evolved from the state $|\mathbf{k}\rangle$, while $|\mathbf{k}^-\rangle$ is the state that will evolve to $|\mathbf{k}\rangle$ so that $\theta|\mathbf{k}^+\rangle$ is the state that will evolve to $|-\mathbf{k}\rangle = \theta|\mathbf{k}\rangle$. In other words, the final state to which $\theta|\mathbf{k}^+\rangle$ will evolve is the time-reversed initial state of $|\mathbf{k}^+\rangle$, while the final state to which $|\mathbf{k}^+\rangle$ will evolve is the time reverse of the initial state of $\theta|\mathbf{k}^+\rangle$.

It is convenient to characterize a general initial state $|\mathbf{k}, a\rangle$ by a set of quantum numbers a which are eigenvalues of operators A such that eqn (8.27) holds. Thus
$$\theta|\mathbf{k}, a\rangle = |\bar{\mathbf{k}}, \bar{a}\rangle = C_a|-\mathbf{k}, a'\rangle. \tag{8.30}$$
Since by assumption the internal Hamiltonians of the isolated molecules are invariant under θ, the set of operators A can include h, and so
$$\theta|\mathbf{k}, a^+\rangle = C_a|-\mathbf{k}, a'^+\rangle. \tag{8.31}$$

We now see the significance of reciprocity,
$$\langle f|T|i\rangle = \langle \bar{i}|T|\bar{f}\rangle. \tag{8.32}$$
In the time-reversed system $|\bar{f}\rangle$ is the initial state that corresponds to

---
§ Recall that l is really an eigenvalue of $\mathbf{l}^2$, which is invariant under θ,
$$\mathbf{l}^2|l, m_l\rangle = l(l+1)|l, m_l\rangle.$$

the final state $|f\rangle$ in the original system. Equation (8.32) is thus the statement that the transition amplitude between the states $|i\rangle$ and $|f\rangle$ is independent of the direction of transition. In general, however, it is not true that $|\bar{f}\rangle$ and $|f\rangle$ are the same state. Explicitly,

$$\langle E,\mathbf{m}|T|E,\mathbf{n}\rangle = \langle E,\mathbf{n}'|T|E,\mathbf{m}'\rangle C_\mathbf{m} C_\mathbf{n}, \tag{8.33}$$

where the Cs are phase factors and $\mathbf{n}$ is a set of quantum numbers of the suitable set of operators A. For the scattering amplitude we thus have

$$|f_{m,n}(\mathbf{k}',\mathbf{k})|^2 = |f_{n',m'}(-\mathbf{k},-\mathbf{k}')|^2. \tag{8.34}$$

In terms of eqns (8.33) and (4.9) we can also write a reciprocity theorem for transition rates, namely,

$$\dot{N}(f \leftarrow i) = (2\pi/\hbar)\delta(E_f - E_i)|\langle f|T|i\rangle|^2 = \dot{N}(\bar{\imath} \leftarrow \bar{f}). \tag{8.35}$$

Equation (8.35) can be considered as the basic statement of microscopic reversibility. Using eqn (4.31) we can transform this equation to statements about cross-sections. If we are not concerned with the direction of the relative momenta we can use eqns (8.33) and (4.58) to write microscopic reversibility as

$$\frac{\mathrm{d}\dot{N}(m \leftarrow n)}{\mathrm{d}E} = \frac{\mathrm{d}\dot{N}(n' \leftarrow m')}{\mathrm{d}E}. \tag{8.36}$$

If the sets of quantum numbers n and m are all eigenvalues of such observables that satisfy $A = \bar{A}$ (and this excludes angular momenta), $n' = n$ and $m' = m$ and eqn (8.36) connects the rates of the direct and reverse transition.

To apply microscopic reversibility to states with non-vanishing angular momenta, we shall write for the initial state $|\mathbf{k},j,m_j\rangle$, where the set j refers to 'even time' observables ($A = \bar{A}$) and the set m_j to 'odd time' observables ($A = -\bar{A}$). Then, from eqns (8.33) and (8.25),

$$|\langle \mathbf{k}',j',m_{j'}|T|\mathbf{k},j,m_j\rangle|^2 = |\langle -\mathbf{k},j,-m_j|T|-\mathbf{k}',j',-m_{j'}\rangle|^2. \tag{8.37}$$

If we now define

$$\frac{\mathrm{d}\dot{N}(j' \leftarrow j)}{\mathrm{d}E} = g_j^{-1} \sum_{m_j m_{j'}} \frac{\mathrm{d}\dot{N}(j'm_{j'} \leftarrow jm_j)}{\mathrm{d}E}, \tag{8.38}$$

where g_j is the number of states m_j that are summed over, then from eqns (8.37) and (8.38) it follows that

$$g_j \frac{\mathrm{d}\dot{N}(j' \leftarrow j)}{\mathrm{d}E} = g_{j'} \frac{\mathrm{d}\dot{N}(j \leftarrow j')}{\mathrm{d}E}. \tag{8.39}$$

(We have used this result in section 2.3.3.) From eqn (4.59) we obtain a similar statement about the 'degeneracy-averaged' cross-section

$$k^2 g_j \sigma_{j',j} = k'^2 g_{j'} \sigma_{j,j'}, \qquad (8.40)$$

where
$$\sigma_{j',j} = g_j^{-1} \sum_{m_j m_{j'}} \sigma_{j'm_{j'},jm_j}. \qquad (8.41)$$

An alternative derivation of these results, which is applicable also to cases where the initial state is not an isotropic mixture of all possible orientations $\hat{\mathbf{k}}$, is given later (eqn (8.61)).

Our proof that the transition operator is an even-time observable is valid only for the one-channel case, when T is L-conjugation invariant. However, we have seen (section 2.6.2) that on the energy shell the transition matrix elements between states of different channels do satisfy the consequences of L-conjugation invariance, so that eqn (8.32) is of general validity on the energy shell ($E_i = E_f$). A formal proof can also be given by noting that the operators

$$[H_{0,i}+T_{if}]^{\ddagger} = H_{0,f}+T_{fi} \qquad (8.42)$$

are both invariant under θ. On the energy shell we thus have

$$\langle E, \mathbf{m}, f | T_{fi} | E, \mathbf{n}, i \rangle = \langle E, \bar{\mathbf{n}}, i | T_{if} | E, \bar{\mathbf{m}}, f \rangle. \qquad (8.43)$$

If we denote the initial state by $|i\rangle$ then $\theta S|i\rangle$ is the initial state for $\theta|i+\rangle$, where S is the S operator (section 2.7.2). If our description of the collision is invariant under time reversal then $\theta|i\rangle$ is the final state of $\theta|i+\rangle$, or
$$\theta|i\rangle = S\theta S|i\rangle = S\theta S\theta^{-1}\theta|i\rangle. \qquad (8.44)$$

Finally, we should point out that reciprocity does not imply that our description of the collision is invariant under time reversal. Our description may be at fault if, for example, we do not observe some of the collision products.§ Thus the observed final states will not, after time inversion, evolve to the original initial state. Even so, the transition amplitude between the initial state and an observed final state will satisfy reciprocity. Another example is the Born approximation. Transition amplitudes computed by the Born approximation (to any order) satisfy reciprocity. In particular, in the first Born approximation, for a self-adjoint V,
$$\langle f|V|i\rangle = \langle V^{\dagger}f|i\rangle = \langle i|V|f\rangle^*, \qquad (8.45)$$
so that in this approximation
$$\dot{N}(f \leftarrow i) = \dot{N}(i \leftarrow f), \qquad (8.46)$$

§ To see that reciprocity and conservation of probability imply eqn (8.44), we recall that the conservation of probability is formally expressed by the unitarity of S, $S^{\dagger} = S^{-1}$. Coupled with reciprocity, $S = \theta S^{\dagger} \theta^{-1}$, we have $S = \theta S^{-1} \theta^{-1}$, which is equivalent to eqn (8.44), since $\theta^2 = \pm 1$.

and microscopic reversibility holds between the rates of the direct and reverse reaction. (For rearrangement collisions the validity of this conclusion follows from eqn (6.33).)

In practice, microscopic reversibility is used to obtain information about rates and cross-sections that are not easily accessible to measurement, notably the recombination reactions

$$A+B \to AB+h\nu \qquad (8.47\,\text{a})$$

(Terenin and Prilezhaeva 1932, Kondratiev 1935) and

$$A+B+M \to AB+M \qquad (8.47\,\text{b})$$

(Keck and Carrier 1965 and references therein) where $h\nu$ is a photon and M is a third body, and in unimolecular breakdown. Equation (8.39) (or (8.35)) is also useful as a guideline for the formulation and checking of the internal consistency of model theories (Pechukas and Light 1965).

2.8.3. *Constants of motion*

A self-adjoint operator Λ is referred to as a constant of motion when it commutes with the total Hamiltonian

$$[\Lambda, H] = 0 \qquad (8.48)$$

so that $\langle \dot{\Lambda} \rangle = 0$. We denote a normalized set of eigenfunctions of Λ by $\{|\lambda_j\rangle\}$,

$$\Lambda |\lambda_j\rangle = \lambda_j |\lambda_j\rangle. \qquad (8.49)$$

For computations on the energy shell we can write for the transition operator

$$T(E^+) = (H-E)+(H-E)G(E^+)(H-E) \qquad (8.50)$$

so that

$$[\Lambda, T(E^+)] = 0 \qquad (8.51)$$

or, taking the off-diagonal elements of eqn (8.51) we conclude that

$$\langle \lambda_{j'} | T(E^+) | \lambda_j \rangle = 0, \quad \lambda_{j'} \neq \lambda_j. \qquad (8.52)$$

The basis $\{|\lambda_j\rangle\}$ is thus suitable for a spectral analysis of the transition operator,

$$T(E^+) = \sum_j |\lambda_j\rangle\langle\lambda_j|T(E^+)|\lambda_j\rangle\langle\lambda_j|. \qquad (8.53)$$

In general, a particular initial (or final) state $|i\rangle$, $(|f\rangle)$, will not be an eigenstate of Λ. We will thus obtain a resolution of $\langle f|T|i\rangle$ as

$$\langle f|T|i\rangle = \sum_j \langle f|\lambda_j\rangle\langle\lambda_j|T(E^+)|\lambda_j\rangle\langle\lambda_j|i\rangle. \qquad (8.54)$$

The physical significance of the resolution is based on eqn (8.52). Each component evolves independently of the others. From a computational point of view the expansion is particularly useful if the matrix elements $\langle \lambda_j | T | \lambda_j \rangle$ are easier to evaluate than $\langle f | T | i \rangle$; the expansion coefficients $\langle f | \lambda_j \rangle$ are easily obtainable and only a few terms in the expansion are required. In most problems involving rotational invariance of molecular systems (section 2.3.1) the last two conditions are not always satisfied.

If the operator Λ is unitary (as is the case for symmetry operations) we note that
$$\langle f|T|i\rangle = \langle f|\Lambda^\dagger \Lambda T|i\rangle = \langle f|\Lambda^\dagger T\Lambda|i\rangle. \tag{8.55}$$
For example, if neither the internal Hamiltonians for the colliding molecules nor their mutual interaction depends on the absolute position of the total centre of mass we can take for Λ the translational operator§ for the centre of mass
$$T_\rho = \exp(i\boldsymbol{\varkappa}\cdot\boldsymbol{\varkappa}), \qquad \boldsymbol{\varkappa} = -i\frac{\partial}{\partial\boldsymbol{\rho}}, \tag{8.56}$$
where $\boldsymbol{\rho}$ is the centre of mass position.

We have previously used eqn (8.56) with $\Lambda = T_\rho$ to derive eqn (4.20), which implies that the momentum of the total centre of mass is conserved.

We can combine reciprocity with invariance under a unitary operator Λ to write
$$\langle f|T|i\rangle = \langle \bar{\imath}|\Lambda^\dagger T\Lambda|\bar{f}\rangle. \tag{8.57}$$
Equation (8.57), which connects the transition amplitude for the process $i \to f$ to that of $\Lambda\bar{f} \to \Lambda\bar{\imath}$, has several important applications.

As an example we consider the parity (space reflection) operator P:
$$P\mathbf{r}P^{-1} = -\mathbf{r},$$
$$P\mathbf{p}P^{-1} = -\mathbf{p}, \tag{8.58}$$
so that
$$P\mathbf{j}P^{-1} = \mathbf{j}, \tag{8.59}$$
where $\mathbf{j}$ is an angular momentum. We normally assume that P commutes with the total Hamiltonian. Combining eqns (8.59), (8.58), and (8.57) and using eqn (8.37) we have a stronger reciprocity statement,
$$\langle \mathbf{k}',j',m_{j'}|T|\mathbf{k},j,m_j\rangle = \langle \mathbf{k},j,-m_j|T|\mathbf{k}',j',-m_{j'}\rangle C, \tag{8.60}$$
where C is a phase factor. Equation (8.60) can also be written as a statement of microscopic reversibility,
$$\sum_{m_j m_j} |\langle \mathbf{k}',j',m_{j'}|T|\mathbf{k},j,m_j\rangle|^2 = \sum_{m_j m_j} |\langle \mathbf{k},j,m_j|T|\mathbf{k}',j',m_{j'}\rangle|^2. \tag{8.61}$$

If we denote by $|T_{j',j}|^2$ the left-hand side of the equation above we can write a degeneracy averaged rate as
$$\mathsf{V}^{-1}R(j' \leftarrow j) = g_j^{-1}|T_{j',j}|^2 \rho(E-E_{j'}) \tag{8.62}$$

§ T_ρ is defined by $\qquad T_\rho F(\boldsymbol{\rho}') = F(\boldsymbol{\rho}'-\boldsymbol{\rho})T_\rho.$ (T)

Expanding the exponential in eqn (8.56) and using the definition of $\hbar\boldsymbol{\varkappa}$ as the centre of mass momentum, we see that equation (T) is just a compact notation for Taylor's theorem.

so that
$$\frac{R(j' \leftarrow j)}{R(j \leftarrow j')} = \frac{g_{j'} \rho(E-E_{j'})}{g_j \rho(E-E_j)}, \quad (8.63)$$

where $\rho(E-E_j)$ is the density of translational states of energy $E-E_j$ and g_j is the density of internal states at energy E_j.§

It is important to remember that all our symmetry theorems are correct on the energy shell, namely, when all the states concerned are taken at the same *total* energy. From eqn (8.63) we have, for example,

$$\frac{d\sigma(j' \leftarrow j)}{d\sigma(j \leftarrow j')} = \frac{v_j^{-1} R(j' \leftarrow j)}{v_{j'}^{-1} R(j \leftarrow j')} = \frac{g_{j'} \mu_f^2 v_{j'}^2}{g_j \mu_i (\mu_f v_{j'}^2 - 2\Delta E)}, \quad (8.64)$$

where μ is the reduced mass and at the same total energy

$$2\Delta E = 2(E_j - E_{j'}) = \mu_f v_{j'}^2 - \mu_i v_j^2$$

with ΔE the difference in the internal energy. If $E_j > E_{j'}$ then for E just above E_j, $\rho(E-E_j)$ is very small and the rate for loss of internal energy is much higher than the rate of the reverse process.

2.8.4. *Permutation symmetry in collisions*

We return in this section to the symmetry associated with the presence of identical‖ particles. Generalizing the discussion of section 3.2.0 we require the wavefunction to satisfy

$$Q\psi_0 = \delta_Q \psi_0, \quad (8.65)$$

where Q is a permutation of identical particles,

$$\delta_Q = \begin{cases} 1 & \text{(bosons)}, \\ (-1)^q & \text{(fermions)}, \end{cases} \quad (8.66)$$

and q is the parity§§ of Q.

A wavefunction of the required symmetry can be projected from an arbitrary wavefunction of N identical particles by the operator

$$O = (N!)^{-1} \sum_Q \delta_Q Q, \quad (8.67)$$

that satisfies
$$QO = \delta_Q O \quad (8.68)$$

and hence is a projection operator, $O^2 = O$.

§ In other words,
$$g_j \rho(E-E_j) = \int g_j \, \delta(E_{\text{int}} - E_j) \rho(E - E_j - E_{\text{int}}) \, dE_{\text{int}}$$

is the total density of states when the internal state j is occupied. For an application see Klots (1967).

‖ We consider two particles as identical if the total Hamiltonian H is invariant under the exchange of the two particles. In the coordinate representation each particle is labelled by its position and (if any) spin, and exchange of particles means exchange of both space and spin coordinates.

§§ Any permutation Q can be written as a product of permutations that exchange one pair at a time. The parity q is the (unique) number of pair permutations in the product. Thus, if $QQ' = Q''$ then $\delta_Q + \delta_{Q'} = \delta_{Q''}$.

SYMMETRY

If there is more than one type of particle (say electrons and nuclei), a separate projection operator must be introduced for each type of particle.

We shall write
$$|Oi\rangle = M_i O|i\rangle, \tag{8.69}$$
where M_i is determined by the condition
$$\langle Oi|Oi\rangle = \langle i|i\rangle = M_i^2 \langle i|O|i\rangle. \tag{8.70}$$

Since the interchange of an identical particle between two molecules is experimentally indistinguishable, we can regard M_i^2 as the number of 'indistinguishable' channels. For composite systems M_i^2 is not necessarily equal to $N!$, where N is the total number of identical particles. For example, in an initial state $a+b$ the wavefunction for a alone does have the proper exchange symmetry, and similarly for b. In this case O merely exchanges particles between a and b, and M_i^2 is the number of possible exchanges. If there are N identical particles, n in a and $N-n$ in b, then $M_i^2 = N!/n!(N-n)!$, or the total number of possible exchanges divided by the number of exchanges that do not permute a particle between a and b. To construct $M_i O$ we note that there are $n!$ different arrangements of the particles in a properly symmetrized wavefunction for a and $(N-n)!$ arrangements for b. For a given arrangement of a and b there are§
$$\binom{N-n}{m}\binom{n}{m} \tag{8.71}$$
ways of exchanging m particles between a and b.

Thus if we denote by $|i(m)\rangle$ the result of permuting m particles between a and b, we have that
$$|Oi\rangle = \binom{N}{n}^{-\frac{1}{2}} \sum_m \binom{N-n}{m}\binom{n}{m} |i(m)\rangle, \tag{8.72}$$
where m is restricted to the range 0 to the smaller of n or $N-n$. If the dependence on m is immaterial for a given purpose we can perform the summation∥ to get
$$|Oi\rangle = \binom{N}{n}^{\frac{1}{2}} |i\rangle = M_i |i\rangle. \tag{8.73}$$

§ $\binom{n}{m} = n!/m!(n-m)!$ is the binomial coefficient, giving the number of ways of selecting m particles from a set of n particles, all particles being alike.

∥ We are using the identity
$$\binom{N}{n} = \sum_{m=0}^{n} \binom{N-m}{m}\binom{n}{m}.$$
If $n > N-n$ we simply replace the upper limit by $N-n$.

For example, in computing the S matrix between distinguishable channels we have to evaluate $\langle Of^-|Oi^+\rangle$ or, since O is a projection,

$$\langle Of^-|Oi^+\rangle = M_f M_i \langle f^-|O|i^+\rangle = M_f\langle f^-|Oi^+\rangle. \tag{8.74}$$

The computations of rates via Ehrenfest's theorem is now based on defining

$$\langle A \rangle = \langle Oi^+|A|Oi^+\rangle. \tag{8.75}$$

It is important, however, to define A properly. For example, in computing rates for collision between two identical atoms we would normally take

$$|On_\alpha n_\beta\rangle = 2^{-\frac{1}{2}}|n_\alpha n_\beta + n_\beta n_\alpha\rangle, \tag{8.76}$$

where the first index is the internal state of the first atom, and put

$$A = |On_\alpha n_\beta\rangle\langle On_\alpha n_\beta|. \tag{8.77}$$

However, when $\alpha = \beta$ we are counting twice as many final states since, in this case only, $M = 1$ rather than $2^{+\frac{1}{2}}$.

In the general case the transition amplitude can be written

$$\langle Of|T|Oi\rangle = \frac{M_f M_i}{N!}\sum_Q \delta_Q \langle f|H-E|Qi^+\rangle. \tag{8.78}$$

In the particular case when $|f\rangle$ and $|i\rangle$ are completely unsymmetrized $M_f = M_i = (N!)^{\frac{1}{2}}$ and only the sum remains. More often than not the initial states are already symmetrized with respect to some of the equivalent particles. For molecular collisions the initial states are usually symmetrized with respect to all electrons beforehand, and each molecule has the proper symmetry with regard to its own nuclei. The number of indistinguishable channels is then determined by the possibility of exchange of identical nuclei between the two molecules.

The matrix elements in eqn (8.78) need not be of equal magnitude. As an example consider the collision

$$H+H_2 \to H+H_2.$$

In chemical kinetics one is primarily interested in the situation where the nuclear spin of the molecule is changed during the collision (*ortho–para* transitions). Equation (8.78) can be written

$$\langle Of|T|Oi\rangle = \langle f|H-E|i^+\rangle + 2\langle f|(H-E)Q|i^+\rangle, \tag{8.79}$$

where Q permutes the incident atom with one atom of the molecule. If the initial and final states correspond to H_2 with different values of the nuclear spin, and if we neglect magnetic interactions in our Hamiltonian, the first term does not contribute due to the orthogonality of the spin functions. The transition amplitude is thus twice the value predicted for 'distinguishable' atoms, and the reaction takes place due to the exchange of the incident atom with either one of the atoms of the molecule.

The L.S. equation for $Q|i^+\rangle$ is obtained from the unsymmetrized L.S. equation, as
$$Q|i^+\rangle = Q|i\rangle + G(E^+)QVQ^{-1}Q|i\rangle, \tag{8.80}$$
since Q commutes with H. QVQ^{-1} is the mutual interaction when the initial state is $Q|i\rangle$. The Schrödinger equation for the initial state is obtained from
$$(E-H_0)|i\rangle = 0 \tag{8.81}$$
as
$$(E-QH_0Q^{-1})Q|i\rangle = 0, \tag{8.82}$$
so that an alternative form of eqn (8.80) is
$$Q|i^+\rangle = Q|i\rangle + (E^+ - QH_0Q^{-1})^{-1}QVQ^{-1}Q|i^+\rangle. \tag{8.83}$$
Combining these states to form $|Oi^+\rangle$ we can re-derive our original conclusion that $|Oi^+\rangle$ satisfies the L.S. equation
$$|Oi^+\rangle = |Oi\rangle + G(E^+)(H-E)|Oi\rangle. \tag{8.84}$$

If the two molecules possess internal spin, the over-all symmetry of the wavefunction is reflected by the possible combination of the space and spin portions of the wavefunction. For example, in the collision of two hydrogen atoms (where the nuclei are fermions of spin $\frac{1}{2}$), a space-symmetric wavefunction must be associated with a spin-antisymmetric wavefunction and vice versa. When we perform a partial wave analysis, a space-symmetric wavefunction will consist of even-l components only. Identical effects are, of course, well known in the spectrum of molecular hydrogen.§ If there are more than two identical particles it is no longer possible to factor the total wavefunction into a simple product of space and spin parts. It can still be written, however, as a sum of such products, cf. Wigner (1959).

§ In molecular hydrogen, l, the angular momentum of relative motion is the rotational quantum number j.

APPENDICES 2

2.A. *Normalization of the solutions of the L.S. equation*

Our proof of the normalization of the solutions of the L.S. equation follows Friedman (1956) and avoids any assumptions about the properties of the solutions. We denote by a single letter the whole set of quantum numbers necessary to specify an initial state, so that the L.S. equation is

$$|i^+\rangle = |i\rangle + G_0(E_i^+) V |i^+\rangle, \qquad (A.1)$$

where $G_0(E_i^+) = (E_i - H_0 + i\epsilon)^{-1}$, and at the moment ϵ is still finite (since we have not taken any matrix elements of $G_0(E_i^+)$ yet). The adjoint L.S. equation is

$$\langle f^+| = \langle f| + \langle f^+| V G_0(E_f^-). \qquad (A.2)$$

(Recall that $G_0^\dagger(E_f^+) = G_0(E_f^-)$.)

Thus

$$\langle f^+|i^+\rangle = \langle f|i\rangle + \langle f^+|V G_0(E_f^-)|i\rangle + \langle f|G_0(E_i^+)V|i^+\rangle + \\ + \langle f^+|V G_0(E_f^-) G_0(E_i^+) V|i^+\rangle. \qquad (A.3)$$

The second term above can be written as

$$\langle f^+|V(E_f - H_0 - i\epsilon)^{-1}|i\rangle = (E_f - E_i - i\epsilon)^{-1}\langle f^+|V|i\rangle \\ = (E_f - E_i - i\epsilon)^{-1}[\langle f^+|V|i^+\rangle - \langle f^+|V G_0(E_i^+) V|i^+\rangle],$$

where we have substituted for $|i\rangle$ the L.S. equation (A.1). A similar transformation can be made on the third term, so that the sum of the second and third terms is

$$(E_f - E_i - i\epsilon)^{-1}[\langle f^+|V G_0(E_f^-) V|i^+\rangle - \langle f^+|V G_0(E_i^+) V|i^+\rangle] \\ = (E_f - E_i - i\epsilon)^{-1}[\langle f^+V|G_0(E_f^-) - G_0(E_i^+)|V i^+\rangle] \\ = \frac{E_i - E_f + 2i\epsilon}{E_f - E_i - i\epsilon} \langle f^+|V G_0(E_f^-) G_0(E_i^+) V|i^+\rangle,$$

where we have written

$$G_0(E_f^-) - G_0(E_i^+) = [(E_i + i\epsilon) - (E_f - i\epsilon)] G_0(E_f^-) G_0(E_i^+).$$

Summing up the last three terms in equation (A.3), we have

$$\langle f^+|i^+\rangle = \langle f|i\rangle + \frac{i\epsilon}{E_f - E_i - i\epsilon} \langle f_s|i_s\rangle \qquad (A.4)$$

where we used the notation $|i_s\rangle$ for the 'scattered' wave, $|i^+\rangle = |i\rangle + |i_s\rangle$. In the limit $\epsilon \to +0$

$$\langle f^+|i^+\rangle = \langle f|i\rangle. \qquad (A.5)$$

The actual value of the normalization integral depends on the normalization convention chosen for the initial states.

The evaluation of $\langle f^-|i^+\rangle$ proceeds in identical steps. When we make an expansion analogous to eqn (A.3) the third term is modified as follows:

$$\langle f|G_0(E_i^+)V|i^+\rangle = (E_i - E_f + i\epsilon)^{-1}\langle f|V|i^+\rangle \\ = [(E_i - E_f - i\epsilon)^{-1} - 2\pi i \delta(E_i - E_f)]\langle f|V|i^+\rangle,$$

with the result that

$$\langle f^-|i^+\rangle = \langle f|i\rangle - 2\pi i \delta(E_i - E_f)\langle f|V|i^+\rangle. \qquad (A.6)$$

2.B. On the convolution theorem and linear systems

LET $g(\omega)$ be the Fourier transform of $G(t)$,

$$g(\omega) = \int_{-\infty}^{\infty} G(t)\exp(i\omega t)\,dt, \tag{B.1}$$

and similarly let $h(\omega)$ be the Fourier transform of $H(t)$. The convolution theorem (Morse and Feshbach 1953) states that $g(\omega)h(\omega)$ is the Fourier transform of the convolution of G and H, or

$$g(\omega)h(\omega) = \int_{-\infty}^{\infty} \exp(i\omega t)\left[\int_{-\infty}^{\infty} G(\tau)H(t-\tau)\,d\tau\right] dt, \tag{B.2}$$

where H and G are square integrable. A simplified proof is obtained by changing the order of integration in eqn (B.2),

$$\int_{-\infty}^{\infty} G(\tau)\left[\int_{-\infty}^{\infty} H(t-\tau)\exp(i\omega t)\,dt\right] d\tau$$

$$= \int_{-\infty}^{\infty} G(\tau)\exp(i\omega\tau)h(\omega)\,d\tau$$

$$= g(\omega)h(\omega) \tag{B.3}$$

where, to get the second line, we put $\exp(i\omega t) = \exp[i\omega(t-\tau)\exp(i\omega\tau)]$ and integrate over the new variable $z = t-\tau$.

We can rewrite eqn (B.2) as

$$\int_{-\infty}^{\infty} g(\omega)h(\omega)\exp(-i\omega t)\,d\omega = 2\pi \int_{-\infty}^{\infty} G(\tau)H(t-\tau)\,d\tau, \tag{B.4}$$

where we have used eqn (7.40) to invert the Fourier transforms (B.2).

Parseval's formula is essentially eqn (B.4) for $t = 0$. The usual form of the theorem is obtained when we note that eqn (B.1) implies that the Fourier transform of $H^*(-t)$ is $h^*(\omega)$, where the star denotes complex conjugation, so that

$$(2\pi)^{-1}\int_{-\infty}^{\infty} g(\omega)h^*(\omega)\,d\omega = \int_{-\infty}^{\infty} G(\tau)H^*(\tau)\,d\tau. \tag{B.5}$$

We now take $H(t) = \theta(t)$, where $\theta(t)$ is the step function. Then

$$h(\omega) = \int_{-\infty}^{\infty} \exp[i\omega t - \epsilon t]\theta(t)\,dt,$$

where we have introduced a convergence factor, $\exp(-\epsilon t)$, to ensure the existence of the integral ($\theta(t)$ is not square integrable). Putting $z = \epsilon - i\omega$,

$$h(\omega) = \int_0^{\infty} \exp(-zt)\,dt = z^{-1} = i(\omega + i\epsilon)^{-1}. \tag{B.6}$$

Using the complex conjugate of eqn (B.4),

$$\lim_{\epsilon \to +0} \int_{-\infty}^{\infty} g(\omega)\frac{\exp(i\omega t)}{\omega - i\epsilon}\,d\omega = 2\pi i \int_{-\infty}^{\infty} G(\tau)\theta(t-\tau)\,d\tau = 2\pi i \int_{-\infty}^{t} G(\tau)\,d\tau. \tag{B.7}$$

Using eqn (B.1), we obtain the limit

$$\lim_{t \to \pm\infty} (\text{B.7}) = \begin{cases} 2\pi i g(\omega = 0) \\ 0. \end{cases} \tag{B.8}$$

By a linear system we mean the transformation

$$L\{G(t)\} = F(t), \tag{B.9}$$

where $G(t)$ is a given function and L is a linear operation. A system is invariant under time translation if

$$L\{G(t-t')\} = F(t-t'), \tag{B.10}$$

so that only time intervals are meaningful. In particular, for such systems we define

$$L\{\delta(t-t')\} = K(t-t'). \tag{B.11}$$

Given the function $K(t)$ we can evaluate $F(t)$ by putting

$$G(t) = \int_{-\infty}^{\infty} G(t')\delta(t-t') \, dt' \tag{B.12}$$

and operating with L on both sides. Using eqn (B.11),

$$F(t) = \int_{-\infty}^{\infty} G(t')K(t-t') \, dt'. \tag{B.13}$$

If $f(\omega)$ is the Fourier transform of $F(t)$ we have, from eqn (B.2), that

$$f(\omega) = g(\omega)k(\omega). \tag{B.14}$$

As examples of the use of eqn (B.14) we find

$$L\{\exp(-i\omega_0 t)\} = k(\omega_0)\exp(-i\omega_0 t) \tag{B.15}$$

where we have used eqn (7.40) to evaluate $g(\omega)$,

$$\Phi(t) = L\{\theta(t)\} = \int_0^t K(\tau) \, d\tau, \tag{B.16}$$

where we have used eqn (B.6), and

$$\Phi^R(t) = \lim_{\epsilon \to +0} L\{\theta(-t)\exp(\epsilon t)\} = \lim_{\epsilon \to +0} \int_t^{\infty} K(t')\exp(-\epsilon t') \, dt', \tag{B.17}$$

where we have changed the sign of t' in the integrand.

Another useful form is obtained by noting the identities

$$G(t) = G(t_0) + \int_{t_0}^{t} \dot{G}(\tau) \, d\tau = G(t_0) + \int_{t_0}^{\infty} \dot{G}(\tau)\theta(t-\tau) \, d\tau,$$

where t_0 is a constant. Using eqns (B.15) (with $\omega_0 = 0$) and (B.16),

$$F(t) = K(0)G(t_0) + \int_{t_0}^{\infty} \dot{G}(\tau)\Phi(t-\tau) \, d\tau. \tag{B.18}$$

In view of eqn (B.16) we often impose the condition

$$\Phi(t-\tau) = 0, \quad t < \tau. \tag{B.19}$$

To investigate the Fourier transforms of functions that satisfy eqn (B.19) (known as 'causal' functions) we note that we can write

$$\Phi(t) = \theta(t)\Phi(t), \tag{B.20}$$

or
$$\varphi(\omega) = (2\pi)^{-1}h(\omega) \otimes \varphi(\omega) = (2\pi i)^{-1}\int_{-\infty}^{\infty}\frac{\varphi(\omega')}{\omega'-\omega^+}\,d\omega', \tag{B.21}$$

where $h \otimes g$ denotes the convolution of h and g.

By definition of the principal value integral we can rewrite eqn (B.21) as

$$\varphi(\omega) = (\pi i)^{-1}Pv\int_{-\infty}^{\infty}\frac{\varphi(\omega')}{\omega'-\omega}\,d\omega' \tag{B.22}$$

or, taking the real part of eqn (B.22),

$$\mathrm{re}[\varphi(\omega)] = \pi^{-1}Pv\int_{-\infty}^{\infty}\frac{\mathrm{im}[\varphi(\omega')]}{\omega'-\omega}\,d\omega'. \tag{B.23}$$

An opposite relation holds on taking the imaginary part. Equation (B.23) is known as a Hilbert transform.

If $\varphi(\infty)$ does not vanish our manipulations are clearly invalid. One can, however, consider $\varphi(\omega) - \varphi(\infty)$ and establish eqns (B.22) and (B.23) for it.

Our discussion of causal functions can be summarized by quoting two theorems (Titchmarsh 1948, theorems 93 and 95). (1) Alternative necessary and sufficient conditions that $\varphi(E)$ be the limit as $\lambda \to E$ of an analytic function $\varphi(\lambda)$ such that

$$\int_{-\infty}^{\infty}|\varphi(E+i\epsilon)|^2\,dE < \infty \tag{B.24}$$

are (i) the real and imaginary parts of $\varphi(E)$ are mutual Hilbert transforms and (ii) $\Phi(t)$ is null for $t < 0$. (2) Let $\varphi(\lambda)$ be analytic and let (B.24) hold for every positive ϵ, then (i) as $\epsilon \to +0$ $\varphi(E+i\epsilon) \to \varphi(E)$ for almost all ϵ, (ii) the real and imaginary parts of $\varphi(E)$ are mutual Hilbert transforms, and (iii) $\Phi(t)$ is null for $t < 0$.

2.C. *Intertwining and the S operator for reactive collisions*

THE Møller wave operator Ω_i^+ generates (in the implied limit $\epsilon \to +0$) a stationary state of H from a stationary state of $H_{0,i}$. In a formal fashion this statement is written as the 'intertwining property'

$$\Omega_i^+ \exp(iH_{0,i}t/\hbar)C_i = \exp(iHt/\hbar)\Omega_i^+. \tag{C.1}$$

The result can be extended to other definable functions of H and H_0 (for example, by expanding them in a Fourier-like series) and, in particular,

$$\begin{aligned}\Omega_i^+\delta(E-H_{0,i})C_i &= \delta(E-H)\Omega_i^+ \\ &= \int dE'\,\Omega_i(E'^+)\delta(E'-H_{0,i})\delta(E-H_{0,i})C_i \\ &= \Omega_i(E^+)\delta(E-H_{0,i})C_i,\end{aligned} \tag{C.2}$$

where in the second line we have used eqn (5.6).

From the result (cf. eqn (6.28))

$$\Omega_f(E^+)C_f = [I + G(E^+)V_f]C_f$$

and the identity
$$G(E^+) - G(E^-) = -2\pi i\delta(E-H)$$

we have that

$$\begin{aligned}C_f[\Omega_f^- - \Omega_f^+]^\dagger &= \int C_f\delta(E-H_{0,f})[\Omega_f(E^-) - \Omega_f(E^+)]^\dagger\,dE \\ &= -2\pi i\int C_f\delta(E-H_{0,f})V_f\delta(E-H)\,dE.\end{aligned} \tag{C.3}$$

For $f \neq i$, in view of the orthogonality relation (7.52),
$$\Omega_f^{-\dagger}\Omega_i^+ = [\Omega_f^- - \Omega_f^+]^\dagger \Omega_i^+ \tag{C.4}$$
and, using eqns (C.4), (C.3), and (C.2),
$$\begin{aligned}\Omega_f^{-\dagger}\Omega_i^+ &= -2\pi i \int C_f \delta(E-H_{0,f}) V_f \Omega_i(E^+) \delta(E-H_{0,i}) C_i \, dE \\ &= -2\pi i \int C_f \delta(E-H_{0,f}) T_{fi}(E^+) \delta(E-H_{0,i}) C_i \, dE. \end{aligned} \tag{C.5}$$

2.D. *On the long-time behaviour and adiabatic switching in collisions*

OUR discussion of time-dependent collision theory was based on the long-time behaviour of the wavefunction $\Psi_i^+(t)$. In particular, we required that
$$\lim_{t \to -\infty} \Psi_i^+(t) = \lim_{t \to -\infty} \Phi_i(t), \tag{D.1}$$
thus identifying $\Psi_i^+(t)$ as the wavefunction that evolved from the initial state Φ_i. Since we are dealing with functions (and more generally with vectors in Hilbert space) we must be more precise in defining the 'meaning' of the equality sign in eqn (D.1). We shall demand that
$$\lim_{t \to -\infty} \|\Psi_i^+(t) - \Phi_i(t)\|^2 = 0, \tag{D.2}$$
where $\|f\|$ is the norm of the function f,
$$\|f\|^2 = \int f^* f \, d\tau. \tag{D.3}$$
Putting $\Psi_i^+(t) = U^+(t)\Psi_i^+(0)$, and similarly for $\Phi_i(t)$, we have
$$\|\Psi_i^+(t) - \Phi_i^+(t)\|^2 = \|\exp(-iHt/\hbar)[\Psi_i^+(0) - \exp(iHt/\hbar)\exp(-iH_0 t/\hbar)\Phi_i(0)]\|^2 \tag{D.4}$$
or, eqn (D.2),
$$\lim_{t \to -\infty} \|\Psi_i^+(0) - \exp(iHt/\hbar)\exp(-iH_0 t/\hbar)\Phi_i(0)\|^2 = 0. \tag{D.5}$$
We can write this condition as
$$\Psi_i^+(0) = \underset{t \to -\infty}{\text{l.i.m.}} \exp(iHt/\hbar)\exp(-iH_0 t/\hbar)\Phi_i(0), \tag{D.6}$$
where l.i.m. stands for 'limit in the mean'. We shall thus understand the limit in eqn (D.1) as limit in the mean.

To show that the norm of $\Psi_i^+(t)$ is conserved for all times t, we use Schwartz's inequality
$$\|f\|^2 \|g\|^2 \geq \|fg\|^2 \tag{D.7}$$
with f an arbitrary function of finite norm and $g = \Psi_i^+(t) - \Phi_i(t)$, so that, from eqns (D.7) and (D.2),
$$\lim_{t \to -\infty} \langle f | \Psi_i^+(t) - \Phi_i(t) \rangle = 0 \tag{D.8}$$
or, taking $f = \Phi_i(t)$,
$$\lim_{t \to -\infty} \langle \Phi_i(t) | \Psi_i^+(t) \rangle = \lim_{t \to -\infty} \langle \Phi_i(t) | \Phi_i(t) \rangle = \|\Phi_i(0)\|^2. \tag{D.9}$$
Thus
$$\begin{aligned}\lim_{t \to -\infty} &\|\Psi_i(0) - \exp(iHt/\hbar)\exp(-iH_0 t/\hbar)\Phi_i(0)\|^2 = 0 \\ &= \lim_{t \to -\infty} [\|\Psi_i(0)\|^2 + \|\Phi_i(0)\|^2 - \langle\Psi_i(t)|\Phi_i(t)\rangle - \langle\Phi_i(t)|\Psi_i(t)\rangle] \\ &= \|\Psi_i(0)\|^2 - \|\Phi_i(0)\|^2. \end{aligned} \tag{D.10}$$

The present proof of conservation of probability is complementary to that of Appendix 2.A, which used stationary methods.

Jauch (1958) was able to show that provided the limit in equation (D.6) exists§ we can write

$$\Psi_i^+(0) = \lim_{\epsilon \to +0} (\epsilon/\hbar) \int_{-\infty}^{0} dt\, \exp(\epsilon t/\hbar)\exp(iHt/\hbar)\Phi_i(t). \quad (D.11)$$

This limiting procedure is sometimes referred to in the literature as 'the adiabatic switching'. (Lippmann and Schwinger 1950, Gell-Mann and Goldberger 1953, and also Snyder 1951 and Moses 1955.) Essentially the idea was to consider $\Phi_i(t')$ in the remote past and let it evolve from that time onwards under the evolution operator $f(t)\exp(iHt/\hbar)$ (recall that $t < 0$), with $f(t)$ slowly increasing from zero (at $t \to -\infty$) to one (at $t = 0$). Indeed, if we take

$$\Phi_i(t) = \exp(-iE_i t/\hbar)|i\rangle \quad (D.12)$$

we have

$$\lim_{\epsilon \to +0} (\epsilon/\hbar) \int_{-\infty}^{0} dt\, \exp(\epsilon t/\hbar)\exp[i(H-E_i)t/\hbar]|i\rangle$$
$$= \lim_{\epsilon \to +0} i\epsilon(E_i + i\epsilon - H)^{-1}|i\rangle = |i\rangle + G(E^+)(H-E_i)|i\rangle, \quad (D.13)$$

which is the L.S. equation.

From our present point of view eqn (D.11) is just a particular method for performing the limiting operation (eqn (D.6))

$$\Psi_i^+(0) = \lim_{t \to -\infty} \exp[iHt/\hbar]\Phi_i(t)$$

since, for any function $g(t)$ such that $g(t \to -\infty)$ exists, we find by integrating by parts

$$\lim_{\epsilon \to +0} \epsilon \int_{-\infty}^{0} \exp(\epsilon t)g(t) = g(t \to -\infty). \quad (D.14)$$

In our discussion in section 2.7.0, we achieved the same aims by using the forward and backward evolution operators. Their use has enabled us to indicate explicitly the inherent 'time ordering' of our iteration scheme and to develop the theory in close parallel with stationary methods.

2.E. *The transition amplitude density method*

As another example of the use of eqn (5.89) we consider the derivation of the method of variable phase (Calogero 1967) which is a particular case of the method of (transition) amplitude density. Our derivation is based on the concept of imbedding. (See, for example, Bellman et al. 1960.) To find the phase shift for the potential $V(R)$, we consider first the equation for $\delta(S)$, the phase shift for the potential $V(R)\theta(S-R)$, where $\theta(S)$ is the unit step function. Recalling that the derivative of the step function is the delta function, we obtain from eqn (5.89)

$$\frac{d\delta(S)}{dS} = -(2\mu/\hbar^2)k|U_{kl}(S)|^2 V(S). \quad (E.1)$$

The potential that determines $\delta(S)$ has the finite range S, so that, for $R \geqslant S$, $U_{kl}(R)$ can be replaced by its asymptotic value, i.e.

$$U_{kl}(R) = k^{-1}[\cos\delta(S)j_l(kR) - \sin\delta(S)n_l(kR)]. \quad (E.2)$$

§ The problem here is what restrictions on the Hamiltonians H and H_0 are necessary in order that the limit does exist. The interested reader will find some answers in Jordan (1962), Van Winter (1965), and Kato (1966, chapter 10).

Substituting this value in eqn (E.1) yields a non-linear first-order differential equation for $\delta(S)$, with the obvious boundary condition $\delta(0) = 0$. The phase shift δ for the physical potential can then be obtained as

$$\delta = \lim_{S \to \infty} \delta(S).$$

As is clear, similar ideas can be used in connection with eqn (5.86) leading to the method of (transition) amplitude density. From a numerical point of view, the resulting non-linear equations provide an efficient method of computation (cf. Johnson and Secrest 1966).

If for $U_{kl}(R)$ we use the exact solutions for a potential $V_1(R)$, we need only truncate the potential $V_2(R)$ to solve for the scattering by V_1+V_2. This modification can be used to advantage in computations. (In fact it is already used in eqn (E.2) in that the centrifugal potential has been incorporated exactly.)

PART 3
Molecular Rate Processes

Introduction

A QUALITATIVE understanding of various aspects of molecular collision theory can be provided either by the use of exact relations (i.e. the generalized Ehrenfest theorem) or exact parametrizations for observed quantities or by the use of models.

A useful approach in the first category has been the use of the partitioning technique. By centring the attention on the particular component of the wavefunction that is of interest, one can obtain an exact equation for that component, where the actual potential is replaced by an (energy-dependent, non-local) equivalent potential. The coupling to the other portion of the wavefunction is taken into implicit (but exact) account through the equivalent potential. The partitioning technique is reviewed in Chapter 1 and its application to molecular collisions is discussed in Chapter 2. One can also apply it to operators as is discussed in Chapter 3.

The models of collision theory can be classified according to the way the dynamical aspects of the problem are handled. In the most fundamental level one obtains a model by the use of an approximate Hamiltonian, for which the problem is solved exactly. One such example is the adiabatic approximation (sections 3.2.1 and 3.4.3). Another example is the impact parameter method (section 3.4.2) where the relative motion is treated classically. The model Hamiltonians employed can sometimes be justified on the basis of a variational principle (section 3.3.1).

A further degree of approximation is obtained when we take into account only the conservation of probability (and any other conservation laws) but otherwise make some model assumptions regarding the dynamics. In the simplest case this can be formulated in terms of a reaction criterion. An extension of this approach leads to the optical model (Chapter 4) and to the statistical theories (Chapter 5).

Part 3 concludes with a discussion of unimolecular reactions and relaxation processes.

3.1. The partitioning technique

THE partitioning technique is introduced to provide a compact formalism for the discussion of collisions of systems with internal degrees of freedom. It is primarily a perturbation theoretic approach based on the early work of Friedrichs (1948), Heitler and Ma (1949), Schönberg 1951), Van Hove (1955, 1956), and Zumino (1956). Like all general schemes it can be specialized in various ways to deal with particular problems. In common with other branches of perturbation theory, the particular selection of a starting point can have a considerable effect on the results of approximate computations.

The partitioning procedure can also be introduced from the point of view of time-dependent perturbation theory, and many of the earlier papers on the subject were written from this point of view. In several of these papers the initial state was specified at $t = 0$ rather than at $t \to -\infty$ as was done in section 2.7.0. The applications to time-dependent theory are discussed in section 3.6.0.

The partitioning technique, applied to the solution of the L.S. equation, consists in the introduction of two projection operators, P and Q,

$$P + Q = I, \qquad P^2 = P = P^\dagger, \tag{1.1}$$

so that
$$PQ = 0, \tag{1.2}$$

such that $P\psi$ is the component of the wavefunction which is of interest. To illustrate our discussion we consider the collision of an atom and a diatomic molecule (section 2.3.0). We can require that $P\psi$ and ψ have the same *asymptotic* behaviour (Feshbach 1962). At a given energy E the asymptotic form of $\psi_{\mathbf{k},n}^+(\mathbf{R}) = \langle \mathbf{R} | \mathbf{k}, n^+ \rangle$ is (eqn (2.3.15 b))

$$\psi_{\mathbf{k},n}^+(\mathbf{R}) \to \sum_m |m\rangle [f_{m,n}(\mathbf{k}', \mathbf{k}) R^{-1} \exp(ik'R) + \delta_{mn} \exp(i\mathbf{k}.\mathbf{R})], \tag{1.3}$$

where $|m\rangle$ is an internal state of the diatomic and the summation over m is restricted to those states $|m\rangle$ for which $E_m < E$, where E_m is the internal energy. As eigenstates of the internal Hamiltonian the states $|m\rangle$ are orthogonal and we can put

$$P = \sum_{m=1}^{M} |m\rangle\langle m|, \tag{1.4}$$

with M restricted by $E_M < E$, $E_{M+1} > E$. Using the expansion (eqn (2.3.8)),

$$\psi_{\mathbf{k},n}^+(\mathbf{R}) = \sum_m |m\rangle F_{\mathbf{k},n}^m(\mathbf{R}), \tag{1.5}$$

we have
$$P\psi_{\mathbf{k},n}^+ = \sum_{m=1}^{M} |m\rangle F_{\mathbf{k},n}^m(\mathbf{R}), \qquad (1.6)$$

which asymptotically reduces to eqn (1.3) (cf. eqn (2.3.14)).

Equation (1.4) is clearly only one possible choice of P. A projection that is R-dependent but reduces asymptotically to eqn (1.4) will also lead to the specified form of ψ, eqn (1.3). Moreover, we can enlarge the subspace on to which P projects (namely increase M in eqns (1.4) and (1.6)) without modifying the asymptotic form of eqn (1.6). This extensive freedom in the explicit construction of P can be used to advantage in particular problems. The merit of eqn (1.4) is its simplicity and the fact that P commutes with H_0. (If P depends on R it will not commute with the kinetic energy operator for relative motion.) A disadvantage, however, is the fact that as we change the energy the value of M changes whenever a new internal state can be asymptotically observed.

Alternatively, we can demand that $P\psi$ represents asymptotically the elastic scattering only:
$$P\psi_{\mathbf{k},n}^+ \xrightarrow[R\to\infty]{} |n\rangle[\exp(i\mathbf{k}\cdot\mathbf{R}) + R^{-1}\exp(ikR)f_{n,n}(\mathbf{k}',\mathbf{k})] \qquad (1.7)$$

(Watson 1957, Mittleman and Watson 1959). In terms of eqns (1.5) and (1.7) the simplest selection for P is
$$P = |n\rangle\langle n|. \qquad (1.8)$$

It is equally clear that the definition above is but one possibility consistent with eqn (1.7). In particular, if we put
$$|n_a\rangle = \sum_m C_m^n(\mathbf{R})|m\rangle \qquad (1.9)$$

with
$$C_m^n(\mathbf{R}) \xrightarrow[R\to\infty]{} \delta_{m,n} + O(R^{-2}), \qquad (1.10)$$

then $P_a\psi_{\mathbf{k},n}^+$, where
$$P_a = |n_a\rangle\langle n_a|, \qquad (1.11)$$

will also satisfy eqn (1.7).

The derivation of the L.S. equation for $P\psi$ is simplest when P commutes with H_0 (say eqns (1.4) or (1.8)) and $Q\phi = 0$ where ϕ is the initial state. Using eqn (1.1) we write the L.S. equation as
$$(E^+ - H)(P+Q)\psi^+ = (E^+ - H_0)\phi. \qquad (1.12)$$

Pre-operating on both sides by P and Q respectively, we get (recall that $P^2 = P$)
$$(E^+ - PHP)P\psi^+ - PHQ\psi^+ = (E^+ - H_0)P\phi \qquad (1.13)$$
and
$$(E^+ - QHQ)Q\psi^+ - QHP\psi^+ = 0. \qquad (1.14)$$

Equation (1.14) can be solved immediately to give§

$$Q\psi^+ = (E^+ - QHQ)^{-1}QHP\psi^+. \tag{1.15}$$

Substituting (1.15) in (1.13) we can write

$$(E^+ - P\mathcal{H}P)P\psi^+ = (E^+ - H_0)P\phi, \tag{1.16}$$

an L.S. equation with the effective Hamiltonian $P\mathcal{H}P$,

$$P\mathcal{H}P = PHP + PHQ(E^+ - QHQ)^{-1}QHP. \tag{1.17}$$

Equation (1.16) is an exact equation for the P component of ψ, which is of interest. It can be formally solved as follows. Let $P\psi_1^+$ be the solution of
$$(E^+ - PHP)P\psi_1^+ = (E^+ - H_0)P\phi, \tag{1.18}$$

which we regard as known. We now obtain from eqn (1.13)

$$P\psi^+ = P\psi_1^+ + (E^+ - PHP)^{-1}PHQ\psi^+ \tag{1.19}$$

and, using eqn (1.19) in eqn (1.14),

$$[E^+ - QHQ - QHP(E^+ - PHP)^{-1}PHQ]Q\psi^+ = QHP\psi_1^+, \tag{1.20}$$

we obtain an equation for $Q\psi^+$ that does not involve the unknown $P\psi^+$. Inverting eqn (1.20) yields a solution for $Q\psi^+$ that can be substituted in eqn (1.19). We re-derive these results below (eqn (1.36)), using the theory of scattering by two potentials.

If we define P by $P = |n\rangle\langle n|$, where n is the lowest internal state of h, and approximate ψ by $P\psi_1^+$, we obtain the so-called 'static' approximation. In this approximation

$$\psi_{\mathbf{k},n}^+|_{\text{static}} = \phi_n(\mathbf{r})L_{\mathbf{k}}^n(\mathbf{R}) \tag{1.21}$$

for any R, and the internal state of the molecules is unmodified throughout the collision. When inelastic processes are possible we often select

$$P = \sum_{m=1}^{N} |m\rangle\langle m|, \tag{1.22}$$

where m enumerates the initial internal state and those final internal states that are of interest (not necessarily all the possible ones). The approximation of ψ^+ by $P\psi_1^+$ is then known as the close-coupling approximation,

$$\psi_{\mathbf{k},n}^+|_{\text{c.c.}} = \sum_{m=1}^{N} \phi_m(\mathbf{r})F_{\mathbf{k},n}^m(\mathbf{R}), \tag{1.23}$$

since this approximation is equivalent to the set of N coupled equations

$$[E - E_m - K - V_{m,m}(\mathbf{R})]F_{\mathbf{k},n}^m(\mathbf{R}) = \sum_{m'}' V_{m,m'}(\mathbf{R})F_{\mathbf{k},n}^{m'}(\mathbf{R}). \tag{1.24}$$

§ Equation (1.15) implies that as long as E is below the continuous spectrum of QHQ, the asymptotic behaviour of $Q\psi^+$ in the coordinate representation is that of $\exp(-\kappa R)$, cf. eqns (2.2.18) and (2.3.14).

The partitioning technique can also be viewed as an example of the theory of scattering by two potentials with $V_1 = PVP$ and $V_2 = V - PVP$. In contrast with the example discussed in section 2.5.2, V_1 can induce inelastic transitions. For example, with P defined by eqn (1.4)

$$V_1 = PVP = \sum_{m,m'}^{M} |m'\rangle V_{m',m}(\mathbf{R})\langle m| \qquad (1.25)$$

and the close-coupling approximation corresponds to neglecting V_2. $P\psi_1^+$ is the solution of the L.S. equation (1.18) with the potential $V_1 = PVP$ and so eqn (2.5.72) can be written as

$$\begin{aligned}\psi^+ &= (1+GV_2)P\psi_1^+ \\ &= P\psi_1^+ + GQHP\psi_1^+ \\ &= P\psi_1^+ + (PG_1PH+1)QGQHP\psi_1^+,\end{aligned} \qquad (1.26)$$

since $PV_2P = 0$ and $QV_2P = QVP = QHP$, when H_0 and P commute. Since Q annihilates the possible final states we need to evaluate only PTP and, from eqns (2.5.66) and (2.5.65),

$$PTP = T_1 + \Omega_1^\dagger PHQGQHP\Omega_1. \qquad (1.27)$$

It is also possible to define P to simplify the relevant perturbation expansions (Watson 1957, Riesenfeld and Watson 1956). In certain instances it is convenient to define P as the complement of Q (namely $P\psi = (I-Q)\psi$). In particular, one often selects Q to project on a subspace of states that would be stable, but for the coupling between P and Q (Friedrichs 1948, Heitler and Ma 1949, Zumino 1956, Newton and Fonda 1960, Fano 1961, Harris 1963, and MacDonald 1964). This point of view is often useful towards a qualitative understanding and the construction of model theories. There are several ways in which we can apply this idea to our example. Regarding the atom and the diatomic molecule as a triatomic molecule we can perform a normal-mode analysis of the vibrational motion of the triatomic and introduce harmonic oscillator wavefunctions to describe the vibrational motion of each of the two modes. If we denote by $|s\rangle$ a particular vibrational state so defined we have

$$Q = |s\rangle\langle s| \qquad (1.28)$$

and the Hamiltonian QHQ will have a purely discrete spectrum. The physical situation is not described entirely by QHQ since above a certain energy the triatomic can dissociate. (For a qualitative discussion see Herzberg (1966), p. 430, and for some model computations see Rosen (1933).) Those states $|s\rangle$ for which E_s is above the dissociation energy are referred to as 'discrete states in the continuum'. Their temporary existence is attributed physically to the fact that the relative coordinate for the dissociative motion (say R of Fig. 3.1) is not a normal mode of the triatomic and it requires a certain time-lag before the internal energy distribution in the molecule is such that sufficient energy is localized in the relative coordinate (Polanyi and Wigner 1928, Slater 1959). This idea is closely related to Bohr's (1936) concept of the compound nucleus. (See also Rice and Ramsperger (1927), Kassel (1932), and sections 3.5.0 and 3.6.0.)

Other constructions of Q with the same aim are possible. In general, we look for an approximate Hamiltonian H_0' that can support bound states above the start of the continuous spectrum. A somewhat different approach is based on explicitly ensuring that the subspace on which Q projects consists of states that are localized in (ordinary) space. Essentially, one considers a sphere inside which these states are localized, and diagonalizes the total Hamiltonian inside the sphere. There are basically two different approaches depending on the handling of the boundary conditions on the sphere. These are associated with Kapur and Peierls (1938; see also Herzenberg and Mandl 1963, 1966) and Wigner and Eisenbud (1947; see also Lane and Thomas 1958. The R-matrix theory has also been considered for molecular problems, see Blum (1966) and Eu and Ross (1966).) For a comparison of different theories see Lane and Robson (1966).

It is also possible to partition Schrödinger's time-dependent equation to obtain

$$i\hbar \frac{\partial P\psi(t)}{\partial t} = PH\psi(t) = PHP\psi(t) + PHQ\psi(t) \tag{1.29}$$

and

$$i\hbar \frac{\partial Q\psi(t)}{\partial t} = QH\psi(t) = QHQ\psi(t) + QHP\psi(t). \tag{1.30}$$

The solutions of these equations depend on the initial conditions imposed. If we select Q to annihilate the initial state then (cf. Appendix 2.D)

$$\lim_{t \to -\infty} \|Q\psi(t)\| \to 0, \tag{1.31}$$

while the initial state $\phi(t)$ is defined by

$$\lim_{t \to -\infty} \|\psi(t) - \phi(t)\| \to 0. \tag{1.32}$$

It can be shown directly (see, for example, Coester and Kümmel (1958)) that the solutions of eqn (1.29), subject to the boundary conditions (1.31) and (1.32) and for an initial state of well-defined energy, satisfy

$$i\hbar \frac{\partial P\psi(t)}{\partial t} = P\mathcal{H}P\psi(t) \tag{1.33}$$

with $P\mathcal{H}P$ defined by eqn (1.17).

Instead of partitioning Schrödinger's time-dependent equation and considering separately the boundary conditions, it is often more convenient to partition the integral equations obtained with the use of the evolution operator (as discussed in Section 2.7.0). Using the discussion in section 2.7.2, in particular eqn (2.7.30), we can use the results of partitioning the L.S. equation in the time-dependent theory as well. Thus, for example, from eqn (2.7.30)

$$Q\psi^+(t) = \sum_{\mathbf{n}} \int dE\, A(\mathbf{n}, E; t) Q|E, \mathbf{n}^+\rangle, \qquad (1.34)$$

where the initial state is

$$\phi(t) = \sum_{\mathbf{n}} \int dE\, A(\mathbf{n}, E; t)|E, \mathbf{n}\rangle, \qquad (1.35)$$

and similarly for $P\psi^+(t)$. We discuss these ideas explicitly in sections 3.6.1 and 3.6.2.

To conclude, the aim of the partitioning technique is to obtain an exact equation for the portion of the wavefunction that contains the information of interest. The effect of the remaining portion is taken into implicit but exact account via the modified Hamiltonian (eqns (1.16) and (1.33)). Similar methods can also be applied to the density matrix (Zwanzig 1960), to other functions of the Hamiltonian (using eqn (3.30) for example), and to other observables. (For example, if Q annihilates the asymptotic states only PAP can be observed after a collision event.) We turn now to specific applications of partitioning the L.S. equation.

3.2. Molecular encounters

3.2.0. CONSIDER the collision of two molecules with internal degrees of freedom (section 2.3.0). During the collision the kinetic energy of relative motion can be converted into internal excitation. This migration of energy can lead to a temporary formation of bound states, when the energy of internal excitation exceeds the initial relative energy. Eventually the energy migrates back to the coordinate of relative motion and the two molecules can separate.

In principle the formalism of section 2.3.0 accounts for the phenomena just described. Indeed, we have shown (eqn (2.3.14)) the presence of internally excited states of the molecules that are associated with bounded relative motion. The (first) Born approximation neglects those contributions that are at least of the second order in the mutual interaction (one excitation, one de-excitation). In other words, the probability density to observe the two molecules at small relative separations can be much enhanced with respect to the density predicted by the (first) Born approximation. Under these circumstances it is more reasonable to try to obtain a formalism that exhibits such features explicitly, so that in the computational stage these features are not lost, rather than go to higher-order Born approximations. In particular, we have already questioned the convergence of the Born approximation under such circumstances. To see this explicitly let us provide a crude estimate for the transition amplitude in the second Born approximation (eqn (2.2.93)) for elastic scattering

$$\langle \mathbf{k}, n | T | \mathbf{k}, n \rangle_{\mathrm{2ndB}} = \langle \mathbf{k}, n | V | \mathbf{k}, n \rangle + \langle \mathbf{k}, n | V G_0(E^+) V | \mathbf{k}, n \rangle. \quad (2.1)$$

The second term can be written exactly as

$$\sum_m \langle \mathbf{k}, n | V | m \rangle (E^+ - E_m - K)^{-1} \langle m | V | \mathbf{k}, n \rangle,$$

where we have used eqn (2.3.7). Even if $E < E_m$ (all $m \neq n$) the denominator can be quite significant when the internal state $|m\rangle$ is associated with a bounded relative motion. Thus we make an Unsöld (1927)-like approximation for the second term, replacing it by

$$\langle \mathbf{k}, n | V^2 | \mathbf{k}, n \rangle / \Delta E_n, \quad (2.2)$$

where ΔE_n is an 'average excitation energy' out of the state $|n\rangle$ and can be quite small.

At sufficiently low energies only elastic collisions can take place. If $|n\rangle$ is the lowest internal state we take
$$P = |n\rangle\langle n| \qquad (2.3)$$
and premultiply eqn (1.16) by $\langle n|$ to obtain an L.S. type equation for $\langle n|\psi\rangle$,
$$[E^+ - E_n - K - \langle n|\mathscr{V}|n\rangle]\langle n|\psi\rangle = (E^+ - E_n - K)\langle n|\phi\rangle \qquad (2.4)$$
where, using eqn (1.17), we have put
$$\langle n|\mathscr{H}|n\rangle = E_n + K + \langle n|\mathscr{V}|n\rangle$$
$$\langle n|\mathscr{V}|n\rangle = \langle n|V + VQ(E^+ - QHQ)^{-1}QV|n\rangle.$$

Due to the presence of the Green's function $(E^+ - QHQ)^{-1}$ the 'potential' $\langle n|\mathscr{V}|n\rangle$ is non-local in the coordinate representation. We can write
$$\langle n|\mathscr{V}|n\rangle\langle n|\psi\rangle = \int \langle n|\mathscr{V}|n, \mathbf{R}'\rangle\langle n, \mathbf{R}'|\psi\rangle\,\mathrm{d}\mathbf{R}'$$
and, putting $\quad\langle \mathbf{R}, n|\phi\rangle = \langle \mathbf{R}, n|\mathbf{k}, n\rangle = \exp(\mathrm{i}\mathbf{k}.\mathbf{R})$
and $\quad\langle \mathbf{R}, n|\psi\rangle = \langle \mathbf{R}, n|\mathbf{k}, n^+\rangle = F_{\mathbf{k},n}^n(\mathbf{R}),$

we obtain an equation for the relative motion in the coordinate representation by preoperating on eqn (2.4) by $\langle \mathbf{R}|$,
$$[E^+ - E_n - K]F_{\mathbf{k},n}^n(\mathbf{R}) - \int \langle \mathbf{R}|\mathscr{V}_{n,n}|\mathbf{R}'\rangle F_{\mathbf{k},n}^n(\mathbf{R}')\,\mathrm{d}\mathbf{R}'$$
$$= (E^+ - E_n - K)\exp(\mathrm{i}\mathbf{k}.\mathbf{R}). \qquad (2.5)$$
Equation (2.5) can now be written as an integral equation,
$$F_{\mathbf{k},n}^n(\mathbf{R}) = \exp(\mathrm{i}\mathbf{k}.\mathbf{R}) +$$
$$+ \iint \mathrm{d}\mathbf{R}'\,\mathrm{d}\mathbf{R}''\langle \mathbf{R}|(E^+ - E_n - K)^{-1}|\mathbf{R}'\rangle\langle \mathbf{R}'|\mathscr{V}_{n,n}|\mathbf{R}''\rangle F_{\mathbf{k},n}^n(\mathbf{R}''). \qquad (2.6)$$

Equation (2.6) is an exact L.S. equation for the relative motion. Since $F_{\mathbf{k},n}^n(\mathbf{R})$ is the only component of the relative motion that is asymptotically observed, the solution of eqn (2.6) will account for the observed scattering. We have thus replaced the set of coupled equations (2.3.14) by a single equation, where, however, the effective potential is non-local and as yet not explicitly known. The effective potential is often referred to as an 'optical potential' or a 'pseudo potential'. In view of the other definitions sometimes associated with these terms (cf. section 3.5.2) we prefer the label 'equivalent potential', since the solution of eqn (2.6) is equivalent to solving a set of coupled equations.

3.2.1. *The adiabatic approximation*

The current experiments on the low-energy elastic collisions of atoms (and small diatomics) are often conveniently accounted for by the introduction of a local potential in eqn (2.6).

From eqns (1.5) and (2.3)
$$\langle \mathbf{R}|Q\psi\rangle = \sum_{m\neq n} |m\rangle F_{\mathbf{k},n}^m(\mathbf{R}), \tag{2.7}$$
so that, from eqns (2.5), (2.4), and (1.18),
$$\int d\mathbf{R} \langle n, \mathbf{R}'|\mathscr{V}|n, \mathbf{R}\rangle F_{\mathbf{k},n}^n(\mathbf{R}) = \sum_m \langle n, \mathbf{R}'|V|m, \mathbf{R}'\rangle F_{\mathbf{k},n}^m(\mathbf{R}'). \tag{2.8}$$
When the equivalent potential is local,
$$\mathscr{V}_{n,n}(\mathbf{R}', \mathbf{R}) = \mathscr{V}_{n,n}(\mathbf{R})\delta(\mathbf{R}'-\mathbf{R}), \tag{2.9}$$
we can define a local operator F by its matrix elements
$$F_{\mathbf{k},n}^m(\mathbf{R}) = F_{m,n}(\mathbf{R})F_{\mathbf{k},n}^n(\mathbf{R}) \tag{2.10}$$
or, from eqns (2.9) and (2.8),
$$\mathscr{V}_{n,n}(\mathbf{R}) = V_{n,n}(\mathbf{R}) + \sum_m{}' V_{n,m}(\mathbf{R})F_{m,n}(\mathbf{R}). \tag{2.11}$$
If we define
$$F_{n,n}(\mathbf{R}) = I, \tag{2.12}$$
the total wavefunction can be written, using eqns (2.10) and (1.5),
$$\langle \mathbf{R}|\psi\rangle = \left[\sum_m |m\rangle F_{m,n}(\mathbf{R})\right] F_{\mathbf{k},n}^n(\mathbf{R}). \tag{2.13}$$
We can regard the term in the square brackets above as the 'internal wavefunction' of the system. The internal energy $E_{\text{int}}(R)$ of this state is determined by
$$[E_{\text{int}}(\mathbf{R}) - (h+V)]\left[\sum_m F_{m,n}(\mathbf{R})|m\rangle\right] = 0. \tag{2.14}$$
Premultiplying by $\langle n|$ and using eqns (2.12) and (2.11),
$$E_{\text{int}}(\mathbf{R}) = E_n + \mathscr{V}_{n,n}(\mathbf{R}). \tag{2.15}$$

The approximation of the total wavefunction as in eqn (2.13) is known as the adiabatic approximation. The coefficients $F_{m,n}(\mathbf{R})$ are determined by solving eqn (2.14) subject to eqn (2.12). It can then be shown that, provided $E_m \neq E_n$,
$$F_{m,n}(\mathbf{R}) \xrightarrow[R\to\infty]{} O(R^{-2}), \quad m \neq n. \tag{2.16}$$

Equation (2.16) is subject to the assumption that $V_{m,n}(R)$ decreases faster than R^{-1} at large separations. To prove equation (2.16) we compute $F_{m,n}(\mathbf{R})$ from eqn (2.14) as
$$F_{m,n}(\mathbf{R}) = [E_{\text{int}}(\mathbf{R}) - E_m]^{-1} \langle m|V \sum_p F_{p,n}|n\rangle. \tag{2.17}$$

At large, relative separation first-order perturbation theory should be sufficient, so that
$$F_{m,n} \sim (E_n - E_m)^{-1} V_{m,n}(R). \tag{2.18}$$

Since eqn (2.14) is independent of the total energy the asymptotic conditions (2.16) are energy independent, and in the adiabatic approximation no permanent internal excitation can take place.

Equation (2.18) is the lowest approximation for eqn (2.14). It does incorporate some distortion of the relative motion. For a collision of two atoms the internal states are the electronic states and, using eqn (2.18) in eqn (2.11), we find to lowest order

$$\mathscr{V}_{n,n}(R) = V_{n,n}(R) + \sum_m |V_{m,n}(R)|^2 (E_n - E_m)^{-1}. \quad (2.19)$$

By expanding the interatomic interactions§ in powers of R^{-1} (see, for example, Dalgarno and Davison 1966) one can show that for closed-shell atoms the leading contribution to $\mathscr{V}_{n,n}$ is the London (1937) term proportional to R^{-6}.

The validity criterion of the adiabatic approximation is that it is essentially a medium-range, low relative velocity, high internal velocity type of approximation. A qualitative discussion can be based on the uncertainty relations (cf. the discussion of eqn (2.3.83)). If τ_c is the duration of the collision, the total energy during the collision is imprecise by $\hbar/\tau_c$ and, for an ineffective internal excitation, this energy uncertainty has to be less than the energy ΔE necessary for the transition. The adiabatic region is thus characterized by

$$\tau_c \gg \hbar/\Delta E. \quad (2.20)$$

In the adiabatic approximation the relative motion is governed by the average (over internal degrees of freedom) interaction. The relative motion should not induce internal transition so that the internal wavefunction, at any R (the term in the square brackets in eqn (2.13)) can be computed by 'clamping' the relative separation at R (eqn (2.14)). In order for such a description to be valid the average mutual interaction should vary sufficiently slowly with R with respect to the wavelength of relative motion, which should be much smaller than the wavelength associated with the internal motion. These conditions are required by the position–momentum uncertainty principle, in order that we may restrict the relative motion to a small interval around R, over which the variation of the potential is negligible‖ and that the resulting uncertainty

---

§ Strictly speaking, R is the relative separation of the two centres of mass. In the multipole expansion R is taken as the distance between the two nuclei. The two distances are only equal to the order of m/M where m is the electronic mass and M the nuclear mass. $\mathscr{V}$ is now a function of $|R|$ only since the electronic coordinates are referred to the atoms, that is, to a system which is rotating with **R**. In this system the kinetic energy operator should be modified (Kronig 1930, Thorson 1961, 1965) leading to rotational terms that couple nuclear and electronic rotations.

‖ This restriction is particularly severe near the steep repulsive part of most intermolecular interactions.

in the relative momentum be small compared with the relative momentum and with the change in the relative momentum due to internal excitation.§

Qualitatively, the low internal velocity and small energy differences between rotational states suggest that the adiabatic approximation should not be invoked for rotational transitions (cf. eqn (2.3.83)). Vibrational transitions usually satisfy eqn (2.20) at moderate temperature with ΔE the asymptotic energy difference. However, the effect of distortion during the collision is normally such that the force constants are weakened and so the relevant ΔE may be much reduced. Even for electronic excitations the considerable distortion during the collision may decrease ΔE sufficiently. Recent experimental evidence (Dworetsky, Novick, Smith, and Tolk 1967, also Lipeles, Novick, and Tolk 1965) indicates significant transition probabilities, for excitation of He, by He$^+$ ($\Delta E \sim 22$ eV), already near the apparent threshold ($E \sim 60$ eV). Equation (2.20) predicts significant transition probabilities at about $E \sim 10^5$ eV, and these are indeed also observed.

From a computational point of view the validity of the adiabatic approximation depends on the consistency of eqn (2.14) and the Schrödinger equation
$$[E-K-(h+V)]\psi = 0, \qquad (2.21)$$
with ψ defined by eqn (2.13). Substituting the definition of ψ into the Schrödinger equation (2.21) it is seen that the two are consistent if
$$[K, F_{m,n}(\mathbf{R})]F^n_{\mathbf{k},n}(\mathbf{R}) = 0 \quad (\text{all } m), \qquad (2.22)$$
so that
$$[E-K-E_n-\mathscr{V}_{n,n}(\mathbf{R})]F^n_{\mathbf{k},n}(\mathbf{R}) = 0, \qquad (2.23)$$
which is eqn (2.4) with a local equivalent potential.

Equation (2.22) is the quantitative statement of the adiabatic approximation. The relative motion cannot cause any internal excitation. The only way $F_{m,n}$ can change with R is due to the mutual interaction V. As far as $F_{m,n}$ is concerned it is permissible to consider the relative kinetic energy as a function of R, say $p^2(R)/2\mu$, where $p(R)$ is the classical momentum at the relative separation R.

In actual computations the local potential $\mathscr{V}_{n,n}(R)$ is a given function and one simply solves eqn (2.23) for $F^n_{\mathbf{k},n}$, and regards it as 'the' nuclear wavefunction. We should remember, however, that in principle $F_{m,n}$ is an operator on the nuclear coordinates, and that only in the adiabatic approximation does $F_{m,n}$ become a function of R. We can therefore

§ In particular, states that are degenerate at very large separations (say when the molecules possess internal angular momentum) will be strongly coupled.

continue to regard $F_{m,n}(\mathbf{R})F^n_{\mathbf{k},n}(\mathbf{R})$ as the nuclear wavefunction associated with the state $|m\rangle$.

The adiabatic theory of chemical reactions is discussed in section 3.4.3.

In practice, only a finite number (say N) internal states of h are included in the expansion of ψ (eqn (2.13) with $m = 1,...,N$). Equation (2.14) can thus be converted to an $N \times N$ secular equation with N eigenvalues $E_m(R)$ and N orthogonal solutions depending parametrically on R,

$$\chi_m(\mathbf{r}|\mathbf{R}) = \sum_{m'} F_{m',m}(\mathbf{R})\phi_{m'}(\mathbf{r}), \qquad (2.24)$$

which differ by their asymptotic behaviour in that

$$F_{m',m}(R) \xrightarrow[R\to\infty]{} \delta_{m'm}. \qquad (2.25)$$

In principle, eqn (2.14) should be solved for each value of R. An important property of the eigenvalues $E_m(\mathbf{R})$ was pointed out by von Neumann and Wigner (1929). The general statement of their result is that two eigenvalues, belonging to eigenfunctions of the same symmetry cannot be equal on a whole surface in $\mathbf{R}$ space. A more familiar version of this rule applies when we consider the interaction between two atoms in a coordinate system fixed to the internuclear axis. In this case the internal energy is a function of R only, and two eigenvalues of the same symmetry cannot cross, but for an accidental degeneracy.

In the adiabatic approximation the set of wavefunctions

$$\psi_m = \chi_m(\mathbf{r}|\mathbf{R})F^m_{\mathbf{k},m}(\mathbf{R}) \qquad (2.26)$$

are orthogonal, and non-interacting with respect to the total Hamiltonian H since, by assumption,

$$H\psi_m = E_m(\mathbf{R})\psi_m + \chi_m K F^m_{\mathbf{k},m}, \qquad (2.27)$$

so that

$$\langle \psi_{m'}|H|\psi_m\rangle = 0 \quad (m' \neq m) \qquad (2.28)$$

due to the orthogonality of the internal wavefunctions. The failure of the adiabatic approximation is the failure of eqn (2.27) since the correct result is

$$K\psi_m = -(\hbar^2/2\mu)\left[2\frac{\partial\chi_m(\mathbf{r}|\mathbf{R})}{\partial\mathbf{R}}\frac{\partial}{\partial\mathbf{R}} + \frac{\partial^2\chi_m(\mathbf{r}|\mathbf{R})}{\partial\mathbf{R}^2}\right]F^m_{\mathbf{k},m} + \chi_m K F^m_{\mathbf{k},m}. \qquad (2.29)$$

Corrections to the adiabatic approximation can be introduced by using the set of states χ_m to expand the total wavefunction,

$$\psi = \sum_m \chi_m(\mathbf{r}|\mathbf{R})G^m_n(\mathbf{R}), \qquad (2.30)$$

where the subscript on G is a reminder that in the initial state the internal state n (of h) is occupied. In view of eqns (2.24) and (2.25) the transition amplitude $n \to m$ is determined by the asymptotic behaviour

of $G_n^m(\mathbf{R})$. At sufficiently low energies only $G_n^n(\mathbf{R})$ is asymptotically outgoing, and one can take
$$P = |\chi_n\rangle\langle\chi_n|. \tag{2.31}$$
It is important to note, however, that this P does not commute with H_0 (in view of eqn (2.29)).

A set of coupled differential equations for the functions G_n^m is obtained by substituting ψ into the Schrödinger equation and noting that as eigenfunctions of eqn (2.14),
$$\langle\chi_{m'}|(h+V)|\chi_m\rangle = \delta_{m,m'}E_m(\mathbf{R}), \tag{2.32}$$
so that
$$[E-K-E_m(\mathbf{R})]G_n^m(\mathbf{R}) = \sum_{m'} C_{m,m'}(\mathbf{R})G_n^{m'}(\mathbf{R}), \tag{2.33}$$
where
$$C_{m,m'}(\mathbf{R}) = \int d\mathbf{r}\, \chi_m^* K \chi_{m'}$$
$$= -(\hbar^2/2\mu)\int d\mathbf{r}\left[2\left(\chi_m^*\frac{\partial}{\partial\mathbf{R}}\chi_{m'}\right)\frac{\partial}{\partial\mathbf{R}} + \chi_m^*\frac{\partial^2}{\partial\mathbf{R}^2}\chi_{m'}\right]. \tag{2.34}$$

It is essentially an exercise in transformation theory and the use of Hellmann–Feynman type identities§ to show that, for the same N states of h, eqns (2.33) and (1.24) are equivalent. The difference between them is that in (1.24) the kinetic energy operator is diagonal and the transitions are induced by the mutual interaction, while in (2.33) the mutual interaction is diagonalized at every R and the transitions are induced by the relative motion. An equivalent potential can be introduced to reduce the computation of G_n^n to a single equation in a manner analogous to eqn (2.4). The lowest (static) approximation to this equivalent potential is to neglect all coupling, which leads back to eqn (2.23). (Note that $|\chi_n\rangle E_n(\mathbf{R})\langle\chi_n| = P(h+V)P$ with P defined by eqn (2.31).) As is well known, $E_n(R)$ is the lowest local equivalent potential for a given N basis function. This is essentially the variational theorem for $h+V$. We shall argue below that the phase shifts computed using $E_n(R)$ as a potential, provide a useful bound to the true phase shift up to an energy E that equals the lowest eigenvalue of QHQ with $Q = I - |\chi_n\rangle\langle\chi_n|$. (This energy may be lower than the threshold for inelastic transitions.) An explicit discussion of the coupled equations (2.33) of the adiabatic theory is given by Nielsen and Dahler (1966).

The corrections to the adiabatic approximation are determined by the magnitude of the coupling terms $C_{m',m}$. As is evident from eqn (2.35) these terms can be quite significant when two 'potential energy curves',

§ For $m \neq m'$, $\int d\mathbf{r}\left(\chi_m^*\frac{\partial}{\partial R}\chi_{m'}\right) = [E_{m'}(R)-E_m(R)]^{-1}\int d\mathbf{r}\, \chi_m^*\frac{\partial(h+V)}{\partial R}\chi_{m'}.$ (2.35)

$E_n(R)$ and $E_m(R)$, approach one another. Indeed, a semi-classical analysis of the relative motion problem indicates that ΔE in eqn (2.20) should be the minimal value of $E_n(R)-E_m(R)$. (See Stueckelberg 1932, Bates, Massey, and Stewart 1953, section 2.3, and section 3.4.2.) The computation of transition amplitudes using the solution of eqns (2.33) is known as the 'perturbed stationary states' method. In practice, the coupling terms $C_{m,n}$ are not easy to evaluate, particularly near those regions of R where they are significant, since it is in these regions that the functions $\chi_m(\mathbf{r}|\mathbf{R})$ vary rapidly as functions of R. To see this we consider a simple one-dimensional model obtained by considering only two spherically symmetrical internal states. If $|m\rangle$ and $|n\rangle$ are two internal states of h we have to diagonalize

$$\mathbf{h}+\mathbf{V} = \begin{pmatrix} (h+V)_{n,n} & V_{n,m} \\ V_{m,n} & (h+V)_{m,m} \end{pmatrix}. \quad (2.36)$$

Introducing a unitary matrix

$$\mathbf{U} = \begin{pmatrix} \cos\theta & \sin\theta \\ -\sin\theta & \cos\theta \end{pmatrix}, \quad (2.37)$$

where θ is a function of R, we obtain two eigenvalues, $E_n(R)$ and $E_m(R)$

$$\mathbf{h}+\mathbf{V} = \begin{pmatrix} (h+V)_{n,n}+V_{n,m}\tan\theta & 0 \\ 0 & (h+V)_{m,m}-V_{m,n}\cot\theta \end{pmatrix} \quad (2.38)$$

and two eigenvectors

$$\chi_n = \cos\theta|n\rangle+\sin\theta|m\rangle \quad \chi_m = \cos\theta|m\rangle-\sin\theta|n\rangle, \quad (2.39)$$

where $\quad \tan 2\theta = 2V_{n,m}/[(h+V)_{n,n}-(h+V)_{m,m}]. \quad (2.40)$

As $R \to \infty$, $\theta \to 0$ as expected, the non-crossing rule is also apparent. The behaviour of θ for intermediate values of R is particularly interesting when, at a particular R,

$$(h+V)_{n,n} = (h+V)_{m,m}, \quad R = R_c. \quad (2.41)$$

Equation (2.41) is by no means uncommon and is particularly frequent in interactions of atoms of different polarity (London 1932, Mulliken 1937, Berry 1957; see also Levine 1967a). It is also common in vibronic coupling; for example, see Donath (1965) and Gouterman (1965). From eqns (2.36) and (2.38) we have that

$$(h+V)_{n,n} = E_n(R)\cos^2\theta+E_m(R)\sin^2\theta,$$
$$(h+V)_{m,m} = E_m(R)\cos^2\theta+E_n(R)\sin^2\theta, \quad (2.42)$$

and $\quad V_{n,m} = [E_n(R)-E_m(R)]\cos\theta\sin\theta, \quad (2.43)$

so that $\quad \cos^2\theta = \sin^2\theta, \quad R = R_c. \quad (2.44)$

For $R > R_c$, $\cos^2\theta > \sin^2\theta$ and conversely for $R < R_c$. Thus over the crossing region $R = R_c$ the two internal adiabatic states exchange their 'character', in that for $R > R_c$, χ_n resembles more the state $|n\rangle$, while for $R < R_c$ it has a larger overlap with $|m\rangle$ and conversely for χ_m.

During an actual collision with a finite relative velocity the internal motion may fail to 'readjust' itself sufficiently readily during the passage via the crossing region. In other words, the same internal state $|n\rangle$ may be occupied after the relative separation decreased beyond R_c. From the point of view of the adiabatic picture this corresponds to a non-adiabatic transition from χ_n to χ_m. There are thus two alternative paths to observe an elastic event. Either the collision is adiabatic or it involves two non-adiabatic transitions (the first as R decreases from infinity through R_c to about the turning point and the second as R increases from the turning point via R_c to infinity). It is important to remember, however, that the two alternatives need not be mutually exclusive or, in other words, the amplitudes for the two events can interfere. In the same fashion the inelastic transition $n \to m$ can proceed by two paths, in that a non-adiabatic transition can occur on the first passage via R_c but not on the second, or vice versa.

Around R_c, $\quad \Delta E(R) = E_n(R) - E_m(R) \simeq V_{n,m}(R)$

and so the probability of a non-adiabatic transition is determined by comparing $h/V_{n,m}$ with the time of passage through the crossing region $a/v(R_c)$, where a is the range of the crossing region

$$a \simeq \Delta E(R_c) \bigg/ \left|\left(\frac{\partial \Delta E}{\partial R}\right)_{R=R_c}\right| \approx V_{n,m}(R_c) \bigg/ \left|\left[\frac{\partial}{\partial R}(V_{n,n} - V_{m,m})\right]_{R=R_c}\right|.$$

The probability of a non-adiabatic transition is thus expected to decrease with the increasing magnitude of

$$\frac{2\pi |V_{n,m}(R_c)|^2}{|\hbar v(R_c)[\partial(V_{n,n} - V_{m,m})/\partial R]_{R=R_c}|}. \tag{2.45}$$

A quantitative discussion leading to this conclusion was first given by Landau (1932) and Zener (1932). If we denote the above expression by v_0/v these authors showed that the probability of a non-adiabatic transition is $\exp(-v_0/v)$. It should be stressed, however, that this result was obtained under restrictive assumptions (cf. Bates 1960, Nikitin 1962): in particular, taking proper account of the nuclear motion in the crossing region may modify the functional form of the result (Coulson and Zalewski 1962, Levine 1967a). See also section 3.4.2, eqn (4.107) for a derivation of eqn (2.45).

Our discussion has thus far neglected to include the possibility of exchange of identical particles. This is particularly important in the present context since the set of states $\{O|n\rangle\}$ is no longer necessarily a set of orthonormal states. This raises some problems regarding the explicit realization of P and Q. One possible solution is to perform a Schmidt orthogonalization procedure on the basis set, starting with $O|n\rangle$, where $|n\rangle$ is the internal state that is initially occupied. In a two-state problem we can thus take

$$P = |On\rangle\langle On|,$$
$$Q = [|Om\rangle - |On\rangle\langle On|Om\rangle][\langle Om|On\rangle\langle On| - \langle Om|], \quad (2.46)$$

so that P and Q are orthogonal and Hermitian. This definition is by no means unique. (One could perform a symmetric or canonical orthogonalization for example.)

The advantage of the adiabatic basis is that it takes proper account of the considerable distortion of the relative motion during the collision. However, once non-adiabatic transitions become important, the required coupling terms $C_{n,m}$ are awkward to evaluate, and the physical significance of the basis set is no longer very useful. (In a sense, a non-adiabatic transition is a non-event, in that the term implies the absence of abrupt changes in the character of the internal wavefunction.) On the other hand, a basis constructed from the eigenstates of h incorporates only the diagonal matrix elements of V as the distortion of the relative motion, and consequently the matrix elements $V_{n,m}$ can be quite large, but these operators are well-behaved functions of R. It is clearly desirable to have a basis (sometimes labelled 'diabatic', cf. Lichten 1963) which will incorporate the advantage of the adiabatic basis in incorporating a considerable portion of the mutual distortion, and which will lead to reasonably simple coupling terms in the set of coupled differential equations. Moreover, the formulation of the asymptotic boundary conditions is simpler in the H_0 representation. This is associated with the use of a rotating coordinate system, to describe the internal motion, in the conventional adiabatic theory. At large relative separation the internal angular momentum is not coupled to the relative angular momentum, and the adiabatic states that rotate with R have to be decoupled to properly represent the limit $R \to \infty$. A discussion of a suitable basis that corrects for the last point was given by Thorson (1961, 1963, 1965). However, no unique prescription for a basis that simplifies the problem of the crossing region is currently available.

In practice, the currently available accurate adiabatic wavefunctions for the collision of two atoms involve values of N up to 100 to describe the electronic motion. This is, of course, a reflection of the considerable distortion of the electronic motion during the collision. Even so, the crossing region primarily involves rapid changes only in the amplitudes ($F_{m,n}$) of very few states of h. Thus we should seek a description that will treat adiabatically those states not directly involved in the crossing, and centre attention on the exact dynamics of the few states problem under the average (adiabatic) influence of the remaining states. Towards this aim we continue, in the following section, our study of the effective potential in the H_0 representation.

3.2.2. *Internal excitation in collisions*

To discuss the migration of energy during the collision it is convenient to consider a model where only two states of h ($|m\rangle$ and $|n\rangle$) are introduced. If $|m\rangle$ is the state of higher energy, we take

$$Q = |m\rangle\langle m| \qquad (2.47)$$

and $\qquad \langle m|\psi^+\rangle = [E^+ - K - E_m - V_{m,m}]^{-1}\langle m|V|n\rangle\langle n|\psi^+\rangle, \qquad (2.48)$

where we have used eqn (1.15), and

$$\langle m|H|m\rangle = K + \langle m|h+V|m\rangle = K + E_m + V_{m,m}. \qquad (2.49)$$

If $E < E_m$, the Green's operator in eqn (2.48) implies that

$$\langle \mathbf{R}, m|\psi^+\rangle \xrightarrow[R\to\infty]{} \exp(-\kappa R)f, \quad \hbar^2\kappa^2 = 2\mu(E_m - E), \qquad (2.50)$$

provided $V_{m,m}(\mathbf{R}) \to 0$ as $R \to \infty$. If the potential $V_{m,m}(\mathbf{R})$ is attractive, the Hamiltonian $K + \langle m|h+V|m\rangle$ may be able to support bound states§

$$(E_i - K - \langle m|h+V|m\rangle)g_i(\mathbf{R}) = 0 \quad (E_i < E_m) \qquad (2.51)$$

of energies below E_m. These correspond to a bounded relative motion of the molecules, internally excited to the state m. Apart from these bound states there will be unbound solutions to eqn (2.5) with energies above E_m, which we denote by $g_\mathbf{k}(\mathbf{R})$,

$$[(\hbar^2k^2/2\mu) - K - \langle m|h+V|m\rangle]g_\mathbf{k}(\mathbf{R}) = 0. \qquad (2.52)$$

§ The representation of the bound states of QHQ as $g_i(\mathbf{R})\phi_m(\mathbf{r})$ is done purely for convenience. We can use any complete set of states in the $\mathbf{R}, \mathbf{r}$ space to diagonalize QHQ. In particular, in terms of the model discussed in the introduction, the normal modes of the triatomic (which are linear combinations of r and R) can be used as the independent variables.

In terms of these solutions we can expand the Green's operator in eqn (2.48) as

$$\langle \mathbf{R}|(E^+ - K - \langle m|h+V|m\rangle)^{-1}|\mathbf{R}'\rangle$$
$$= \sum_i g_i(\mathbf{R})(E^+ - E_i)^{-1} g_i^*(\mathbf{R}') + \int d\mathbf{k}\, g_\mathbf{k}(\mathbf{R})[E^+ - E(k)]^{-1} g_\mathbf{k}^*(\mathbf{R}')$$

and obtain

$$F^m_{\mathbf{k},n}(\mathbf{R}) = \langle \mathbf{R}, m|\psi^+\rangle$$
$$= \sum_i g_i(\mathbf{R})(E^+ - E_i)^{-1} \int d\mathbf{R}'\, g_i^*(\mathbf{R}') V_{m,n}(\mathbf{R}') F^n_{\mathbf{k},n}(\mathbf{R}') +$$
$$+ \int d\mathbf{k}\, g_\mathbf{k}(\mathbf{R})[E^+ - E(k)]^{-1} \int d\mathbf{R}'\, g_\mathbf{k}^*(\mathbf{R}') V_{m,n}(\mathbf{R}') F^n_{\mathbf{k},n}(R'). \quad (2.53)$$

Equation (2.53) provides a formal statement of our qualitative picture. During an elastic collision ($E < E_m$) the amplitude ($F^m_{\mathbf{k},n}(\mathbf{R})$) to observe the two molecules internally excited can attain significant values, particularly when the interaction between the excited states ($V_{m,m}(\mathbf{R})$) is attractive. When $E \sim E_i$ the main contribution to $F^m_{\mathbf{k},n}$ is from the bound state $g_i(\mathbf{R})$. In other words, when $E \sim E_i$,

$$\mathscr{V}_{n,n}(\mathbf{R},\mathbf{R}') \simeq V_{n,n}(\mathbf{R})\delta(\mathbf{R}-\mathbf{R}') +$$
$$+ V_{n,m}(\mathbf{R})g_i(\mathbf{R})(E^+ - E_i)^{-1} g_i^*(\mathbf{R}') V_{m,n}(\mathbf{R}'). \quad (2.54)$$

The second term in eqn (2.54) is a separable potential of the type discussed in section 2.2.2.

Using eqns (1.19) and (1.15) we can write

$$P\psi^+ = P\psi_1^+ + (E^+ - PHP)^{-1} PHQ(E^+ - QHQ)^{-1} QHP\psi^+, \quad (2.55)$$

where $P\psi_1^+$ is the solution of the scattering problem with the interaction PVP
$$(E^+ - PHP) P\psi_1^+ = (E^+ - H_0)\phi. \quad (2.56)$$

In the present model $P\psi_1^+$ is that component of the total wavefunction that takes no account of the internal excitation, since the mutual interaction PVP cannot lead to a transition to the Q component of ψ. Equation (2.53) can be iterated by substituting $P\psi_1^+$ in the second term on the right, and so on. Thus

$$\langle n|\psi^+\rangle = \langle n|\psi_1^+\rangle + g_{n,n} V_{n,m} g_{m,m} V_{m,n} \langle n|\psi_1^+\rangle +$$
$$+ g_{n,n} V_{n,m} g_{m,m} V_{m,n} g_{n,n} V_{n,m} g_{m,m} V_{m,n} \langle n|\psi_1^+\rangle + \dots, \quad (2.57)$$

where $$g_{n,n} = (E^+ - H_{n,n})^{-1}.$$

Successive terms in the iteration differ by the factor $g_{n,n} V_{n,m} g_{m,m} V_{m,n}$ which (in contrast to the Born iteration) is of the second order in the perturbation. The first term in the iteration corresponds to the collision without any internal excitation. The mutual interaction is $V_{n,n}$, the interaction between the unexcited molecules. The significance of the higher-order terms is best analysed from the point of view of time-ordered iterations, section 2.7.1. Thus in the second term proceeding from right to left, the mutual interaction $V_{m,n}$ leads to an excitation of the internal state m. The system then evolves ($g_{m,m}$) under the mutual interaction $V_{m,m}$. Finally, the interaction $V_{n,m}$ leads to a de-excitation back to the state n. In each successive term the system undergoes yet another excitation–de-excitation stage. It is important to bear in mind that the Green's functions $g_{n,n}$ and $g_{m,m}$ are operators on the coordinates of relative motion, and that in the coordinate representation they are non-local. For example, the coordinate representation of the second term in the iteration is, for $E \sim E_i$,

$$\int \langle \mathbf{R}|g_{n,n}|\mathbf{R}'\rangle V_{n,m}(\mathbf{R}')(E^+ - E_i)^{-1} g_i(\mathbf{R}') \, d\mathbf{R}' \times$$
$$\times \int g_i^*(\mathbf{R}'') V_{m,n}(\mathbf{R}'') \langle \mathbf{R}'', n|\psi_1^+\rangle \, d\mathbf{R}'', \quad (2.58)$$

where for $E \sim E_i$ we approximated $g_{m,m}$ by one term in the spectral resolution. Note that in principle the transition $n \to m$ can occur for any value of $\mathbf{R}''$, and that the integrations over $\mathbf{R}''$ and $\mathbf{R}'$ are independent. In practice, the main contributions to such matrix elements come from regions of constructive interference between $g_i(\mathbf{R})$ and $\langle \mathbf{R}, n|\psi_1^+\rangle$, provided $V_{m,n}(\mathbf{R})$ is a reasonably smooth function. The function $\langle \mathbf{R}, n|\psi_1^+\rangle$ is a wavefunction for the unbounded relative motion, and so is rapidly oscillating as a function of R (Fig. 3.1). There will be extensive cancellation in the integrand apart from those regions of R where the bound state function and $g_i(\mathbf{R})$ have a significant overlap. Qualitatively, this conclusion is equivalent to the Franck–Condon principle. Quantitative estimates are usually based on a semi-classical approximation to the continuum wavefunction (Zener 1932, Landau and Lifshitz 1965, Coulson and Zalewski 1962, and section 3.4.2).

To sum the series (2.57) we define the Green's function $G_{m,m}$ by§

$$(E^+ - H_{m,m} - V_{m,n} g_{n,n} V_{n,m}) G_{m,m} = I. \quad (2.59)$$

Then, from eqn (1.20),

$$\langle m|\psi^+\rangle = G_{m,m} V_{m,n} \langle n|\psi_1^+\rangle. \quad (2.60)$$

Equation (2.60) is the solution for the amplitude $\langle m|\psi^+\rangle$ in terms of the

§ In general, eqn (2.59) reads
$$[E^+ - QHQ - QHP(E^+ - PHP)^{-1}PHQ]QGQ = Q,$$
see eqn (3.13).

(assumed) known $\langle n|\psi_1^+\rangle$. Using this amplitude in eqn (2.55),

$$\langle n|\psi^+\rangle = \langle n|\psi_1^+\rangle + g_{n,n}V_{n,m}G_{m,m}V_{m,n}\langle n|\psi_1^+\rangle. \quad (2.61)$$

In equation (2.61) the effects of internal excitation are contained in the factor $V_{n,m}G_{m,m}V_{m,n}$. To the right and to the left of this factor the

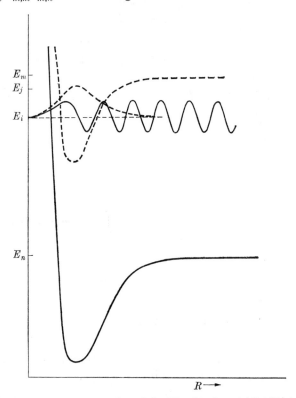

FIG. 3.1. Schematic representation of the Hamiltonians $\langle n|(h+V)|n\rangle$ (bottom curve) and $\langle m|(h+V)|m\rangle$ as functions of R. The second Hamiltonian can support two bound states at energies E_i and E_j. Shown also is the profile of $g_i(R)$ and the continuum s-wave function of the Hamiltonian $\langle n|(h+V)|n\rangle$ at the energy E_i. The s-wave phase shift, with and without coupling to the discrete bound states, is shown in Fig. 3.4.

molecules interact subject to $V_{n,n}$ only. We shall show below that $G_{m,m}$ takes exact account of the evolution of the two molecules in the excited state m, including the possibility of de-excitation. We thus expect that under suitable (namely non-stationary) experimental conditions the collision mechanism represented by the second term will lead to a time delay in the appearance of the final states, when compared to the first term (see section 3.6.1).

When the total energy E is near E_i it is convenient to centre attention on the excitation of the bound state $g_i(\mathbf{R})$, which corresponds to the bound relative motion of the two excited molecules at the total energy E_i, $E_i < E_m$.

To this end we put
$$Q = |m,i\rangle\langle m,i|, \tag{2.62}$$
where
$$g_i(\mathbf{R}) = \langle \mathbf{R}|i\rangle,$$
and define§ $P = I - Q$. With this definition $P\psi_1^+$ includes all collision events in which the state $|m,i\rangle$ does not participate as an intermediate.

From eqn (2.59),
$$\langle m,i|G|m,i\rangle = [E^+ - E_i - \langle m,i|H(E^+ - PHP)^{-1}H|m,i\rangle]^{-1},$$
so that
$$Q\psi^+ = |m,i\rangle\langle m,i|G|m,i\rangle\langle m,i|H|P\psi_1^+\rangle. \tag{2.63}$$

Using eqn (2.1.9) we can write
$$\langle m,i|G|m,i\rangle = (E^+ - E_i - \Delta_i - i\Gamma_i)^{-1}, \tag{2.64}$$
where
$$\Delta_i = \langle m,i|HP_v(E - PHP)^{-1}H|m,i\rangle \tag{2.65}$$
and∥
$$\Gamma_i = \pi\langle m,i|H\delta(E - PHP)H|m,i\rangle = \pi\int d\hat{\mathbf{k}} \ |\langle m,i|H|E,\hat{\mathbf{k}},n_1^+\rangle|^2, \tag{2.66}$$
with
$$(E - PHP)|E,\hat{\mathbf{k}},n_1^+\rangle = 0. \tag{2.67}$$

The asymptotic form of $(E^+ - PHP)^{-1}$ in the coordinate representation is (cf. eqn (2.5.73))
$$\langle n, \mathbf{R}|(E^+ - PHP)^{-1} \xrightarrow[R\to\infty]{} (-\mu/2\pi\hbar^2 R)\exp(ikR)\langle n,\mathbf{k}_1^-| \tag{2.68}$$
with $\hat{\mathbf{k}} = \hat{\mathbf{R}}$.

Let $f_d(\mathbf{k}',\mathbf{k})$ be the elastic scattering amplitude for collisions that do not proceed via the state $|m,i\rangle$ as intermediate,
$$\langle \mathbf{R},n|\psi_1^+\rangle \xrightarrow[R\to\infty]{} \exp(i\mathbf{k}\cdot\mathbf{R}) + f_d(\mathbf{k}',\mathbf{k})\exp(ikR)/R. \tag{2.69}$$

Then, from eqns (2.69), (2.68), (2.64), and (2.61),
$$\langle \mathbf{R},n|\psi^+\rangle \xrightarrow[R\to\infty]{} \exp(i\mathbf{k}\cdot\mathbf{R}) + R^{-1}\exp(ikR) \times$$
$$\times \left[f_d(\mathbf{k}',\mathbf{k}) - (\mu/2\pi\hbar^2)\frac{\langle n,\mathbf{k}_1^-|H|m,i\rangle\langle m,i|H|n,\mathbf{k}_1^+\rangle}{E - E_i - \Delta_i - i\Gamma_i}\right]. \tag{2.70}$$

Equation (2.70) identifies the scattering amplitude, where the second term is the contribution from the collisions that did involve $|m,i\rangle$.

§ Of course, with this definition $P|m\rangle \neq 0$. This does not matter in principle since $P\psi^+$ does equal ψ^+ asymptotically. We can, if we wish, retain the previous definition $(P = I - |m\rangle\langle m| = |n\rangle\langle n|)$ and put $Q = Q_i + Q^i$, $Q_i Q = Q_i$, $Q_i = |m,i\rangle\langle m,i|$ and proceed as in eqns (1.26) and (1.27). One can show that the results are the same.

∥ Note that Γ_i is non-negative.

To perform a partial wave expansion of eqn (2.70), we simplify the problem by assuming that $g_i(\mathbf{R})$ (eqn (2.62)) can be written as

$$g_i(\mathbf{R}) = g_i(R) Y_l^{m_l}(\hat{\mathbf{R}}) \tag{2.71}$$

and that both H and H_0 commute with $\mathbf{l}$, the angular momentum of relative motion. In this case only the lth partial wave contributes to the third term in eqn (2.70). To see this explicitly we put, following eqn (2.2.68),

$$\langle \mathbf{R}, \mathbf{r} | \psi_1^+ \rangle = \sum_{l, m_l} \psi_{klm_l}^+(R) Y_l^{m_l}(\hat{\mathbf{R}}) Y_l^{m_l *}(\hat{\mathbf{k}}) \phi_n(\mathbf{r}) \tag{2.72}$$

so that, since H and $\mathbf{l}$ commute,

$$\langle m, i | H | n, \mathbf{k}_1^+ \rangle = Y_l^{m_l}(\hat{\mathbf{k}}) \int g_i^*(R) V_{m,n}(R) \psi_{klm_l}^+(R) R^2 \, \mathrm{d}R, \tag{2.73}$$

where we used the orthogonality of the spherical harmonics over $\hat{\mathbf{R}}$. To handle the second matrix element we note from eqns (2.8.29) and (2.2.82) that§ with δ_l^0 as the phase shift induced by PVP,

$$\psi_{klm_l}^+ = \exp(2i\delta_l^0) \psi_{klm_l}^-. \tag{2.74}$$

The rest of the calculation is identical to that carried out in section 2.2.3. We finally have

$$f(\mathbf{k}', \mathbf{k}) = \sum_{l'} \frac{(2l'+1)}{k} P_{l'}(\cos\theta) \bigg[\sin\delta_{l'}^0 \exp(i\delta_{l'}^0) + \\ + \delta_{l,l'} \exp(2i\delta_l^0) \frac{\Gamma_i^l}{E - E_i - \Delta_i^l - i\Gamma_i^l} \bigg], \tag{2.75}$$

where the first contribution is from the 'direct collision', and

$$\Gamma_i = \pi \int \mathrm{d}\hat{\mathbf{k}}\, \rho(k) |\langle m, i | H | n, \mathbf{k}_1^+ \rangle|^2 \\ = \pi \rho(E - E_n) \sum_{m_l} \bigg| \int g_i(R) V_{m,n} \psi_{klm_l}^+(R) R^2 \, \mathrm{d}R \bigg|^2 \\ = \Gamma_i^l. \tag{2.76}$$

Introducing the resonance phase shift δ_l^r by

$$\tan\delta_l^r = \Gamma_i^l / (E - E_i - \Delta_i^l), \tag{2.77}$$

we can write (cf. eqn (2.5.54))

$$\tau_l = -\pi^{-1} \bigg[\sin\delta_l^0 \exp(i\delta_l^0) + \exp(2i\delta_l^0) \frac{\Gamma_i^l}{E - E_i - \Delta_i^l - i\Gamma_i^l} \bigg] \\ = -\pi^{-1} \sin(\delta_l^0 + \delta_l^r) \exp[i(\delta_l^0 + \delta_l^r)]. \tag{2.78}$$

The cross-section for the lth partial wave is

$$\sigma_l = (4\pi/k^2)(2l+1) \sin^2(\delta_l^0 + \delta_l^r). \tag{2.79}$$

§ In principle, this conclusion follows from eqns (2.5.30) and the normalization of the solutions, so that $\psi_j^+ = S_j \psi_j^-$, where $S_j = \exp(2i\delta_j)$.

Only if $\delta_l^0 = 0$ we can put
$$\sigma_l = (4\pi/k^2)(2l+1)\sin^2\delta_l^r = (4\pi/k^2)(2l+1)\frac{(\Gamma_i^l)^2}{(E-E_i-\Delta_i^l)^2+(\Gamma_i^l)^2}. \quad (2.80)$$

In general, following Fano (1961), we can rewrite eqn (2.79) as
$$\sigma_l = (4\pi/k^2)(2l+1)\sin^2\delta_l^0 \frac{(q+\epsilon)^2}{1+\epsilon^2} \quad (2.81)$$

with $q = -\cot\delta_l^0$, $\epsilon = -\cot\delta_l^r$. In general, the energy dependence of σ_l will deviate from the symmetric form of eqn (2.80) since ϵ is an odd function of $E-E_i-\Delta_i$. The asymmetry is due to interference between the two outgoing waves $P\psi_1^+$ and $(E^+-PHP)^{-1}PHQ\psi^+$. To see this explicitly we use eqns (2.75), (2.72), (2.2.80), and (2.2.82) to write

$$\psi_{kl}^+ \xrightarrow[R\to\infty]{} (kR)^{-1}\sin(kR-\tfrac{1}{2}l\pi+\delta_l^0)\exp(i\delta_l^0)+$$
$$+(kR)^{-1}\exp[i(kR-\tfrac{1}{2}l\pi)]\exp(2i\delta_l^0)\sin\delta_l^r\exp(i\delta_l^r), \quad (2.82)$$

where the first term is the contribution of $P\psi_1^+$ to the lth partial wave and the second term is the contribution of the resonance. The resonance term has a component of the same spatial dependence as the direct term. We can rewrite eqn (2.82) as

$$\psi_{kl}^+ \to (kR)^{-1}[\sin(kR-\tfrac{1}{2}l\pi+\delta_l^0)\cos\delta_l^r+$$
$$+\cos(kR-\tfrac{1}{2}l\pi+\delta_l^0)\sin\delta_l^r]\exp[i(\delta_l^0+\delta_l^r)], \quad (2.83)$$

where the interference is explicit. At the resonance energy, $E = E_i+\Delta_i$, $\cos\delta_l^r$ vanishes and only the second term remains. We can sum the two components to

$$\psi_{kl}^+ \to (kR)^{-1}\sin(kR-\tfrac{1}{2}l\pi+\delta_l^0+\delta_l^r)\exp[i(\delta_l^0+\delta_l^r)] \quad (2.84)$$

as expected (cf. eqn (2.2.82)). We turn now to a generalization of this problem.

The line shape $A_i(E)$ is defined as the density of the state $|m,i\rangle$ at the energy E,
$$A_i(E) = \langle m,i|\delta(E-H)|m,i\rangle = \sum_{\mathbf{n}} |\langle m,i|E,\mathbf{n}^+\rangle|^2, \quad (2.85)$$

where the summation over $\mathbf{n}$ stands for an integration over all directions of the relative momentum. (Recall that $\mathbf{n} = (n,\hat{\mathbf{k}})$.) Using eqns (2.1.9) and (2.64),
$$A_i(E) = -\pi^{-1}\mathrm{im}\langle m,i|G(E^+)|m,i\rangle = \pi^{-1}\frac{\Gamma_i}{(E-E_i-\Delta_i)^2+(\Gamma_i)^2}. \quad (2.86)$$

The same conclusion can also be obtained by evaluating explicitly eqn (2.85), using eqns (2.63) and (2.66).

The line shape is not directly measurable. As we shall see (section

3.7.1) the measurable quantity is the line profile $S_z(E)$, which we can define in general as
$$S_z(E) = B\langle z|\delta(E-H)|z\rangle \tag{2.87}$$
with some general coefficient B. If $P|z\rangle = 0$ (and this is an exception in practice) $Q|z\rangle = |z\rangle$ and
$$S_z(E) = B|\langle z|m,i\rangle|^2 A_i(E). \tag{2.88}$$

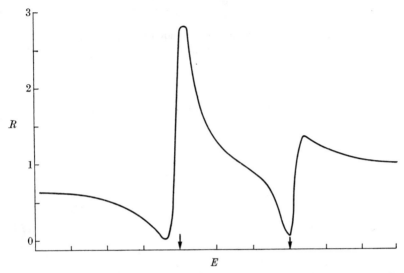

FIG. 3.2. The ratio of the line profile $S_z(E)$ to the line profile in the absence of the coupling to the discrete states of QHQ, $S_z^0(E) = B\langle z|\delta(E-PHP)|z\rangle$, from Levine (1967b). The two discrete eigenvalues of QHQ are indicated by the arrows. For the cross-section, with one resonance, eqns (2.81) and (2.91) give
$$R = (q+\epsilon)^2/(1+\epsilon^2).$$

In general,§
$$|E, \mathbf{n}^+\rangle = |E, \mathbf{n}_1^+\rangle + W|m,i\rangle\langle m,i|E, \mathbf{n}^+\rangle, \tag{2.90}$$
in which case S_z will contain interference terms between the two components (see Fig. 3.2 and, for example, Levine 1967b). Equation (2.83) is a particular case of this conclusion. To see this we recall the result
$$\sigma(E) = -(2/\hbar v)\mathrm{im}\langle \mathbf{k},n|T(E^+)|\mathbf{k},n\rangle \tag{2.91}$$
and eqn (2.5.27). Thus taking $|z\rangle = V|E,n\rangle$ we can bring the cross-section to the form of eqn (2.87).

Equation (2.70) is an exact equation for any energy E, $E < E_m$. However, when $E \sim E_j$ we expect that the first term will vary rapidly with E due to the excitation of the state $|m,j\rangle$. Provided the energy

§ To obtain the analogue of eqn (2.83) the wave operator has to be written as
$$W = I - i\pi\delta(E-PHP)P(H-E) + P_v(E-PHP)^{-1}P(H-E). \tag{2.89}$$
(See for details, Levine 1967b, Appendix A.)

difference $|E_i+\Delta_i-E_j-\Delta_j|$ is larger than $\Gamma_i+\Gamma_j$ we can, when $E \sim E_j$, replace eqn (2.70) by an equation derived with $Q = |m,j\rangle\langle m,j|$. Under these conditions it is meaningful to centre attention on each energy interval separately, a situation which is commonly referred to as 'isolated resonances'. When the regions of rapid energy variations due to different states overlap, one should consider Q as the projection on all the relevant states:

$$Q = \sum_i |m,i\rangle\langle m,i|. \qquad (2.92)$$

The summation over i is over all states with energy E_i in the specified interval (since *a priori* we know neither the magnitude nor the sign of Δ_i).

In general, using eqn (1.26),

$$P\psi^+ = P\psi_1^+ + (E^+ - PHP)^{-1} PHQGQHP\psi_1^+, \qquad (2.93)$$

so that, using eqn (2.92),

$$P\psi^+ = P\psi_1^+ + (E^+ - PHP)^{-1} PH \sum_{i,j} |m,i\rangle\langle m,i|G|m,j\rangle\langle m,j|H|P\psi_1^+\rangle. \qquad (2.94)$$

The Green's operator QGQ,

$$QGQ = [E^+ - QHQ - QHP(E^+ - PHP)^{-1}PHQ]^{-1}$$
$$= (E^+ - Q\mathcal{H}Q)^{-1}, \qquad (2.95)$$

is not diagonal in the QHQ representation. The reason can be seen by iterating QGQ in terms of $(E^+ - QHQ)^{-1}$,

$$QGQ = (E^+ - QHQ)^{-1} +$$
$$+ (E^+ - QHQ)^{-1} QHP(E^+ - PHP)^{-1} PHQ(E^+ - QHQ)^{-1} + \ldots. \qquad (2.96)$$

When we substitute the iteration (2.96) in the equation for $P\psi^+$, the first term is diagonal. This term corresponds to a single excitation–de-excitation stage. The next term, which corresponds to two excitation–de-excitation stages in succession, is no longer diagonal. For example, one contribution will be

$$(E^+ - PHP)^{-1} PH|m,i\rangle(E^+ - E_i)^{-1}\langle m,i|HP(E^+ - PHP)^{-1}PH|m,j\rangle \times$$
$$\times (E^+ - E_j)^{-1}\langle m,j|H|P\psi_1^+\rangle.$$

As is clear, the term is not diagonal because the first excitation–de-excitation stage can be followed by a second stage with respect to a different state. One can bring QGQ to a diagonal form by introducing the eigenstates of the operator $Q\mathcal{H}Q$, which is not Hermitian but is invariant under L-conjugation.

Let $|s\rangle$ be an eigenvector of $Q\mathcal{H}Q = QHQ + QHP(E^+ - PHP)^{-1}PHQ$,

$$(Q\mathcal{H}Q - \lambda_s)|s\rangle = 0. \qquad (2.97)$$

From eqn (2.95)
$$Q\mathcal{H}(\lambda^*)Q = Q\mathcal{H}^*(\lambda)Q, \qquad (2.98)$$
so that
$$[Q\mathcal{H}(\lambda^*)Q - \lambda_s^*]|s^*\rangle = 0 \qquad (2.99)$$
and, taking the Hermitian conjugate of this equation,
$$\langle s^*|(Q\mathcal{H}^\ddagger Q - \lambda_s) = 0$$
or, since $Q\mathcal{H}Q$ is L-conjugation invariant,
$$\langle s^*|(Q\mathcal{H}Q - \lambda_s) = 0. \qquad (2.100)$$

Equations (2.97) and (2.100) show that the eigenvectors of $Q\mathcal{H}Q$ form a bi-orthogonal set. (See, for example, Morse and Feshbach 1953, p. 884.) The bi-orthogonality relation follows from the L-conjugation invariance of $Q\mathcal{H}Q$,
$$\langle s'^*|Q\mathcal{H}Q|s\rangle - \langle s'^*|Q\mathcal{H}^\ddagger Q|s\rangle = 0 = (\lambda_s - \lambda_{s'})\langle s'^*|s\rangle = 0. \qquad (2.101)$$

We can thus put
$$\langle s'^*|s\rangle = \delta_{s's} \qquad (2.102)$$
and
$$Q = \sum_s |s\rangle\langle s^*|. \qquad (2.103)$$

In practice, we shall solve eqn (2.97) by expanding
$$|s\rangle = \sum_i B_{i,s}|m,i\rangle. \qquad (2.104\,\text{a})$$

If we denote $\mathcal{H}_{i,j} = \langle m,i|Q\mathcal{H}Q|m,j\rangle$, we have to solve the secular equations
$$(\mathcal{H} - \lambda\mathbf{I})\mathbf{B} = 0 \qquad (2.105\,\text{a})$$
and
$$(\mathcal{H}^* - \lambda^*\mathbf{I})\mathbf{C} = 0, \qquad (2.105\,\text{b})$$
where
$$|s^*\rangle = \sum_i C_{i,s}|m,i\rangle. \qquad (2.104\,\text{b})$$

Equation (2.102) can now be written as
$$\sum_i C_{i,s'}^* B_{i,s} = \delta_{s's} \qquad (2.106\,\text{a})$$
and the orthogonality relation $\langle m,i|m,i'\rangle = \delta_{ii'}$ as
$$\sum_s C_{i',s}^* B_{i,s} = \delta_{ii'}. \qquad (2.106\,\text{b})$$

We can now write
$$QGQ = \sum_s \frac{|s\rangle\langle s^*|}{E^+ - \lambda_s} \qquad (2.107)$$
and, using eqns (2.97) and (2.100), confirm that
$$(E^+ - Q\mathcal{H}Q)QGQ = Q = QGQ(E^+ - Q\mathcal{H}Q). \qquad (2.108)$$

Using eqn (2.93) we obtain for $P\psi^+$,
$$P\psi^+ = P\psi_1^+ + (E^+ - PHP)^{-1}PH\sum_s \frac{|s\rangle\langle s^*|H|P\psi_1^+\rangle}{E^+ - \lambda_s} \qquad (2.109\,\text{a})$$
and
$$Q\psi = \sum_s |s\rangle\frac{\langle s^*|H|P\psi_1^+\rangle}{E^+ - \lambda_s}. \qquad (2.109\,\text{b})$$

The scattering amplitude is given, by analogy to eqn (2.70), as

$$f_d(\mathbf{k}', \mathbf{k}) - (\mu/2\pi\hbar^2) \sum_s \frac{\langle n, \mathbf{k}_1^-|H|s\rangle\langle s^*|H|n, \mathbf{k}_1^+\rangle}{E-\lambda_s}. \quad (2.110)$$

In view of eqns (2.97) and (2.100) it is not possible to write $\mathrm{im}\,\lambda_s$ in terms of $\langle s^*|Q\mathcal{H}Q|s\rangle$. Rather, by premultiplying eqn (2.97) by $\langle s|$ we obtain

$$\Gamma_s = -\mathrm{im}\,\lambda_s = \pi\langle s|QHP\delta(E-PHP)PHQ|s\rangle/\langle s|s\rangle. \quad (2.111)$$

We can write, however, the invariance of $\mathrm{Tr}\,Q\mathcal{H}Q$ under the transformation (2.104) as

$$\sum_s \lambda_s = \sum_i \mathcal{H}_{i,i}, \quad (2.112\,\mathrm{a})$$

so that

$$\sum_s \Gamma_s = -\pi \sum_i \mathrm{im}\,\mathcal{H}_{i,i}. \quad (2.112\,\mathrm{b})$$

Also, using eqn (2.106),

$$\sum_s \langle n, \mathbf{k}_1^-|H|s\rangle\langle s^*|H|n, \mathbf{k}_1^+\rangle = \sum_i \langle n, \mathbf{k}_1^-|H|m, i\rangle\langle m, i|H|n, \mathbf{k}_1^+\rangle. \quad (2.113)$$

Finally, we note Hadamard's inequalities, for every eigenvalue λ of $Q\mathcal{H}Q$

$$\min\left(|\mathcal{H}_{i,i}| - \sum_{j=1}^N |\mathcal{H}_{i,j}|\right) \leqslant \lambda \leqslant \max \sum_{j=1}^N |\mathcal{H}_{i,j}|,$$

where $1 \leqslant i \leqslant N$, and that $\mathrm{im}\,\lambda$ is between the smallest and largest eigenvalue of $(2i)^{-1}(Q\mathcal{H}Q - Q\mathcal{H}^\dagger Q)$. Now

$$-(2i)^{-1}(Q\mathcal{H}Q - Q\mathcal{H}^\dagger Q) = \pi QHP\delta(E-PHP)PHQ \quad (2.114)$$

is non-negative, thus confirming that Γ_s is non-negative.

Our discussion also shows explicitly that the states we are interested in are the eigenvectors of $Q\mathcal{H}Q$ (and not QHQ). If the interaction QHP is large, it is possible that QHQ will have no discrete eigenvalues, and yet $Q\mathcal{H}Q$ will lead to typical resonance behaviour in eqns (2.110) and (2.109).

Equation (2.110) has primarily a formal significance, in showing that the scattering amplitude for overlapping resonances can be written in a form analogous to the isolated resonance case. Thus one can fit the parameters Γ_s to the results of experimental observations (see, for example, Chen 1964). From a computational standpoint if the states $|s\rangle$ are evaluated as linear combinations of the eigenstates of QHQ then one can equally well by-pass the introduction of the basis $\{|s\rangle\}$ and work with QGQ in the QHQ representation.

When a very large number of states contribute in a given energy interval one can begin to consider them as a continuum and introduce statistical arguments, as is discussed in section 3.5.3.

The present discussion has an obvious extension to the case of non-reactive collisions. The simplest choice of P (Feshbach 1962) is a projection on all the internal states of h that can be asymptotically observed

$$P = \sum_m |m\rangle\langle m| \quad (E_m < E).$$

The scattering amplitude into a particular final internal state is obtained by considering the asymptotic limit of $\langle m|P\psi^+\rangle$ by analogy to eqn (2.70), so that for an initial state $|\mathbf{k}, n\rangle$

$$f_{m,n}(\mathbf{k}', \mathbf{k}) = (f_{m,n}(\mathbf{k}', \mathbf{k}))_a +$$
$$+ (-\mu/2\pi\hbar^2) \sum_s \frac{\langle m, \mathbf{k}_1'^-|H|s\rangle\langle s^*|H|n, \mathbf{k}_1^+\rangle}{E - \lambda_s}, \quad (2.115)$$

where
$$(E^- - PHP)|m, \mathbf{k}_1^-\rangle = (E^- - H_0)|m, \mathbf{k}\rangle$$

and the first term is obtained from the asymptotic behaviour of $\langle m|n, \mathbf{k}_1^+\rangle$.

It is important to observe that since (cf. eqn (2.7.50))

$$\langle E, \mathbf{n}_1^+|E, \mathbf{m}_1^+\rangle = 0 \quad (m \neq n), \quad (2.116)$$

we have
$$-\mathrm{im}\, Q\mathcal{H}Q = \pi QHP\delta(E - PHP)PHQ$$
$$= \pi \sum_m QHP|E, \mathbf{m}_1^+\rangle\langle E, \mathbf{m}_1^+|PHQ, \quad (2.117)$$

so that
$$\Gamma_s = \sum_m \Gamma_s^m, \quad (2.118)$$

where Γ_s^m refers to a particular (non-reactive) channel of relative motion. Equation (2.118) is the optical theorem for the rate operator.

It is often convenient to distinguish between the elastic and inelastic collisions by redefining P as

$$P = |n\rangle\langle n|,$$

where $|n\rangle$ is the internal state of the molecules before the collision. In this case, $Q\psi^+$ need not vanish asymptotically, and the asymptotic behaviour of $\langle m|Q\psi^+\rangle$ and $\langle n|P\psi^+\rangle$ determines the inelastic $(n \to m)$ and elastic scattering amplitudes respectively.

The equivalent potential $\mathscr{V}$ then determines the elastic scattering amplitude, but is no longer Hermitian. Due to the inelastic processes the norm of $P\psi^+$ need not be conserved in time. To see this we note from eqns (1.17) and (1.12) that

$$V|\psi^+\rangle = \mathscr{V}P|\psi^+\rangle, \quad (2.119)$$

so that we can use Ehrenfest's theorem with $A = P$ (eqn (2.4.5)) and, recalling that $P^2 = P = P^\dagger$,

$$\frac{d}{dt}\langle P\psi^+|P\psi^+\rangle = \frac{d}{dt}\langle \psi^+|P|\psi^+\rangle$$

$$= (i\hbar)^{-1}\langle\psi^+|PV-VP|\psi^+\rangle$$

$$= (i\hbar)^{-1}\langle\psi^+|P\mathscr{V}P-P\mathscr{V}^\dagger P|\psi^+\rangle$$

$$= (2/\hbar)\text{im}\langle P\psi^+|\mathscr{V}|P\psi^+\rangle, \qquad (2.120)$$

where we have used eqn (2.119) and its Hermitian conjugate. From eqn (1.17),

$$\mathscr{V}-\mathscr{V}^\dagger = -2\pi i PHQ\delta(E-QHQ)QHP,$$

provided E is in the continuous spectrum of QHQ or, in other words, provided inelastic transitions can take place. The decrease in norm of $P|\psi^+\rangle$ is due to the increase in norm of $Q|\psi^+\rangle$ (which was, by definition of Q, zero before the collision, and is non-vanishing after the collision if inelastic transitions are energetically allowed). Since $P+Q = I$ we have that

$$\frac{d}{dt}\langle\psi^+|Q|\psi^+\rangle = -(2/\hbar)\text{im}\langle P\psi^+|\mathscr{V}|P\psi^+\rangle. \qquad (2.121)$$

The left-hand side is the rate of inelastic transitions. Using eqns (2.4.35a) and (2.4.38) we obtain for the rate per unit volume and unit density

$$\mathsf{V}^{-1}\sum_m{}' R(m \leftarrow \mathbf{n}) = -(2/\hbar)\text{im}\langle n,\mathbf{k}^+|P\mathscr{V}P|n,\mathbf{k}^+\rangle,$$

or

$$\sigma_{in} = \sum_m{}' \sigma_{m,n} = \mathsf{V}^{-1}\sum_m{}' v_n^{-1} R(m \leftarrow \mathbf{n})$$

$$= -(2/\hbar v_n)\text{im}\langle n,\mathbf{k}^+|P\mathscr{V}P|n,\mathbf{k}^+\rangle. \qquad (2.122)$$

Equations (2.121) and (2.122) can also be derived using the optical theorem (see Lax 1950, Francis and Watson 1953). The problem of an equivalent potential for elastic scattering (when inelastic collisions can take place) has been extensively discussed in the literature, particularly in connection with nuclear scattering. For recent reviews see Fetter and Watson (1965), Mittleman (1965), Goldberger and Watson (1964).

The present discussion can be extended to reactive collisions by selecting P such that $P\psi^+$ determines asymptotically the scattering amplitude for all non-reactive collisions. The simplest choice of P is obviously

$$P = C_i,$$

(cf. section 2.6.0) when the reactants belong to the (reactive) channel i. If we define

$$V|\psi^+\rangle = \mathscr{V}C_i|\psi^+\rangle, \qquad (2.123)$$

then our previous derivation of eqns (2.120)–(2.122) can be easily modified to obtain an expression for the total rate of reactive collisions.

We can introduce in the projection Q a distinction between the inelastic (or reactive) processes that are energetically allowed and those that are not by putting
$$Q = Q_i + Q^i, \quad Q_i Q = Q_i \qquad (2.124)$$
and taking
$$Q_i = \sum_{m \neq n} |m\rangle\langle m|, \quad E_m < E, \quad P = |n\rangle\langle n|. \qquad (2.125)$$

The L.S. equation can now be partitioned to three components:
$$(E^+ - PHP)P\psi^+ - PHQ_i\psi^+ - PHQ^i\psi^+ = (E^+ - H_0)\phi,$$
$$(E^+ - Q_i H Q_i)Q_i\psi^+ - Q_i H Q^i\psi^+ - Q_i H P\psi^+ = 0,$$
$$(E^+ - Q^i H Q^i)Q^i\psi^+ - Q^i H Q_i\psi^+ - Q^i H P\psi^+ = 0, \qquad (2.126)$$
provided $Q\phi = 0$.

Equation (2.126) is, of course, independent of any interpretation of the projection operators, but for the condition (2.124). Eliminating $Q^i\psi$, we obtain two coupled equations (see also Mittleman 1962):
$$[E^+ - Q_i U Q_i]Q_i\psi^+ = Q_i U P\psi^+ \qquad (2.127\,\text{a})$$
and
$$[E^+ - PUP]P\psi^+ - PUQ_i\psi^+ = (E^+ - H_0)\phi, \qquad (2.127\,\text{b})$$
where
$$U = H + HQ^i(E^+ - Q^i H Q^i)^{-1}Q^i H.$$

But for the form of U these equations are entirely analogous to eqns (1.13) and (1.14). If E is below the continuous spectrum of $Q^i H Q^i$ (so that $Q^i\psi^+$ does not contribute asymptotically) U is Hermitian. Moreover, if E is below any eigenvalue of $Q^i H Q^i$ the adiabatic approximation should provide an accurate approximation for U as a local potential.§

In particular, eqns (2.127) provide one way of incorporating as much as possible of the distortion of the relative motion while retaining an exact description of the dynamics of internal excitation for those states that are strongly coupled. As is well known, a large contribution to the adiabatic equivalent potential is from highly excited states of h. However, these states are not strongly coupled to $P\psi^+$ and their excitation is also unfavoured by the Franck–Condon principle. The present resolution of Q incorporates these states in the place where they matter, namely, in providing an accurate mutual distortion potential. Even when $E < E_m$ the present resolution is useful for improving the two-state approximation by taking $Q_i = |m\rangle\langle m|$. For reactive collisions we can take Q_i to be the projection on the other reactive channels that are possible, and

---

§ Also under this condition the second term in U is necessarily non-positive or, in other words, the second term corresponds to an effective attraction.

Q^i then provides an account of the reaction intermediates. The partitioning approach to inelastic collisions is discussed further by Mittleman (1961, 1962; see also the next section), Coz (1965), Chen (1966b), Chen and Mittleman (1966), Hahn (1966), and Levin (1966). Recently these techniques were used to discuss the processes of electron scattering by molecules (O'Malley 1966, Chen 1966a, Chen 1967, Bardsley and Mandl 1968).

3.2.3. *Theories of direct reactions*

Qualitatively the label 'direct' is applied to those reaction events in which the duration of the actual internal change is fast compared to the periods of the internal motions. As the two molecules approach, their mutual interaction leads first to a distortion of their relative motion. At smaller relative separations an internal change occurs following which the final molecules separate under the influence of a distortive mutual interaction. Qualitatively we consider the reaction as 'direct' when the internal change is rapid and involves little internal rearrangement.

Two limiting cases can be considered and are outlined in Fig. 3.3, for the reaction
$$A+BC \to AB+C.$$
Classically (cf. Fig. 3.3), when the relative energy is high and the effect of the distortion is small, we expect that the stripping limit corresponds to a long-range type of mutual interaction, which affects also the high impact parameter components, so that the motion of C is essentially unperturbed by the reaction. (This limit is sometimes referred to as the 'spectator stripping' model; see Minturn, Datz, and Becker (1966) and Polanyi (1967).) In the other limit most of the reactive collisions are characterized by low impact parameters ('head-on' collisions) and the molecules 'rebound' from the interaction region.

Writing the reaction cross-section as (cf. eqn (2.2.36) and the extensive discussion in Chapter 3.4)
$$\sigma = 2\pi \int bP(b)\,\mathrm{d}b, \qquad (2.128)$$
where $2\pi b\,\mathrm{d}b$ is the fraction of all collisions with an impact parameter b to $b+\mathrm{d}b$, we see that, by virtue of the kinematical factor $2\pi b\,\mathrm{d}b$, reactions which are predominantly of the stripping type are expected to have larger cross-sections than reactions which are predominantly rebound.

Evidence for the experimental validity of these ideas is often obtained from the angular distribution (and energy disposal) of the products (see, for example, Herschbach 1965a,b). These classical ideas are also confirmed by trajectory analysis on simple potential energy surfaces (see section 3.4.3 and Fig. 3.7). The actual internal change is seen to occur within one vibrational period and with little internal rearrangement.

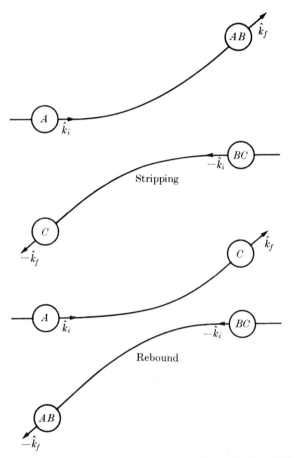

FIG. 3.3. Schematic representation of the stripping and rebound limits in the centre of mass system.

The following discussion is based on the author's thesis (unpublished) and follows well-known models in the theory of nuclear collisions (Tobocman 1961, Banerjee 1960, Butler 1957). Other discussions of direct reactions have been given by Herschbach (1965a,b), by Minturn et al. (1966), by Henglein et al. (1965), by Suplinskas and Ross (1967), and by Karplus and Tang (1967). We shall discuss more background to the concepts discussed here in Chapters 3.4 and 3.5.

From a theoretical point of view we shall restrict ourselves to a particular approach via the partitioning technique, as is outlined below. We begin with a discussion of the theory of scattering by two potentials (following Greider 1959, Bassel and Gerjuoy 1960, and Mittleman 1961, 1962).

Let H be the total Hamiltonian and $H_{0,i}$ be the Hamiltonian for the non-interacting molecules in the initial channel i. Then

$$H = H_{0,i} + V_i. \tag{2.129}$$

We put $\qquad V_i = U_i + V_i',$

where U_i is a central potential $U_i = U_i(R_i)$ or, in general, any potential that does not lead to reactive collisions. If $|i_1^+\rangle$ is the solution of the L.S. equation for the Hamiltonian $H_{0,i} + U_i$,

$$(E^+ - H_{0,i} - U_i)|i_1^+\rangle = (E^+ - H_{0,i})|i\rangle, \tag{2.130}$$

or $\qquad |i_1^+\rangle = |i\rangle + (E^+ - H_{0,i})^{-1} U_i |i_1^+\rangle,$

then our selection for U_i must be such that asymptotically $|i_1^+\rangle$ represents§ non-reactive collisions only.

If U_i is a central potential then $|i_1^+\rangle$ represents asymptotically the elastic scattering by U_i. Other selections for U_i may allow inelastic, but non-reactive, collisions as well. Any selection of U_i must conform to the required asymptotic behaviour of $|i_1^+\rangle$. An improper definition of U_i invalidates the subsequent discussion.

In general, therefore, we can consider U_i as an equivalent potential that leads to a specified asymptotic form for $|i_1^+\rangle$. One way to define such an equivalent potential is to require that

$$|i_1^+\rangle = P|i^+\rangle,$$

where $|i^+\rangle$ is the solution of the L.S. equation for the Hamiltonian H, and P is the projection operator on the initial internal state of the reactants ($P = |n\rangle\langle n|$). In this case U_i is the equivalent potential for elastic scattering, as previously introduced (eqn (2.5) and the subsequent discussion). It thus follows that

$$\langle \mathbf{R}_i, \mathbf{r}_i | i_1^+ \rangle = \langle \mathbf{r}_i | n \rangle \langle n, \mathbf{R}_i | i^+ \rangle = \phi_n(\mathbf{r}_i) F_{\mathbf{k},n}^n(\mathbf{R}_i), \tag{2.131}$$

where $\mathbf{r}_i$ is the set of internal coordinates for the initial internal state and $F_{\mathbf{k},n}^n(\mathbf{R}_i)$ is the solution of eqn (2.6) (which is equivalent, of course, to eqn (2.130) for this particular selection of U_i). A further extension of this idea (Mittleman 1962) is to consider U_i as the equivalent potential for both elastic and inelastic (but non-reactive) collisions by taking $P = C_i$

§ In other words, if R_f is the relative separation of the products,
$$\lim_{R \to \infty} \langle \mathbf{R}_f | i_1^+ \rangle \to O(1/R_f^2).$$

where C_i is the projection on all (or some) of the internal bound states in the initial channel.

The solution of the L.S. equation for the Hamiltonian H,

$$(E^+ - H)|i^+\rangle = (E^+ - H_{0,i})|i\rangle, \qquad (2.132)$$

can be written, using eqn (2.130), as

$$(E^+ - H)|i^+\rangle = (E^+ - H_{0,i} - U_i)|i_1^+\rangle, \qquad (2.133)$$

or
$$|i^+\rangle = |i_1^+\rangle + G(V_i - U_i)|i_1^+\rangle.$$

Only the second term leads asymptotically to reactive scattering and, using eqn (2.6.79), we obtain the transition amplitude to the state $|f\rangle$ in some final reactive channel as

$$\langle f^- |(V_i - U_i)|i_1^+\rangle, \quad f \neq i. \qquad (2.134\,\mathrm{a})$$

If we select U_i such that $|i_1^+\rangle = P|i^+\rangle$ then, since $PU_i = U_i$ (recall eqn (1.17) and that $P^2 = P$), we can write the transition amplitude as

$$\langle f^- |(V_i - U_i)P|i^+\rangle = \langle f^- |[V_i, P]|i^+\rangle, \quad f \neq i, \qquad (2.135)$$

where we have used eqn (2.119).

In a similar fashion we can define U_f as a distortion potential that cannot lead to reactive transitions out of the channel f. Defining $|f_1^+\rangle$ as the scattering solution by U_f we can put

$$|f^-\rangle = |f_1^-\rangle + (E^- - H)^{-1}[V_f - U_f(E^-)]|f_1^-\rangle, \qquad (2.136)$$

where we recognize explicitly that U_f can be energy dependent. Equation (2.135) can now be written as (with $V'_i = V_i - U_i$, $V'_f = V_f - U_f$)

$$\langle f_1^- |V'_i + V'^{\ddagger}_f(E^+ - H)^{-1}V'_i|i_1^+\rangle, \qquad (2.137\,\mathrm{a})$$

where we have used the Hermitian conjugate of eqn (2.136), in eqn (2.134 a). In principle, $U_f(E)$ is L-conjugation invariant (cf. eqn (1.17)) so that $(V'_f)^{\ddagger} = V'_f$; however, approximate equivalent potentials may violate this condition.

In a manner analogous to the derivation of eqn (2.6.10) we can premultiply eqn (2.132) by $(E^+ - H_{0,f} - U_f^{\ddagger})^{-1}$ and obtain for the transition amplitude

$$\langle f_1^- |V_f - U_f^{\ddagger}|i^+\rangle \qquad (2.134\,\mathrm{b})$$

or, using eqn (2.133),

$$\langle f_1^- |V'^{\ddagger}_f + V'^{\ddagger}_f(E^+ - H)^{-1}V'_i|i_1^+\rangle. \qquad (2.137\,\mathrm{b})$$

We also note that in eqn (2.134 a) (and hence in eqn (2.137 a)) the potential U_f is, in principle, arbitrary, while U_i is restricted to be a distorting potential. Conversely, eqns (2.134 b) and (2.137 b) hold for any U_i, provided U_f is a distorting potential.

Equations (2.134) and (2.137) have a general validity (on the energy shell, $E_i = E_f$) for any two different reaction channels and distortion potentials U_i and U_f. To apply it to the theory of direct reactions we select U_i and U_f as the equivalent potentials for non-reactive transitions in the initial and the final channels. The first term in eqn (2.137) represents then a transition from the initial distorted state to the final distorted state. The participation of reaction intermediates is incorporated in the second term.

We shall consider the first term

$$\langle f_1^- | V_i' | i_1^+ \rangle \quad \text{or} \quad \langle f_1^- | V_f' | i_1^+ \rangle, \qquad E_i = E_f \tag{2.138}$$

as representing the direct reaction contribution to the transition amplitude. It is non-vanishing only if the Hamiltonian has non-vanishing matrix elements between the initial and final internal states.

The lowest approximation to eqn (2.138) is obtained by taking U_i (or U_f) as the lowest approximation to the equivalent potential for elastic scattering, namely,

$$U_i = \langle n | V_i | n \rangle, \tag{2.139}$$

where $|n\rangle$ is the initial internal state of the reactants. In this case

$$\langle \mathbf{R}_i, \mathbf{r}_i | i_1^+ \rangle = \phi_n(\mathbf{r}_i) L_\mathbf{k}^n(\mathbf{R}_i), \tag{2.140}$$

where $L_\mathbf{k}^n(\mathbf{R}_i)$ is the solution of eqn (2.6) with $\mathscr{V}_{n,n}(\mathbf{R}_i', \mathbf{R}_i)$ replaced by $V_{n,n}(\mathbf{R}_i)$. Making a similar approximation with respect to U_f, eqn (2.138) reduces to the DWB result for rearrangement collisions, with the perturbing potential $V_i - \langle n | V_i | n \rangle = V_i'$ (or $V_f - \langle m | V_f | m \rangle = V_f'$).

For direct reactions where the actual internal rearrangement is rapid compared to the periods of internal motions we expect that the DWB approximation will provide accurate results for angular distributions and similar kinematical features (see also section 3.4.1, eqns (4.61)–(4.78)). It will, of course, be misleading in the predicted absolute values for the cross-sections. A variational derivation of the DWB approximation is given in section 3.3.2 (in particular, see eqns (3.84)–(3.87) and Rosenberg 1964a).

The failure of the DWB approximation to conserve probability is particularly clear in this formulation. Since inelastic and reactive transitions can take place the equivalent potential $\mathscr{V}_{n,n}$ is not Hermitian and the norm of $|i_1^+\rangle = P|i^+\rangle$ is not conserved in time (see eqns (2.120)–(2.122)). In contrast $V_{n,n}(\mathbf{R}_i)$ is a real potential and the function $L_\mathbf{k}^n(\mathbf{R}_i)$ is a fair approximation to $F_{\mathbf{k},n}^n(\mathbf{R}_i)$ only for large values of R_i. (A better approximation is to take for U_i an adiabatic equivalent potential.)

In the absence of exact results for equivalent potentials many studies in the past have resorted to various empirical means of correcting the DWB approximation. In general, most of these approximations can be schematically considered together as follows. Let $\{|j\rangle\}$ denote a complete set of states for the problem. Then we can write the transition amplitude (2.138) as

$$\sum_{jj'} \langle f_1^- | j' \rangle \langle j' | V_i' | j \rangle \langle j | i_1^+ \rangle. \tag{2.141a}$$

For example, if U_i is the equivalent potential for elastic scattering in the initial channel and similarly for U_f (so that eqn (2.131) holds) we can write eqn (2.141a) as

$$(2\pi)^{-6} \iint d\mathbf{k} d\mathbf{k}' \langle f_1^- | m, \mathbf{k}' \rangle \langle m, \mathbf{k}' | V_i' | n, \mathbf{k} \rangle \langle n, \mathbf{k} | i_1^+ \rangle, \tag{2.141b}$$

where $|n\rangle$ and $|m\rangle$ are the initial and final internal states. One can now try to approximate the form factors $\langle n, \mathbf{k} | i_1^+ \rangle$,

$$\langle n, \mathbf{k} | i_1^+ \rangle = \int d\mathbf{R}_i \langle \mathbf{k} | \mathbf{R}_i \rangle \langle \mathbf{R}_i, n | i_1^+ \rangle = \int \exp(-i\mathbf{k} \cdot \mathbf{R}_i) F_{\mathbf{k},n}^n(\mathbf{R}_i) d\mathbf{R}_i.$$

and $\langle f_1^- | m, \mathbf{k}' \rangle$. The 'Born'-type amplitude in the integral can often be evaluated in a closed analytic form as we show below. The required form factors can, in the high-energy limit, be obtained from the impulse approximation (see Rosenberg 1963, 1964b and Chew 1950).

We consider the collision $A + BC \rightarrow AB + C$ and simplify the problem by taking a 'dumbbell' model for the total Hamiltonian

$$H = K + V_{AB} + V_{BC} + V_{AC},$$

where K is the kinetic energy operator and the interaction is represented as a sum of two-body terms. The initial and final Hamiltonians are

$$H_{0,i} = K + V_{BC} \quad \text{and} \quad H_{0,f} = K + V_{AB}$$

and represent the relative motion of A (or C) and the bound species BC (or AB).

If A is infinitely heavy then, as is clear from Fig. 2.4, we can write $V_{AC} = V_{AC}(\mathbf{R}_f)$ and consider V_{AC} as part of the distorting potential in the final state. In the same fashion, if C is infinitely heavy $V_{AC} = V_{AC}(\mathbf{R}_i)$ and we can include V_{AC} in the distorting potential in the initial state. We are not aware of a rigorous proof that extends these conclusions to an arbitrary situation (although some proofs, which we find unsatisfactory, have been offered). The evaluation of the Born amplitude cannot be done analytically for V_{AC} but can always be done numerically. We shall thus centre attention on the Born amplitude

$$f_{m,n}(\mathbf{k}_f, \mathbf{k}_i)_B = -(\mu_f/2\pi\hbar^2)\langle \mathbf{k}_f, m | V_{BC} | \mathbf{k}_i, n \rangle. \tag{2.142a}$$

We put
$$\langle \mathbf{R}_i, \mathbf{R}_{BC} | \mathbf{k}_i, n \rangle = \varphi_n(\mathbf{R}_{BC}) \exp(i\mathbf{k}_i \cdot \mathbf{R}_i),$$

where φ_n is the initial internal state,

$$[E_n + (\hbar^2/2\mu_{BC})\nabla^2_{\mathbf{R}_{BC}} - V_{BC}(\mathbf{R}_{BC})]\varphi_n(\mathbf{R}_{BC}) = 0, \qquad (2.143)$$

and similarly for the final state, so that

$$f_{m,n}(\mathbf{k}_f, \mathbf{k}_i)_B = -(\mu_f/2\pi\hbar^2) \iint d\mathbf{R}_f d\mathbf{R}_{AB} \times$$
$$\times \varphi_m^*(\mathbf{R}_{AB})V_{BC}(\mathbf{R}_{BC})\varphi_n(\mathbf{R}_{BC})\exp[i(\mathbf{k}_i \cdot \mathbf{R}_i - \mathbf{k}_f \cdot \mathbf{R}_f)]. \quad (2.142\,\text{b})$$

In contrast to the inelastic (but non-reactive) case, eqns (2.3.60)–(2.3.66), here the internal wavefunctions refer to different internal coordinates.

Referring to Fig. 3.3 we can introduce the changes in translational momentum of atoms A and C as follows. Initially the translational momentum of A was $\hbar \mathbf{k}_i$. In the final state A has the fraction $M_A/(M_A+M_B)$ of the translational momentum of AB. Thus, in the stripping picture the change in the translational momentum of A is $\hbar \mathbf{q}_A$,

$$\mathbf{q}_A = \frac{M_A}{M_A + M_B}\mathbf{k}_f - \mathbf{k}_i, \qquad (2.144)$$

and similarly $\quad \mathbf{q}_C = -\mathbf{k}_f + \dfrac{M_C}{M_B + M_C}\mathbf{k}_i.$

Introducing the initial and final relative coordinates

$$\mathbf{R}_i = \mathbf{R}_{BA} + \frac{M_C}{M_B + M_C}\mathbf{R}_{CB}$$

and $\quad \mathbf{R}_f = \mathbf{R}_{CB} + \dfrac{M_A}{M_A + M_B}\mathbf{R}_{BA},$

where $\mathbf{R}_{CB} = \mathbf{R}_C - \mathbf{R}_B$ and $\mathbf{R}_{BA} = \mathbf{R}_B - \mathbf{R}_A$, we obtain the identity

$$\mathbf{k}_i \cdot \mathbf{R}_i - \mathbf{k}_f \cdot \mathbf{R}_f = -\mathbf{q}_C \cdot \mathbf{R}_{BC} + \mathbf{q}_A \cdot \mathbf{R}_{AB}. \qquad (2.145)$$

The integral in eqn (2.142 b) can now be evaluated, since

$$d\mathbf{R}_f\, d\mathbf{R}_{AB} = d\mathbf{R}_{BC}\, d\mathbf{R}_{AB},$$

to give

$$f_{m,n}(\mathbf{k}_f, \mathbf{k}_i)_B = -(\mu_f/2\pi\hbar^2) \int d\mathbf{R}_{BC} V_{BC}(\mathbf{R}_{BC})\varphi_n(\mathbf{R}_{BC})\exp(-i\mathbf{q}_C \cdot \mathbf{R}_{BC}) \times$$
$$\times \int d\mathbf{R}_{AB} \exp(i\mathbf{q}_A \cdot \mathbf{R}_{AB})\varphi_m^*(\mathbf{R}_{AB}). \quad (2.142\,\text{c})$$

The second integral is the form factor for the final state. The first integral can be simplified using the Schrödinger equation for the initial internal state, eqn (2.143), so that

$$\int d\mathbf{R}_{BC} V_{BC}(\mathbf{R}_{BC})\varphi_n(\mathbf{R}_{BC})\exp(-i\mathbf{q}_C \cdot \mathbf{R}_{BC})$$
$$= \int d\mathbf{R}_{BC}[E_n + (\hbar^2/2\mu_{BC})\nabla^2_{\mathbf{R}_{BC}}]\varphi_n(\mathbf{R}_{BC})\exp(-i\mathbf{q}_C \cdot \mathbf{R}_{BC})$$
$$= [E_n + \hbar^2 q_C^2/2\mu_{BC}] \int d\mathbf{R}_{BC}\, \varphi_n(\mathbf{R}_{BC})\exp(-i\mathbf{q}_C \cdot \mathbf{R}_{BC}).$$

Introducing the notation $g_n(\mathbf{q})$ for the form factor of $\varphi_n(\mathbf{R})$,

$$g_n(\mathbf{q}) = \int d\mathbf{R}\exp(-i\mathbf{q}\cdot\mathbf{R})\varphi_n(\mathbf{R}), \tag{2.146}$$

we can write

$$f_{m,n}(\mathbf{k}_f, \mathbf{k}_i)_B = -(\mu_f/2\pi\hbar^2)[E_n + \hbar^2 q_C^2/2\mu_{BC}]g_n(\mathbf{q}_C)g_m^*(\mathbf{q}_A). \tag{2.142 d}$$

The physical interpretation of this Born amplitude depends on the significance of the form factors; $g_n(\mathbf{q})$ is the amplitude for the distribution of internal momentum $\mathbf{q}$ in the initial bound state

$$g_n(\mathbf{q}) = \int \langle \mathbf{q}|\mathbf{R}\rangle\, d\mathbf{R}\, \langle \mathbf{R}|n\rangle = \langle \mathbf{q}|n\rangle.$$

The form factor contributes significantly only for those values of $\mathbf{q}$ that are characteristic of the initial state. If we write $|E_n| = \hbar^2 q_n^2/2\mu_{BC}$ we expect the main contribution to the form factor from the classical region $q_n = q_C$ or

$$q_n = \left| -\mathbf{k}_f + \frac{M_C}{M_B + M_C}\mathbf{k}_i \right|, \tag{2.147}$$

which we can rewrite, on squaring both sides and dividing throughout by $2\mu_f = 2(M_A + M_B)M_C/(M_A + M_B + M_C)$, as

$$\sin^2\beta\, E_n = e_f + \cos^2\beta\, e_i - \hbar^2 \mathbf{k}_f \cdot \mathbf{k}_i \sin^2\beta/M_B, \tag{2.148}$$

where
$$\cos^2\beta = M_A M_C/(M_A + M_B)(M_B + M_C)$$

and e_f and e_i are the initial and final translational energies, $(e_f = \hbar^2 k_f^2/2\mu_f)$.

Equation (2.147) and the analogous relation for $\mathbf{q}_A$ serves as a guide to the distribution of the energy between internal and relative energies of the products, and its dependence on the scattering angle. We should remember, however, that eqn (2.147) is based on the classical correspondence $|E_n| = \hbar^2 q_n^2/2\mu_{BC}$ and that the form factor will also receive contributions from other values of q_n. It is easy to see by evaluating the form factor for a product of a rotor and a harmonic oscillator wavefunction that the significant contributions are, however, from the classical value of q_n. We can regard these conclusions as a semiquantitative statement of the Franck–Condon principle for direct reactions, namely, most of the direct processes proceed with little or no change in the total momentum of each atom. Physically this conclusion follows from the 'direct' nature of the reaction, which is instantaneous in the Born approximation.

The Born scattering amplitude for direct reactions (2.142 d) can be compared to the Born scattering amplitude (2.2.95), which is typical of compound processes (see eqns (2.2.42) and (2.70)). In the direct reaction case the final form factor depends on both $\mathbf{k}_i$ and $\mathbf{k}_f$ (strictly it depends on k_i, k_f, and $\hat{\mathbf{k}}_i \cdot \hat{\mathbf{k}}_f$), while for compound processes the final form factor depends on $\mathbf{k}_f$ only. In this sense one can say that in direct reactions the system 'remembers' the initial direction of the relative motion.

The evaluation of the form factors can be carried out by performing a partial wave analysis of the initial and final internal wavefunctions as discussed in section 2.3.2. Putting

$$\varphi_n(\mathbf{R}_{BC}) = \varphi_n(R_{BC}) Y_j^{m_j}(\hat{\mathbf{R}}_{BC})$$

we obtain

$$g_n(\mathbf{q}) = 4\pi \mathrm{i}^{-j} Y_j^{m_j}(\hat{\mathbf{q}}) g_{n,j}(q),$$

where

$$g_{n,j}(q) = \int_0^\infty R_{BC}^2 \varphi_n(R_{BC}) j_j(q R_{BC}) \, \mathrm{d}R_{BC}.$$

It should be remembered, however, that our discussion from eqn (2.142 a) onward referred to the Born amplitude for V_{BC} and not to the exact result for the scattering amplitude, eqn (2.141b). The Born result should now be averaged over the distribution of $\mathbf{k}_i$ and $\mathbf{k}_f$ as required by eqn (2.141b). Moreover, in order to obtain closed analytic forms we have assumed that $V_f' = V_{BC}$, which is only the lowest possible approximation to V_f'.

3.3. Operator partitioning theory

3.3.0. An abstract operator formalism of the partitioning technique can be obtained by applying the partitioning to the operator equations of section 2.5.0. One can justify this procedure provided a realization of the projection operators P and Q is available in some representation, so that the partitioning can be considered to apply to the matrix representation of the operator in the particular representation.

We define the (Brillouin–Wigner) wave operator W by
$$W(E^+) = I + (E^+ - PHP)^{-1}P(H - E). \tag{3.1}$$
In a perturbation expansion of W there can be no intermediate state which is annihilated by P. Thus, for example, eqn (1.18) can be written
$$P\psi_1^+ = W(E^+)\phi, \tag{3.2}$$
while the results of partitioning the L.S. equation can be written
$$\psi^+ = P\psi_1^+ + W(E^+)Q\psi^+. \tag{3.3}$$

Alternatively, we can introduce the wave operator§ F by
$$F = I + (E^+ - QHQ)^{-1}Q(H - E), \tag{3.4}$$
so that
$$\psi^+ = FP\psi^+. \tag{3.5}$$
If P commutes with H_0 we can write the equivalent potential $P\mathscr{V}P$ as
$$\mathscr{V} = VF. \tag{3.6}$$
In the same way we can put
$$\Gamma = VW \tag{3.7}$$
and write eqn (1.20) as
$$Q\psi^+ = (E^+ - Q\mathscr{H}Q)^{-1}QHP\psi_1^+ = (E^+ - QH_0Q - Q\Gamma Q)^{-1}Q\Gamma\phi, \tag{3.8}$$
where we have used eqn (3.2). To identify the inverse operator in eqn (3.8) we write
$$(E^+ - H)G(E^+) = I \tag{3.9}$$
and preoperate with P and Q respectively to get
$$(E^+ - PHP)PG = PHQG + P, \tag{3.10}$$
$$(E^+ - QHQ)QG = QHPG + Q. \tag{3.11}$$
Operating on eqn (3.10) from the right by Q we have
$$PGQ = PWQGQ, \tag{3.12}$$

§ This operator is a particular realization of the operator F used extensively by Watson (1957) and others.

and writing $QG = QGQ+QGP$, we can rewrite eqn (3.11) as

$$QGQ = (E^+ - QHQ - QHPWQ)^{-1} = (E^+ - QH_0Q - Q\Gamma Q)^{-1}, \quad (3.13)$$

and by a similar manipulation

$$PGP = (E^+ - PHP)^{-1} + PWQGP = (E^+ - PH_0P - P\mathscr{V}P)^{-1}. \tag{3.14}$$

The L-conjugate of eqn (3.12) is

$$QGP = QGQW^{\ddagger}P \tag{3.15}$$

and, adding eqns (3.12)–(3.15) and using eqn (3.15) in eqn (3.14),

$$G = (E^+ - PHP)^{-1} + WQGQW^{\ddagger}. \tag{3.16}$$

A similar argument leads to§

$$G = (E^+ - QHQ)^{-1} + FPGPF^{\ddagger}. \tag{3.17}$$

The transition operator $T(E^+)$

$$T(E^+) = V + VG(E^+)V, \tag{3.18}$$

can be expressed in terms of either Γ or $\mathscr{V}$ using the resolutions (3.16) or (3.17). In terms of eqns (3.12), (3.13), and (3.15),

$$T(E^+)P = \Gamma P + \Gamma QGQ\Gamma P, \tag{3.19}$$

where, when P and H_0 commute,

$$Q\Gamma P = QVWP = QW^{\ddagger}VP. \tag{3.20}$$

The transition amplitude determined by eqn (3.19) corresponds to the scattering amplitude determined by the asymptotic behaviour of the wavefunction (eqns (2.70) or (2.115)) when the same definition of P is adopted. Equation (3.19) also provides a resolution of the Møller wave operator $(T(E^+) = V\Omega(E^+))$ as

$$\Omega(E^+)P = WP + WQGQ\Gamma P. \tag{3.21}$$

In the same fashion we can write

$$T(E^+)P = \mathscr{V}P + \mathscr{V}PGP\mathscr{V}P = \mathscr{V}P + \mathscr{V}P(E^+ - PH_0P)^{-1}P\mathscr{V}P + \dots. \tag{3.22}$$

Both eqns (3.19) and (3.22) are exact; however, they centre attention on complementary aspects of the collision. Equation (3.22) is explicit with

---

§ When P does not commute with H_0 we modify our discussion by redefining a zero order Hamiltonian which satisfies $PH_0Q = 0$. In the theory of rearrangement collisions it is more convenient to put

$$\Gamma = H + (H-E)P(E^+ - PHP)^{-1}P(H-E)$$

and not to specify H_0 in the definition of Γ, although quite often the projection Q will be introduced in terms of some convenient zero order Hamiltonian.

respect to the states on which P projects.§ The operator $\mathscr{V}$ can be considered as the transition operator for collisions that do not involve states in the P subspace as intermediates. From eqns (3.6) and (3.4),

$$\begin{aligned}\mathscr{V}P &= VP+VQ(E^+-QHQ)^{-1}QVP\\ &= VP+VQ(E^+-H_0)^{-1}\mathscr{V}P\\ &= VP+VQ(E^+-H_0)^{-1}VP+VQ(E^+-H_0)^{-1}VQ(E^+-H_0)^{-1}VP+\ldots.\end{aligned} \quad (3.23)$$

The first transition induced by V is into the Q subspace, and from then on, till the final interaction, the mutual perturbation is QVQ (recall that $[H_0, Q] = 0$).

The iteration of T in terms of $\mathscr{V}$ can also be considered as the iteration of the integral equation

$$TP = \mathscr{V}P+\mathscr{V}P(E^+-H_0)^{-1}TP, \qquad (3.24)$$

obtained by summation. We can formally solve this equation to get an analogous relation to (3.19),

$$TP = \mathscr{V}P+VPGPVP. \qquad (3.25)$$

In the adiabatic approximation eqn (3.22) reduces to the ordinary Born iteration for the local equivalent potential.

On the other hand, Γ can be considered as the transition operator for those collisions that proceed via the P subspace only. From eqns (3.2) and (3.7) we can write

$$VP\psi_1^+ = VW\phi = \Gamma\phi \qquad (3.26)$$

and $\quad \psi_1^+ = \phi+(E^+-H_0)^{-1}PVP\psi_1^+ = \phi+(E^+-H_0)^{-1}P\Gamma P\phi. \quad (3.27)$

When $[P, H_0] = 0$ both Γ and $\mathscr{V}$ can be written as

$$A = V+VBG_0 A, \qquad (3.28\,\mathrm{a})$$

where $A = \Gamma$ or $\mathscr{V}$ and $B = P$ or Q respectively, and from eqns (3.6) and (3.7) we can write

$$W = I+PG_0\Gamma$$

or $\quad\qquad F = I+QG_0\mathscr{V}. \qquad (3.28\,\mathrm{b})$

A function of the Hamiltonian H can be defined by eqn (1.2.23 a):

$$F(H) = (2\pi\mathrm{i})^{-1}\int_{\mathrm{sp}} F(\lambda)G(\lambda)\,\mathrm{d}\lambda. \qquad (3.29)$$

If Q is a projection operator, $Q^2 = Q$, we obtain from eqn (3.29)

$$\mathrm{Tr}\{QF(H)\} = (2\pi\mathrm{i})^{-1}\int_{\mathrm{sp}} F(\lambda)\mathrm{Tr}\{QG(\lambda)Q\}\,\mathrm{d}\lambda, \qquad (3.30)$$

provided $\mathrm{Tr}\{QF(H)\}$ exists. In writing the right-hand side we have used

§ We shall refer to the subspace onto which P projects as 'the P subspace'.

the cyclic property of the trace

$$\operatorname{Tr}(ABC) = \operatorname{Tr}(CAB). \tag{3.31}$$

In particular, if $\operatorname{Tr} F(H)$ exists, eqn (3.30) holds with Q being the identity operator. An important application of eqn (3.30) is the evaluation of partition functions for interacting systems (see, for example, Watson 1956). The partition function $Z(\beta)$ is defined by

$$Z(\beta) = \operatorname{Tr}\{\exp(-\beta H)\}. \tag{3.32}$$

Given the diagonal elements of $G(\lambda)$ in some representation we can evaluate the partition function,

$$Z(\beta) = (2\pi i)^{-1} \int_{\mathrm{sp}} \exp(-\beta\lambda)\operatorname{Tr} G(\lambda)\,\mathrm{d}\lambda. \tag{3.33}$$

To derive eqn (3.33) explicitly rather than through eqn (3.29) we recall that the trace of an operator is independent of the representation used to evaluate the trace (consider eqn (3.31) with $A = C^\dagger$, A unitary). In the representation that diagonalizes H,

$$Z(\beta) = (2\pi i)^{-1} \int_{\mathrm{sp}} \exp(-\beta\lambda) \sum_n \frac{g_n}{\lambda - E_n}\,\mathrm{d}\lambda = \sum_n \exp(-\beta E_n)g_n,$$

using the theorem of residues. If H has a continuous spectrum, we note that the contour of integration encircles the spectrum in a counter-clockwise manner, so that the contribution of the integrand is determined by the difference in the values of $G(E^-)$ and $G(E^+)$ (see the derivation of eqn (3.34) below).

In the same fashion we can evaluate $\operatorname{Tr}\{F(H)\} - \operatorname{Tr}\{F(H_0)\}$, and, using the discussion in section 2.5.4, obtain

$$\operatorname{Tr}\{F(H)\} - \operatorname{Tr}\{F(H_0)\} = (2\pi i)^{-1} \int_{\mathrm{sp}} F(\lambda)\,\mathrm{d}\ln D(\lambda)$$

$$= (2\pi i)^{-1} \int_{-\infty}^{\infty} \mathrm{d}E\, F(E) \frac{\mathrm{d}}{\mathrm{d}E} \ln \frac{D(E^-)}{D(E^+)}, \tag{3.34}$$

where $D(E)$ is the Fredholm determinant. To obtain the second line we resolved the counter-clockwise contour of integration over λ to three parts: (a) $\lambda = E + i\epsilon$, $\epsilon > 0$, E goes from ∞ to $-M$; (b) $\lambda = M + iZ$, Z goes from ϵ to $-\epsilon$; and (c) $\lambda = E - i\epsilon$, E goes from $-M$ to ∞. Finally, $M \to \infty$, $\epsilon \to +0$. (This assumes that the spectrum of H is bounded from below.) Using eqns (2.5.95) and (2.5.94) we can write the above as

$$(2\pi i)^{-1} \int_{-\infty}^{\infty} \mathrm{d}E\, F(E)\Delta(E), \tag{3.35}$$

where $\Delta(E)$ is the change in the density of states due to the potential $H-H_0$.

In particular, eqn (3.35) provides a prescription for the difference§ between the partition functions of an interacting and a non-interacting pair of molecules.

We can perform a partial wave resolution of $\Delta(E)$. In the case of a resonance in the lth partial wave (eqns (2.78)) we have that

$$\Delta_l(E) = \Delta_l^0(E)+\Delta_l^r(E), \qquad (3.36)$$

where the first contribution is from δ_l^0, the phase shift in the absence of internal excitation, and the second contribution is due to the resonance phase shift δ_l^r. Using eqns (2.77) and (2.5.97) and assuming that Γ_i^l and Δ_i^l are constants (energy-independent)

$$\Delta_l^r(E) = \pi^{-1}\frac{\mathrm{d}\delta_l^r}{\mathrm{d}E} = \pi^{-1}\frac{\Gamma_i^l}{(E-E_i-\Delta_i^l)^2+(\Gamma_i^l)^2}, \qquad (3.37)$$

in agreement with eqn (2.86).

3.3.1. *Variational principles*

When exact solutions of the L.S. equations are not available, as is almost always the case, the computed results (phase shifts, transition amplitudes, etc.) are necessarily approximate. A variational principle is a prescription for computations of results that are in error only in the second order of the error of the wavefunction or, in other words, results with vanishing first-order corrections. Thus when the available wavefunction is a good approximation, the computed result hardly differs from the exact result.

Under favourable circumstances we can also demand that the error in our result be of a known sign. In this case we have not only a variational principle but also an extremum (maximum or minimum) principle. It is clearly also desirable to be able to improve the result by the introduction of suitable parameters in the wavefunction and to vary these parameters so as to lead to an optimal result. If this can be done we have a variational bound principle.

It is important, however, to bear in mind that for very approximate trial wavefunctions the results of variational calculations may turn out to be numerically inferior to results computed by other means.

For the case of scattering by a local central potential $V(R)$ there exist several formulations of variational and extremum principles for

§ This difference is the 'internal' partition function of Smith (1963, eqn 42) when we recall that by Smith's (1960) definition and eqn (2.5.97) $Q(E,j) = 2\pi\hbar\Delta_j(E)$. For other applications see Levine (1966a).

the phase shift and scattering length. (See Demkov 1963 and Moiseiwitch 1966 and references therein.) From a practical point of view the availability of high-speed computers implies that it is possible to perform a numerical integration of the one-dimensional equation (2.2.84) that determines the phase shift. (For practical details see, for example, Bernstein 1960.) Moreover, it is rapidly becoming possible to provide accurate numerical solutions to a small number (say less than 10) of coupled one-dimensional differential equations. We shall therefore assume that in practice such problems are soluble to a re sonable numerical accuracy, and consider mainly multichannel collisions, and variational principles for the scattering and transition amplitudes.

We begin our discussion with the monotonicity theorem (Spruch 1963), for a single channel, eqn (2.5.89). We recall that for two operators, A, B, we define $A \geqslant B$ when $\langle A \rangle \geqslant \langle B \rangle$ for all diagonal matrix elements in the common domain of definition of A and B.

If a potential V_1 is everywhere more attractive than V_2, then $V_1(R) \leqslant V_2(R)$ and we can put

$$V(R; z) = V_1(R) + z[V_2(R) - V_1(R)], \tag{3.38}$$

where $V(R; 1) = V_2$ and $\dfrac{\partial}{\partial z} V(R; z) \geqslant 0$.

From eqn (2.5.89) we have

$$\delta(z=1) - \delta(z=0) = \int_0^1 \frac{d\delta(z)}{dz}\, dz \leqslant 0. \tag{3.39}$$

Thus the more attractive the potential the larger the phase shift. In multichannel collisions the analogous statement refers to the diagonal elements of the **K** matrix, by replacing the outgoing wave operators in eqn (2.5.86) by their standing wave analogues. (See Hahn, O'Malley, and Spruch 1964, and Percival 1960.) We consider, in particular, the role of the closed channels in a multichannel collision, and assume for simplicity that only one channel is open. The results can, however, be extended to the case of several open channels. The result we wish to prove is

$$\delta \geqslant \delta_{\text{ad}} \geqslant \delta_{\text{st}} \tag{3.40}$$

where δ is the exact phase shift, δ_{ad} the phase in the adiabatic approximation and δ_{st} the phase in the static approximation. In practice one includes only a finite number of closed channels and a subsidiary conclusion is

$$\delta(Q_1) \geqslant \delta(Q_2), \tag{3.41}$$

where the subspace onto which Q_2 projects is contained in (is a subspace of) the subspace onto which Q_1 projects. (These results have a restricted validity when the computed phase shift is only obtained modulo π, so that the inequalities above can always be satisfied by adding multiples of π.)

The exact phase shift δ is determined by the equivalent potential (eqn (2.5)),

$$\mathscr{V}_{n,n} = V_{n,n}(R) + \langle n|HQ(E-QHQ)^{-1}QH|n\rangle, \qquad (3.42)$$

where $|n\rangle$ is the lowest internal state $Q = I - |n\rangle\langle n|$, and if only elastic collisions can take place the total energy E is below the threshold of the continuous spectrum of QHQ. To prove eqn (3.40) we have, by the monotonicity theorem, to show that the static potential is an upper bound to the adiabatic potential, which is, in turn, an upper bound to the exact equivalent potential.

We observe that if E is below the lowest eigenvalue of QHQ the second term in eqn (3.42) is necessarily an attractive§ potential. If we take $|n\rangle$ as the lowest internal state of h then $V_{n,n}(R)$ is the potential in the static approximation and $\delta > \delta_{\mathrm{st}}$. In the adiabatic approximation we can use the Rayleigh–Ritz variational bound principle (using δ to denote a small variation),

$$\delta[F_{\mathrm{adt}}^\dagger H_{\mathrm{int}} F_{\mathrm{adt}} - E_{\mathrm{int}}(R) F_{\mathrm{adt}}^\dagger F_{\mathrm{adt}}] = 0, \qquad (3.43)$$

where F_{adt} is a trial value for the adiabatic wave operator (eqn (2.14)), H_{int} is the internal Hamiltonian, $H_{\mathrm{int}} = h + V$, and $E_{\mathrm{int}}(R)$ is the adiabatic equivalent potential (the 'electronic energy' in atomic collisions). Thus, if we take

$$F_{\mathrm{ad}} = |n\rangle\langle n| + QF_{\mathrm{ad}},$$

it follows that
$$E_{\mathrm{int}}(R) \leqslant V_{n,n}(R) \qquad (3.44)$$

and $\delta_{\mathrm{ad}} \geqslant \delta_{\mathrm{st}}$. Finally, we note that the equivalent of the Rayleigh–Ritz principle holds for the exact wave operator

$$F = P + (E - QHQ)^{-1}QHP, \qquad (3.45)$$

namely,
$$\delta[F_t^\dagger(H-E)F_t] = 0. \qquad (3.46)$$

To prove eqn (3.46) we note the identity

$$Q(H-E)F = QHP + Q(H-E)Q(E-QHQ)^{-1}QHP = 0. \qquad (3.47)$$

If F_t is a trial‖ wave operator,

$$F_t = P + (E - Q_t H Q_t)^{-1} Q_t H P, \quad PQ_t = 0, \qquad (3.48)$$

§ Since all the denominators are negative definite and all numerators are positive definite, a typical term being
$$\frac{\langle n|H|m,i\rangle\langle m,i|H|n\rangle}{E-E_i} \quad m \neq n,\ E < E_i.$$

‖ In general, $F_t = P + QF_t$ with F_t an arbitrary trial value.

we see that, using the substitution $F_t = F + \delta F$ and eqn (3.47),

$$F_t^\dagger(H-E)F_t = F^\dagger(H-E)F + \delta F^\dagger(H-E)\delta F. \tag{3.49}$$

Thus the left-hand side is stationary about the exact solution. In analogy to the adiabatic case eqn (3.49) can provide a bound. Since, by construction, $Q\delta F = \delta F$, the error term can be written

$$\delta F^\dagger (QHQ-E)\delta F. \tag{3.50}$$

For E below the lowest eigenvalue of QHQ the error term is necessarily positive. From eqn (3.47) and the identity $F^\dagger P = P$ it follows that with $\mathscr{H} = PHFP$, eqn (3.2),

$$\mathscr{H} - E = P(H-E)F = F^\dagger(H-E)F. \tag{3.51}$$

Equation (3.51) thus provides an upper bound for $\mathscr{H} - E$. The choice $F_t = P$ corresponds to the static approximation while the choice $F_t = F_{\text{ad}}$ implies $\delta \geqslant \delta_{\text{ad}}$.

Equation (3.41) is valid for both the adiabatic and the exact phase shift. In both cases they follow from the variational principle (3.49). As we enlarge the subspace onto which Q_t projects the error term decreases. Thus, by the monotonicity theorem, the converse is true for the phase shift. The discussion can, in fact, be extended for energies above the lowest discrete eigenvalue of QHQ (if any) up to the threshold for the continuous spectrum of QHQ, by the substitution $Q \to zQ$. $z = 0$ corresponds to the static approximation and $z = 1$ corresponds to the exact Q. The Hamiltonian H can be written as

$$\mathbf{H} = \begin{pmatrix} PHP & zPHQ \\ zQHP & z^2QHQ \end{pmatrix} \tag{3.52}$$

and from eqn (2.5.90) one can show that $(d\delta/dz) \geqslant 0$ up to the threshold. However, at energies in the vicinity of the discrete eigenvalues of QHQ, δ will increase by π (due to the resonance caused by internal excitation, cf. eqn (2.78)), while δ_{st} and δ_{ad} will fail to reproduce this behaviour (cf. Fig. 3.4).

To obtain an upper bound to the phase one should construct an equivalent potential that is more attractive than $\mathscr{V}_{n,n}$. A simple procedure is to replace all the eigenvalues of QHQ by the lowest eigenvalue E_1, so that

$$\mathscr{V}_{n,n} \geqslant V_{n,n} + \frac{\langle n|HQH|n\rangle}{E-E_1}, \tag{3.53}$$

where, with the definition of $Q = I - |n\rangle\langle n|$,

$$\langle n|HQH|n\rangle = \langle n|H^2|n\rangle - |\langle n|H|n\rangle|^2.$$

Most of the results that we discuss below are based on simple operator identities and operator inequalities. (For a review of such inequalities in bound state problems, see Lowdin (1965).) The first identity is

$$A = A + A^{\ddagger} - A^{\ddagger}A^{-1}A, \qquad (3.54)$$

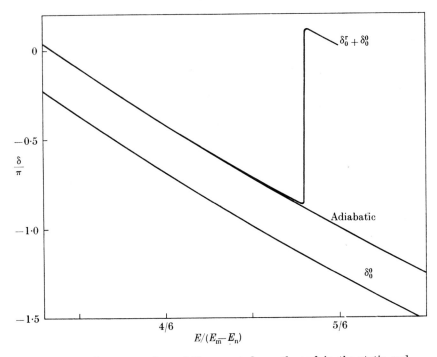

FIG. 3.4. The s-wave phase shift computed exactly and in the static and adiabatic approximations in the vicinity of E_i of Fig. 3.1. (B. R. Johnson, private communication). The slope of the curves is essentially due to the repulsive portion of the potential well. The notation in the figure corresponds to eqns (2.74)–(2.81). It is interesting to note in connection with eqn (2.81) that $\sin^2\delta_0^0$ is also varying with energy. Note that the adiabatic approximation is in close agreement with the exact result, apart from the vicinity of E_i. For more exact computations see Levine et al. (1968).

where A is L-conjugation invariant. If A_t is a trial value for A, $A_t = A + a$ (where a is the error), then the expression

$$[A] = A_t + A_t^{\ddagger} - A_t^{\ddagger} A^{-1} A_t \qquad (3.55)$$

is of second order in a,

$$A = [A] - a^{\ddagger} A^{-1} a. \qquad (3.56)$$

As an example we can consider the transition operator $T = V\Omega$. If we consider a trial value $T_t = V(\Omega + \delta\Omega)$, where Ω is exact,

$$\Omega = I + G_0^+ V\Omega, \qquad (3.57)$$

we find from eqn (3.55)
$$\delta[T] = V\delta\Omega + \Omega^\ddagger(V - VG_0^+ V)\delta\Omega + \ddagger \text{ term}$$
$$= [1 - \Omega^\ddagger(1 - VG_0^+)]V\delta\Omega + \ddagger \text{ term}$$
or $\delta[T] = 0$. In other words, the expression§
$$[T] = V\Omega + \Omega^\ddagger V + \Omega^\ddagger(V - VG_0^+ V)\Omega \tag{3.58}$$
is stationary with respect to variations of Ω and $\Omega^\ddagger$ about their correct value, and reduces to the exact transition operator for exact wave operators.

The stationary character of eqn (3.58) was proved by Lippmann and Schwinger (1950). In terms of matrix elements we have shown that
$$[T]_{f,i} = \langle f|V|i^+\rangle + \langle f^-|V|i\rangle - \langle f^-|V - VG_0^+ V|i^+\rangle \tag{3.59}$$
is stationary about the exact outgoing solutions of the L.S. equation
$$|i^+\rangle = |i\rangle + G_0^+ V|i^+\rangle. \tag{3.60}$$
A similar proof holds for the incoming solutions. In practice one usually restricts the form of the trial wavefunction to some class of functions. As an example, for a collision of an atom and a diatomic we may consider a trial (static) solution
$$\psi^+(\mathbf{R}, \mathbf{r}) = F^+(\mathbf{R})\varphi_n(\mathbf{r}) \tag{3.61}$$
with $F^+(\mathbf{R})$ an arbitrary function of $\mathbf{R}$, provided it satisfies the proper boundary conditions,
$$F^+(\mathbf{R}) \xrightarrow[R\to\infty]{} \exp(i\mathbf{k}\cdot\mathbf{R}) + f_t(\mathbf{k}', \mathbf{k})\exp(ikR)/R \tag{3.62}$$
where $f_t(\mathbf{k}', \mathbf{k})$ is some trial value for the scattering amplitude. Equation (3.59) will then be stationary about the solutions of the static approximation, eqn (1.21). Starting with the approximation $F^+(\mathbf{R}) = \exp(i\mathbf{k}\cdot\mathbf{R})$, ($f_t = 0$), we obtain, in the notation of section 2.3.0,
$$(-2\pi\hbar^2/\mu)[f(\mathbf{k}', \mathbf{k})] = V_{n,n}(\mathbf{q}) + \sum_m \iint d\mathbf{R}' \, d\mathbf{R} \times$$
$$\times \exp(-i\mathbf{k}'\cdot\mathbf{R}')V_{n,m}(\mathbf{R}')\langle\mathbf{R}'|(E^+ - E_m - K)^{-1}|\mathbf{R}\rangle V_{m,n}(\mathbf{R})\exp(i\mathbf{k}\cdot\mathbf{R}), \tag{3.63}$$
where $\mathbf{q} = \mathbf{k} - \mathbf{k}'$, and E_m is the internal energy of the state m. Equation (3.63) clearly corresponds to the second Born approximation, whereas the scattering amplitude computed directly from $T = V\Omega_t$ would be, in this case ($\Omega_t = 1$), the first Born approximation. It is easy to see this directly from eqn (3.59). Also if $\Omega_t = \Omega_{\text{2nd B}} = 1 + G_0^+ V$ is used we get for $[T]$ the fourth Born approximation.

In the same fashion that the static approximation is the optimal solution among all functions of the form (3.61) which satisfy outgoing

§ Note that $T^{-1}T = \Omega^{-1}V^{-1}V\Omega = (1 - G_0^+ V)\Omega$, on using eqn (3.57) to invert Ω.

boundary conditions of the form (3.62), we can show that the adiabatic approximation is the optimal solution among all functions of the form§

$$\psi^+(\mathbf{R}, \mathbf{r}) = \sum_m C_m(\mathbf{R})\varphi_m(\mathbf{r})F^+(\mathbf{R}) \qquad (3.64)$$

such that $C_m(\mathbf{R}) \to \delta_{n,m}$ as $R \to \infty$ and $F^+(\mathbf{R})$ satisfies eqn (3.62).

We note that eqn (3.59) is L-conjugation invariant, so that the computed transition and scattering amplitudes will satisfy the reciprocity theorem (cf. eqns (2.8.13), (2.8.33), and (2.8.34)).

Our results depend on the proper normalization (eqn (3.62)) of the trial wavefunction, and on the assumption that the states of H_0 are known exactly. We can remove the dependence on a specific normalization by writing

$$[T]_{fi} = \frac{\langle f^-|V|i\rangle\langle f|V|i^+\rangle}{\langle f^-|V - V G_0^+ V|i^+\rangle} \qquad (3.65)$$

where we note that

$$\langle f^-|V|i\rangle = \langle f^-|V|\{|i^+\rangle - G_0^+ V|i^+\rangle\}. \qquad (3.66)$$

Equation (3.65) is sometimes referred to as the Schwinger variational principle. (For applications of this principle see Joachain (1965), and references therein.)

When reactive collisions can take place, the formulation should be modified to include explicitly the different possible resolutions of the total Hamiltonian (see Lippmann 1956).

The identity (3.55) can be used for other operators as well. For example, if in eqn (3.58) we replace the outgoing (or incoming) wave operator by a standing wave solution

$$\Omega^\circ = I + P_v G_0 V \Omega^\circ \qquad (3.67)$$

we obtain a stationary expression for K (cf. eqn (2.5.47)). With minor modifications it is possible to derive stationary expressions for the wave operator itself and also for the Green's operator for the Hamiltonian H. In general, the principle can be used for any operator A that satisfies a Fredholm type integral equation

$$A = B + B G_0 A, \qquad (3.68)$$

where B is self adjoint. In particular, the rate operator Γ and equivalent potential $\mathscr{V}$ (eqn (3.28)) fall into this class (see Levine 1966a).

§ An equivalent form is $\psi^+(\mathbf{R}, \mathbf{r}) = \chi_n(\mathbf{R}, \mathbf{r})F^+(\mathbf{R})$
with $\chi_n(\mathbf{R}, \mathbf{r}) \to \varphi_n(\mathbf{r}), \quad R \to \infty.$
See Levine (1968).

A Kohn (1948) type variational expression is
$$[A] = A_t - I_t, \tag{3.69}$$
where I_t, which is a functional of A_t, is defined in a way that ensures that $I_t = 0$ when A_t is the exact value for A and that
$$\delta[A] = \delta A_t - \delta I_t = 0. \tag{3.70}$$
If we are concerned with the scattering (or transition) amplitude we can, following Kohn, take
$$I_t = \langle \psi_{f,t}^- | (E-H) | \psi_{i,t}^+ \rangle, \tag{3.71}$$
where we require that our trial wavefunctions satisfy the proper asymptotic behaviour, with some trial value for the scattering amplitude. For exact wavefunctions $I_t = 0$ since in the limit $\epsilon \to 0$
$$(E-H)\psi_i^+ = 0. \tag{3.72}$$
For the case of scattering by a central potential, for example, we take
$$\psi_{\mathbf{k}}^+(\mathbf{R}) \to \exp(i\mathbf{k} \cdot \mathbf{R}) + f_t(\mathbf{k}', \mathbf{k}) \exp(ikR)/R, \tag{3.73}$$
with $E = \hbar^2 k^2 / 2\mu$, $k' = k$. In this case
$$(\psi_{\mathbf{k}}^-)^* = \psi_{-\mathbf{k}}^+, \tag{3.74}$$
so that
$$I_t(\mathbf{k}', \mathbf{k}) = \int d\mathbf{R}\, \psi_{-\mathbf{k}'}^+(\mathbf{R})[E + (\hbar^2/2\mu)\nabla_{\mathbf{R}}^2 - V(\mathbf{R})]\psi_{\mathbf{k}}^+(\mathbf{R}). \tag{3.75}$$
It can be shown (Kohn 1948), using the methods of section 2.6.4, that
$$\delta I_t(\mathbf{k}', \mathbf{k}) = -(2\pi\hbar^2/\mu)\delta f_t(\mathbf{k}', \mathbf{k}), \tag{3.76}$$
where
$$\delta \psi_{\mathbf{k}}^+(\mathbf{R}) = \delta f_t(\mathbf{k}', \mathbf{k}) \exp(ikR)/R. \tag{3.77}$$
Taking $A_t = f_t(\mathbf{k}', \mathbf{k})$ we obtain from eqn (3.49)
$$[f(\mathbf{k}', \mathbf{k})] = f_t(\mathbf{k}', \mathbf{k}) + I_t(\mathbf{k}', \mathbf{k})(\mu/2\pi\hbar^2). \tag{3.78}$$
It is clear that starting with $f_t = 0$ we obtain for I_t the first Born approximation. Equation (3.78) can be used as a variational principle by requiring that $\delta[f(\mathbf{k}', \mathbf{k})] = 0$.

To prove the Kohn principle it is convenient to make use of a generalization (following Rosenberg 1964) of the Kato (1951) identity.

We write the trial solution of the L.S. equation as
$$|i_t^+\rangle = \Omega_t |i\rangle,$$
$$\Omega_t = I + G_0 T_t$$
$$= I + G_0 T^\ddagger + G_0 (T_t - T^\ddagger). \tag{3.79}$$
Then
$$T^\ddagger - T_t = \Omega^\ddagger (1 - VG_0)(T^\ddagger - T_t)$$
$$= \Omega^\ddagger (E^+ - H) G_0 (T^\ddagger - T_t)$$
$$= \Omega^\ddagger (E^+ - H)(\Omega - \Omega_t). \tag{3.80}$$

OPERATOR PARTITIONING THEORY

Taking the i, f matrix element and then the limit $\epsilon \to 0$ we obtain

$$\langle f|T|i\rangle - \langle f|T_t|i\rangle = \langle f^-|(E-H)|i_t^+\rangle, \qquad (3.81)$$

where $\qquad \langle f^-|(E-H)|i^+\rangle = 0$

for exact solutions in the limit $\epsilon \to +0$. $\langle f|T_t|i\rangle$ is defined from eqn (3.79) in terms of the asymptotic behaviour of $|i_t^+\rangle$. Equation (3.81) is the required identity. For reactive collisions one has to consider matrix elements on the energy shell only.

If on the right-hand side of eqn (3.81) we replace $\langle f^-|$ by $\langle f_t^-|$ we are introducing an error of the second order only. To see this we put $|f^-\rangle = |f_t^-\rangle - |\delta f^-\rangle$ so that

$$\langle f^-|(E-H)|i_t^+\rangle = \langle f_t^-|(E-H)|i_t^+\rangle + \langle \delta f^-|(E-H)|\delta i^+\rangle, \quad (3.82)$$

where in the second matrix element we put $\langle \delta f^-|(E-H)|i^+\rangle = 0$ in the limit $\epsilon \to +0$. Neglecting quantities that are of second order in the error of the wavefunction we obtain

$$[T]_{f,i} = (T_{f,i})_t - I_t,$$
$$I_t = \langle f_t^-|(E-H)|i_t^+\rangle \qquad (3.83)$$

as a Kohn-type variational principle. Note that while $|\delta i^+\rangle$ is restricted by eqn (3.79) to be such that $|i_t^+\rangle$ satisfies eqn (3.79) (namely, $|i\rangle$ as the incident wave, and outgoing boundary conditions), $\langle \delta f^-|$ is arbitrary. For a single-channel collision eqn (3.83) reduces to eqn (3.78).

To derive the distorted-wave Born approximation (eqns (2.5.83)–(2.5.87)) from eqn (3.83) we take as a trial wavefunction the exact solution for the Hamiltonian H_1, $H_1 = H_0 + V_1$, where V_1 is the part of the mutual interaction that cannot induce inelastic (or reactive) transitions. Normally V_1 is a central potential so that the trial wavefunction can be assumed known,

$$|i_t^+\rangle = \Omega_1 |i\rangle. \qquad (3.84)$$

From eqn (2.5.72) the error in the trial wavefunction is given exactly by
$$|\delta i^+\rangle = |i_t^+\rangle - |i^+\rangle = -GV_2|i_t^+\rangle. \qquad (3.85)$$

Thus $\qquad -I_t = -\langle f_t^-|(E-H)|i_t^+\rangle = \langle f_t^-|V_2|i_t^+\rangle, \qquad (3.86)$

while (since $\langle \delta f^-|(E-H)|i^+\rangle = 0$)

$$\langle \delta f^-|(E-H)|\delta i^+\rangle = \langle f_t^- V_2 G(E^-)|E-H|i_t^+\rangle$$
$$= \langle f_t^-|V_2 G(E^+)V_2|i_t^+\rangle$$
$$= \langle f|\Omega_1^\ddagger V_2 G(E^+) V_2 \Omega_1|i\rangle. \qquad (3.87)$$

In this case the Kato identity (3.81) is equivalent§ to eqn (2.5.77). The

---

§ We also see that the limits $\epsilon \to +0$ on the left- and right-hand sides of $\Omega_1^\ddagger (E-H)\Omega_1$ can be taken independently of one another.

distorted-wave Born approximation consists in neglecting the second-order error, eqn (3.87), and is equivalent to eqn (2.5.84). The DWB transition amplitude (3.86) is stationary about the solution of the L.S. equation for H_1.

The close-coupling approximation can also be derived from the Kohn principle (Mittleman and Pu 1962). Moreover, the discussion of eqn (3.46) shows that with the definition $P = \sum_m |m\rangle\langle m|$, where m sums over the coupled states, the close-coupling approximation also leads to a stationary expression for $\mathscr{H}$, the effective Hamiltonian in P space.

In general, the decoupling approximation corresponds to neglecting the coupling between the P and Q components of the total wavefunction. It can be shown using the Lippmann–Schwinger or the Kohn variational principles that the decoupled components should satisfy

$$(E^+ - PHP)P\psi_1^+ = (E^+ - H_0)\phi \qquad (3.88)$$

and
$$(E^+ - QHQ)Q\psi_1^+ = 0, \qquad (3.89)$$

where $Q\psi_1^+$ has no incident wave (cf. eqns (1.13) and (1.14)). Different choices of P lead to different specific approximations. Thus eqns (1.8), (2.31), and (1.4) correspond to the static, adiabatic, and close-coupling approximations respectively. Corrections to the decoupling approximation can be introduced by allowing coupling to first order only. This approximation (analogous to a distorted-wave approximation) can also be derived from the variational principles, and replaces eqn (3.89) by

$$Q\psi^+ = (E^+ - QHQ)^{-1}QHP\psi_1^+. \qquad (3.90)$$

The various choices of P correspond to the various choices of the distorting potential U_i as discussed in section 3.2.3 (i.e. $U_i = PHP$). Thus the transition amplitude corresponding to eqn (3.90) is a distorted-wave type amplitude as in eqn (2.138), as can also be seen by using eqn (2.5.74) to evaluate the asymptotic form of the Green's function ($Q|s\rangle = |s\rangle$),

$$\lim_{R\to\infty} \langle \mathbf{R}, s|(E^+ - QHQ)^{-1} = -(\mu/2\pi\hbar^2)\frac{e^{ik_s R}}{R}\langle \mathbf{k}', s_1^-|, \qquad (3.91)$$

$$|\mathbf{k}', s_1^-\rangle = F(E^-)|\mathbf{k}', s\rangle, \qquad (3.92)$$

$$E = E_s + \hbar^2 k_s^2/2\mu, \qquad \mathbf{k}' = k_s \hat{\mathbf{R}}.$$

Identical conclusions follow also from the use of eqn (2.122).

3.4. Models in collision theory

3.4.0. IN Chapters 3.4 and 3.5 we consider some physical models that have been employed in collision theory. In the present chapter we consider in particular those models that treat the relative motion during the collision in some approximate fashion, either by treating it in a classical (or semi-classical) fashion, or by the introduction of some reaction criteria in terms of the relative motion.

In any microscopic mechanical model of inelastic or reactive collisions we necessarily discuss the relative motion in terms of classical trajectories. Such an approach requires that the wavelength of relative motion $\lambda = k^{-1}$ be small compared to the characteristic distance of the potential,

$$\lambda \ll a \quad \text{or} \quad \mu v a \gg \hbar, \tag{4.1}$$

in order that the motion can be localized. The high atomic mass (in atomic units) ensures that eqn (4.1) is satisfied at low velocities. (Equation (4.1) also implies that many partial waves contribute to any event, since $2ka$ is essentially the number of partial waves that are required in the expansion of the scattering amplitude.)

Since we are dealing with a three-dimensional trajectory, we can also require that the deflexion (determined by the momentum transfer) be well-defined when compared to the uncertainty in the deflection due to δp, the uncertainty in momentum due to the localization to within a, or

$$\Delta p \sim \text{force} \cdot \tau_c \sim \frac{V_0}{a}\frac{a}{v} \gg \delta p \sim \frac{\hbar}{a},$$

where V_0 is the average strength of the interaction. Thus

$$\frac{V_0 a}{\hbar v} \gg 1 \tag{4.2}$$

or
$$\mu v a \gg \hbar(E/V_0).$$

Inverting eqn (4.2) we obtain the conditions of validity of the Born approximation, where the deflection is a single event when viewed in the momentum representation. A classical behaviour on the other hand, corresponds to many small deflections so that the change is continuous rather than abrupt. (For further discussion see Williams 1945, Glauber 1959.)

In what follows we shall mainly use eqn (4.1) and evaluate differential cross-sections using approximations to the summation over many partial wave amplitudes. If ΔR is the uncertainty in position we shall require that $p\Delta R \gg \hbar$ so that we can consider an initial state with a well-defined impact parameter, and for a given transition evaluate $P(b)$, the probability of the transition at an impact parameter b. The differential cross-section can be written as

$$d\sigma = 2\pi P(b) b \, db, \tag{4.3}$$

where, in an initial state of uniform density, $2\pi b \, db$ is the fraction of all collisions with an impact parameter in the range b to $b+db$.

For many purposes it is a useful first approximation to characterize molecules by their size. We can thus try to determine $P(b)$ by the requirement that any relative motion that will bring the centres of the two molecules to within a distance d will lead to reaction (Present 1955). In this ('reactive box') approximation the relative motion is determined by a spherically symmetric interaction $V(R)$ (which we can consider as the true interaction averaged over the internal degrees of freedom).

To reach a mutual separation d the translational energy should exceed not only $V(d)$ but also the centrifugal barrier $\hbar^2 l(l+1)/2\mu d^2$. With the substitution $l(l+1) = b^2 k^2$ we have that for a given energy E the highest impact parameter that will lead to reaction is the root of

$$E = V(d) + E b_m^2/d^2, \tag{4.4}$$

or

$$b_m^2 = d^2 \left[1 - \frac{V(d)}{E}\right], \tag{4.5}$$

and $P(b) = \theta(b_m - b)$, where θ is the step function.

If we regard the two molecules as hard spheres, with d being the sum of their radii, we can reformulate the criterion for reaction as 'reactive collisions will occur whenever, on impact, the energy along the line of centres exceeds E_0', and identify E_0 with $V(d)$.

In several applications we regard $V(R)$ as the long-range part of the mutual interaction. For example, in the recombination of ions (Mahan 1964) $V(d) = -Z_1 Z_2/d$ is the Coulomb potential, and

$$b_m^2 = d^2 \left(1 + \frac{Z_1 Z_2}{Ed}\right). \tag{4.6}$$

In general, we expect the long-range interaction to be a dispersion-type force $V(R) = -aR^{-n}$. If only the short-range interaction can lead to a reaction we can determine d to be the maximum point of the effective potential (Mahan 1960),

$$\frac{\partial}{\partial R}[V(R) + Eb^2/R^2]_{R=d} = 0.$$

Only those molecules that can surmount the centrifugal barrier and reach the repulsive inner core will react. In this model d depends on E

$$d = \left(\frac{na}{2Eb^2}\right)^{1/(n-2)} \tag{4.7}$$

and

$$b_m^2 = \left(\frac{na}{2E}\right)^{2/n}\left(\frac{n+2}{n}\right)^{(2-n)/n}. \tag{4.8}$$

From eqns (4.3) and (4.5),

$$\sigma = \int d\sigma = \pi b_m^2 = \pi d^2\left(1-\frac{E_0}{E}\right), \tag{4.9}$$

the yield function is (cf. eqn (2.6.48 b))

$$Y(E) = \pi^{-1}k^2\sigma(E) = (2\mu/\hbar^2)(E-E_0)d^2, \tag{4.10}$$

and the rate constant in the canonical ensemble

$$k_c = Z_{tr}^{-1}h^{-1}(2\mu/\hbar^2)d^2\int_0^\infty (E-E_0)\exp(-\beta E)\,dE$$

$$= \pi d^2\left(\frac{8kT}{\pi\mu}\right)\exp(-\beta E_0). \tag{4.11}$$

Equation (4.11) is the well-known expression for the hard-sphere collision rate. When the spheres are charged (eqn (4.6))

$$k_c = \pi d^2\left(\frac{8kT}{\pi\mu}\right)^{\frac{1}{2}}\left(1+\frac{Z_1 Z_2}{dkT}\right),$$

while for ion-molecule reactions, when we expect $n = 4$ in eqn (4.8),

$$k_c = 2\pi\left(\frac{2a}{\mu}\right)^{\frac{1}{2}}. \tag{4.12}$$

The maximum value of the impact parameter leading to a reaction at a given energy can be rewritten as a maximum angular momentum l_m. From eqns (4.10), (4.9), and (4.5),

$$Y(E) = l_m(l_m+1). \tag{4.13}$$

It is interesting to note that $Y(E)$ can be rewritten as the number of internal states, in this case the rotational states, of the colliding molecules up to the energy $E-E_0$; $B = \hbar^2/2\mu d^2$

$$Y(E) = \sum_l \int_0^{E-E_0} \delta[E-Bl(l+1)](2l+1)\,dE$$

$$= \sum_{l=0}^{l_m}(2l+1)$$

$$= l_m(l_m+1). \tag{4.14}$$

We have previously remarked that eqn (4.14) has the correct formal form, and so a possible extension of the model is to assume that an analogue of eqn (4.14) will be correct when the colliding molecules have other internal degrees of freedom (Marcus 1966a, Coulson and Levine 1967).

In other words, from the point of view of the collision model
$$Y(E)\exp(-\beta E)\,\mathrm{d}E/Z_i$$
is the number of colliding pairs, drawn from a canonical ensemble, of total energy E and internal energy less than $E-E_0$, at a specified configuration. These considerations imply that

$$Y(E) = \int_0^{E-E_0} \rho_{\ddagger}(E-E_0-e)\,\mathrm{d}e, \qquad (4.15)$$

where $\rho_{\ddagger}$ is the density of internal states of the system at the specified configuration and E_0 is the minimal energy, in the degree of freedom of (one-dimensional) relative motion, necessary for the reaction to occur. Using eqn (4.15) in the expression for the rate in the canonical ensemble (eqn (2.6.49))

$$k_c = \frac{\mathbf{k}T}{h}\frac{Z_{\ddagger}}{Z}\exp(-\beta E_0), \qquad (4.16)$$

where
$$Z_{\ddagger} = \int \rho_{\ddagger}(E)\exp(-\beta E)\,\mathrm{d}E \qquad (4..17)$$

is the 'partition function' for the internal degrees of freedom in the specified configuration. For our simple model (eqn (4.14))

$$Z_{\ddagger} = 2I\mathbf{k}T/\hbar^2, \quad I = \mu d^2, \qquad (4.18)$$

which we recognize as the high-temperature partition function for a rotor of reduced mass μ and separation d. As is clear, neither the simple model nor its extension above invoke the concept of a 'transition state'. The specified configuration (in the simple model this is simply the separation d) enters the theory through our choice of a criterion for reaction. For further discussion see Fowler and Guggenheim (1952, sections 1200–10) and Slater (1959, sections 2.3–2.6 and references therein).

An alternative criterion for reaction, which also leads to eqn (4.15) has been formulated by Eyring (1935). See also Glasstone, Laidler, and Eyring (1941) and Marcus (1965). In this approach the reaction rate is given by the rate of change of the number of systems in a specified configuration (the 'transition state'), which one considers as an intermediate structure between the reactants and products. Almost invariably the transition state is chosen as a saddle point in a one-dimensional relative motion. The number of states, $\mathrm{d}N(E_*)$, at energy E to $E+\mathrm{d}E$ is

$$\mathrm{d}N(E_*) = \rho_{\ddagger}(E-E_*)\,\mathrm{d}E\mathrm{d}p\mathrm{d}R/h, \qquad (4.19)$$

where E_* is the energy of one-dimensional relative motion,
$$E_* = E_0 + p^2/2\mu,$$
and $\mathrm{d}p\,\mathrm{d}R/h$ is the element of phase space for the motion along R. Noting that $p = \mu\,\mathrm{d}R/\mathrm{d}t$, we can write
$$\frac{\mathrm{d}\dot{N}(E_*)}{\mathrm{d}E} = \rho_\ddagger(E-E_*)p\,\mathrm{d}p/\mu h$$
and obtain the total rate at a given energy interval E to $E+\mathrm{d}E$ by integrating over all possible values of p,
$$\frac{\mathrm{d}\dot{N}}{\mathrm{d}E} = \frac{1}{h}\int \rho_\ddagger(E-E_0-p^2/2\mu)\,\mathrm{d}(p^2/2\mu), \qquad (4.20)$$
in agreement with eqn (4.15). The required counting of states that lead to eqn (4.12) was performed by Eyring, Hirschfelder, and Taylor (1936), and the concept of rotational energy barrier in the transition state theory is discussed by Eyring, Gershionowitz, and Sun (1935; see also Gorin 1938).

While the two models discussed differ by their criteria for reaction, they have two assumptions in common. The first and innocuous one, is that the distribution of energies of the (non-interacting) molecules is canonical. The second assumption is essentially the assumption that the distribution of energy between the internal and relative degrees of freedom is that of an equilibrium situation, which enabled us to write eqn (4.19). While this assumption in its simplest form is questionable, it is clear that it can be improved by involving only some degrees of freedom. Moreover, if we are willing to regard $\rho_\ddagger(E)$ more as a free parameter than as a predetermined function then eqn (4.20) is certainly dimensionally valid. Similar models are discussed in section 3.5.0.

An essential difference between the two reaction criteria is the nature of the coordinate R. In Eyring's theory and its generalizations R varies continuously from the reactants to the products (R is the so-called 'reaction coordinate'; see section 3.4.3), while in the collision model R is the relative separation between the reactants.

3.4.1. *The opacity function—the optical model*

To consider the quantum-mechanical extension of these ideas we introduce the opacity function. In a collision that can lead to inelastic and reactive events, let ψ_e^+ be the component of the total wavefunction that represents asymptotically the elastic events,
$$\psi_\mathrm{e}^+ \xrightarrow[R_i\to\infty]{} [\exp(i\mathbf{k}\cdot\mathbf{R}_i) + f(\mathbf{k}',\mathbf{k})\exp(ikR_i)/R_i]\varphi_i, \qquad (4.21)$$

where R_i is the coordinate of relative motion in the initial channel, φ_i is the initial internal state, and $k' = k$. When the orbital angular momentum is conserved we can perform a partial wave resolution, with $\cos\theta = \hat{\mathbf{k}}'.\hat{\mathbf{k}}$,
$$\psi_e^+ = \sum_l i^l(2l+1)\psi_{kl}^+(R_i)P_l(\cos\theta)$$
and (cf. eqn (2.2.80))
$$\psi_{kl}^+ \xrightarrow[R_i\to\infty]{} (kR_i)^{-1}\sin(kR_i - \tfrac{1}{2}l\pi) + i^{-l}k^{-1}\beta_{kl}\exp(ikR_i)/R_i,$$
where
$$f(k,\theta) = k^{-1}\sum_{l=0}(2l+1)\beta_{kl}P_l(\cos\theta). \tag{4.22}$$

We can regard eqn (4.22) as an expansion of the scattering amplitude in a set of orthogonal polynomials, $P_l(\cos\theta)$, with expansion coefficients β_{kl} that can be determined from the orthogonality relation as

$$\beta_{kl} = \tfrac{1}{2}k\int_{-1}^{1} f(k,\theta)P_l(\cos\theta)\,\mathrm{d}\cos\theta. \tag{4.23}$$

In terms of eqn (2.2.35) the cross-section for elastic scattering of the lth partial wave is given by the surface integral of the asymptotic density

$$\sigma_l^e = \int k^{-2}(2l+1)^2|\beta_{kl}|^2 P_l^2(\cos\theta) R^{-2}\,\mathrm{d}S$$
$$= \frac{4\pi}{k^2}(2l+1)|\beta_{kl}|^2. \tag{4.24}$$

The total cross-section for all events is given from the optical theorem (2.6.71) by

$$\sigma_l^t = (4\pi/k)\mathrm{im}[(2l+1)k^{-1}\beta_{kl}P_l(0)] = \frac{2\pi i}{k^2}(2l+1)[\beta_{kl} - \beta_{kl}^*]. \tag{4.25}$$

In contrast to the discussion of section 2.2.3 one cannot equate σ_l^t to σ_l^e. Rather, since for the computations of rates and cross-sections we can superpose probabilities (cf. section 2.4.0) we put

$$\sigma_l^t = \sigma_l^e + \sigma_l^{\mathrm{in}}, \tag{4.26}$$

where σ_l^{in} is the cross-section for all inelastic (whether reactive or not) processes. In view of the optical theorem, if σ_l^e vanishes, so does σ_l^{in}. Any inelastic process is necessarily accompanied by an elastic process. To cast these relations in a quantitative form we put§

$$\beta_{kl} = [\exp(2i\gamma_l) - 1]/2i, \tag{4.27}$$

§ The chosen form for β_{kl} is dictated empirically by the observation that $\sigma_l^t > \sigma_l^e$ whenever σ_l^{in} is non-vanishing. From a theoretical point of view we can regard $1 + 2i\beta_{kl}$ as the diagonal element of the **S** matrix. Since there are inelastic processes, the non-diagonal elements of **S** are non-vanishing and the unitarity of **S** implies that $|1+2i\beta_{kl}| < 1$. Alternatively, if we restrict our attention to the elastic scattering by the introduction of an equivalent potential (sections 3.1.0 and 3.2.2), then γ_l is the phase shift computed with a complex equivalent potential. (Recall that the equivalent

where γ_l is energy dependent and is not necessarily real,
$$\gamma_l = \delta_l + i\eta_l.$$
We thus obtain from eqns (4.24)–(4.26),
$$\sigma_l^t = \frac{2\pi}{k^2}(2l+1)[1-\cos(2\delta_l)\exp(-2\eta_l)],$$

$$\sigma_l^{in} = \frac{\pi}{k^2}(2l+1)[1-\exp(-4\eta_l)] \qquad (4.28)$$

and $\qquad \sigma_l^e = \frac{2\pi}{k^2}(2l+1)\exp(-2\eta_l)[\cosh(2\eta_l)-\cos(2\delta_l)].$

To ensure a positive inelastic cross-section we must take $\eta_l \geqslant 0$. We note that
$$\sigma_l^t \leqslant \frac{4\pi}{k^2}(2l+1), \quad \sigma_l^{in} \leqslant \frac{\pi}{k^2}(2l+1). \qquad (4.29)$$

The inelastic cross-section for all partial waves is
$$\sigma^{in} = \frac{\pi}{k^2}\sum_l (2l+1)[1-\exp(-4\eta_l)] = \frac{\pi}{k^2}\sum_l (2l+1)P_l, \qquad (4.30)$$

which can be bounded only if we truncate the sum over l. It is common to refer to $P_l = 1-\exp(-4\eta_l)$ as the 'opacity function' (see, for example, Greider and Glassgold 1960), since it is a direct measure of the attenuation of the elastic scattering,
$$P_l = (\sigma_l^t - \sigma_l^e)/(\pi/k^2)(2l+1). \qquad (4.31)$$

In the random phase (RPH) approximation (Massey and Mohr 1934) $\langle \sin^2\delta_l \rangle = \tfrac{1}{2}$ where the bracket denotes an average value, so that
$$\langle \sigma_l^t \rangle_{\text{RPH}} = \frac{2\pi}{k^2}(2l+1). \qquad (4.32)$$

In the absence of inelastic processes, the RPH value of the elastic cross-section is also given by eqn (4.32). Thus in the RPH approximation the magnitude of the total cross-section, for a given l is the same, irrespective of whether inelastic processes do or do not take place. Moreover, in the same approximation
$$\langle \sigma_l^{in} \rangle_{\text{RPH}} = \tfrac{1}{2}\langle \sigma_l^t \rangle_{\text{RPH}} P_l, \qquad (4.33)$$
so that $\qquad \langle \sigma_l^{in} \rangle_{\text{RPH}}/\langle \sigma_l^e \rangle_{\text{RPH}} \leqslant 1 \qquad (4.34)$

potential for elastic scattering is complex whenever inelastic transitions can take place, see eqns (2.120)–(2.122).) We can also regard γ_l as the phase shift averaged over an energy interval. This phase shift is determined by an optical potential (section 3.5.2). In this case, however, the interpretation of the elastic and inelastic cross-sections must be modified as described in section 3.5.2.

and (see also Rosenfeld and Ross 1966)
$$P_l = 1 - \langle \sigma_l^e \rangle_{\text{RPH}} / \langle \sigma_l^t \rangle_{\text{RPH}}. \tag{4.35}$$

An alternative approximation is that of 'shadow' scattering in which $\text{re} f(k,\theta) = 0$. In this case β_{kl} is pure imaginary, or, in other words, the observed elastic cross-section is exactly that required by the optical theorem to support a given inelastic cross-section
$$(\sigma_l^e)_s = \frac{4\pi}{k^2}(2l+1)|\text{im}\,\beta_{kl}|^2 = \frac{\pi}{k^2}(2l+1)(1-\zeta_l)^2, \tag{4.36}$$

where $\zeta_l = \exp(-2\eta_l)$. The inelastic cross-section is the same, while
$$(\sigma_l^t)_s = \frac{2\pi}{k^2}(2l+1)(1-\zeta_l). \tag{4.37}$$

In this approximation also the optical theorem is satisfied, however
$$(\sigma_l^{\text{in}})_s / (\sigma_l^e)_s \geqslant 1, \tag{4.38}$$
although an analogue of eqn (4.33) does hold
$$(\sigma_l^{\text{in}})_s = \tfrac{1}{2} P_l (\sigma_l^t)_{\text{RPH}}. \tag{4.39}$$

A further extension of the RPH approximation for inelastic collision is considered in section 3.5.1.

To characterize the interaction region by an effective size we assume that ζ_l is a constant for $l \leqslant L$ and vanishes for $l > L$, or
$$\text{im}\,\beta_{kl} = (1-\zeta)\theta(L-l). \tag{4.40}$$

We then have
$$\sigma^{\text{in}} = \sum_l \sigma_l^{\text{in}} = \frac{\pi}{k^2}(1-\zeta^2)(L+1)^2. \tag{4.41}$$

The results for σ^t and σ^e depend on the model selected for the elastic phase δ_l.
$$(\sigma^t)_s = \frac{2\pi}{k^2}(L+1)^2(1-\zeta) = (\sigma^t)_{\text{RPH}}(1-\zeta). \tag{4.42}$$

Using the semi-classical correspondence $l+\tfrac{1}{2} \leftrightarrow bk$ we can obtain an impact parameter description. Thus eqn (4.41) can be rewritten as
$$\sigma^{\text{in}} = 2\pi \int b P(b)\,\mathrm{d}b \tag{4.43}$$

with $P(b) = P_l$. Equation (4.43) is thus obtained as an approximation. However, since σ^{in} has the dimensions of an area, it can be represented exactly by eqn (4.43) with a suitable definition of $P(b)$. The aim of the impact parameter representation is to obtain such exact relations and to consider their semi-classical limiting form. Our discussion follows Adachi and Kotani (1966 and references therein). We define an impact parameter amplitude, $\beta_k(b)$, by analogy to eqn (4.23) as

$$\beta_k(b) = \tfrac{1}{2}k \int_0^1 f(k,\theta) J_0(2bk \sin \tfrac{1}{2}\theta) \sin \theta \, d\theta, \qquad (4.44)$$

where $J_n(x)$ is the Bessel function of order n and argument x. Using the orthogonality relation

$$\int_0^\infty x^{-1} J_n(x) J_m(x) \, dx = \frac{\delta_{n,m}}{2n} \qquad (4.45)$$

we can invert eqn (4.44) to

$$f(k,\theta) = 2k \int_0^\infty \beta_k(b) J_0(2bk \sin \tfrac{1}{2}\theta) b \, db, \qquad (4.46)$$

while using the representation

$$\sum_l (2l+1) J_{2l+1}(x) P_l(\cos \theta) = \tfrac{1}{2} x J_0(x \sin \tfrac{1}{2}\theta) \qquad (4.47)$$

in eqn (4.44) gives, on comparing with eqn (4.22),

$$\beta_k(b) = (bk)^{-1} \sum_l (2l+1) J_{2l+1}(2bk) \beta_{kl}. \qquad (4.48)$$

For $x \gg 1$ the first maxima of $J_n(x)$ occur at $x \simeq n$, and so with $l' + \tfrac{1}{2} = bk$

$$\beta_k(b) \simeq (bk)^{-1} \beta_{kl'} \sum_l (2l+1) J_{2l+1}(2bk)$$

$$= \beta_{kl'} = [\exp(2i\gamma_{l'}) - 1]/2i, \qquad (4.49)$$

where we have used eqn (4.47) with $P_l(0) = 1$ and $J_0(0) = 1$.

The identification of the impact parameter amplitude with the partial wave amplitude is thus a semi-classical, $b\mu v \gg \hbar$, approximation. In fact, if we regard eqn (4.48) as an 'average' and introduce the notation

$$\beta_k(b) = \{\beta_{kl}\},$$

we can show that $\qquad (bk)^2 = \{l(l+1)\} \qquad (4.50)$

so that our correspondence $l + \tfrac{1}{2} \leftrightarrow bk$ is indeed a semi-classical correspondence.

In terms of the impact parameter amplitudes we obtain the exact relations

$$\sigma^t = \frac{4\pi}{k} \operatorname{im} f(k,0) = 8\pi \int_0^\infty b \operatorname{im} \beta_k(b) \, db, \qquad (4.51)$$

by eqn (4.46). Similarly,

$$\sigma^e = 8\pi \int_0^\infty b |\beta_k(b)|^2 \, db. \qquad (4.52)$$

We define the function $P_k(b)$ by

$$\sigma^{\text{in}} = 2\pi \int_0^\infty b P_k(b)\, db, \tag{4.53}$$

so that, by analogy to eqns (4.51) and (4.48), we can write

$$P_k(b) = (bk)^{-1} \sum_l (2l+1) J_{2l+1}(2bk) P_l \tag{4.54}$$

and identify $P_k(b)$ as the opacity function in the impact parameter representation that is equal to P_l, $l+\tfrac{1}{2} \leftrightarrow bk$ in the semi-classical limit. Using eqn (4.47) for $\theta = 0$ we note that since $|\beta_{kl}| \leqslant 1$ and $P_l \leqslant 1$ it is also true that

$$|\beta_k(b)| \leqslant 1 \quad \text{and} \quad P_k(b) \leqslant 1. \tag{4.55}$$

In contrast to the partial wave amplitudes, we cannot write down an optical theorem for a particular b value, but can only conclude that we can introduce $K_k(b)$ such that

$$\operatorname{im} \beta_k(b) = |\beta_k(b)|^2 + \tfrac{1}{4} P_k(b) + K_k(b) \tag{4.56}$$

where, from eqns (4.51)–(4.53),

$$\int_0^\infty b K_k(b)\, db = 0.$$

Since
$$\operatorname{im} \beta_{kl} = |\beta_{kl}|^2 + \tfrac{1}{4} P_l,$$

we obtain, on taking the impact parameter 'average' of both sides and using eqns (4.54) and (4.48),

$$K_k(b) = \{|\beta_{kl}|^2\} - |\beta_k(b)|^2 = \{|\beta_{kl}|^2\} - |\{\beta_{kl}\}|^2. \tag{4.57}$$

Thus $K_k(b)$ is the measure of the error in replacing $\beta_k(b)$ by β_{kl}. We note, however, that since $J_n(0) = \delta_{n,0}$ it follows that as $k \to 0$, $\beta_k(b)$ and β_{kl} are identical and in this limit $K_k(b) = 0$. Also, from eqn (4.47), $K_k(b)$ is very small in the semi-classical region, tending to zero as $k \to \infty$.

From eqns (4.3), (4.51)–(4.53), and (4.56) we can write

$$d\sigma^t = d\sigma^{\text{el}} + d\sigma^{\text{in}} + 8\pi b K_k(b)\, db. \tag{4.58}$$

In the semi-classical limit the last term is negligible. The semi-classical limit of eqn (4.58) can be considered as the equivalent of $\sigma_l^t = \sigma_l^{\text{el}} + \sigma_l^{\text{in}}$ in the partial wave analysis.

It may seem more natural to introduce a characteristic length directly in terms of the impact parameter amplitude. Thus in the shadow approximation we have $\operatorname{re} \beta_k(b) = 0$, and we can take

$$\operatorname{im} \beta_k(b) = \tfrac{1}{2}(1-\zeta)\theta(d-b). \tag{4.59}$$

However, since $K_k(b) \neq 0$ this equation does not imply that

$$P(b) = (1-\zeta^2)\theta(d-b),$$

although it is true that

$$\sigma^{\text{in}} = 2\pi \int_0^\infty (1-\zeta^2)\theta(d-b)b\,\mathrm{d}b = \pi d^2(1-\zeta^2). \quad (4.60)$$

(Compare with eqns (4.39) and (4.9).) If we compute $\{\beta_{kl}\}$ with β_{kl} given by eqn (4.40) we find no sharp cut-off but only a strong oscillatory decline in $\beta_k(b)$ for $bk > L+1$.

The model can be refined by making ζ a function of other parameters beside d. In particular, it can be a function of the relative orientation of the two molecules during the collisions (see, for example, Herschbach 1965a). Thus one can define $P(b)$ as the fraction of the area of a ring, of radius b and width db, that will lead to a reaction. This fraction is a function of the relative orientation. In this way one can introduce a 'steric factor' into the model.

Quantum-mechanical computations of the opacity function have been mostly carried out for inelastic (but non-reactive) transitions, using an approximate treatment for the relative motion which almost invariably neglects the effects of the inelastic transition on the relative motion.

As an example, we consider the distorted-wave Born approximation in which we have resolved the potential into V_1 and V_2 where V_1 is a central potential and can lead to elastic scattering only. We further assume that the angular momentum of the relative motion is conserved. Strictly speaking, only the total angular momentum is conserved, but in the semi-classical region, $bk \gg 1$, small changes in the internal angular momentum can be neglected.§ The scattering amplitude for the transition $n \to m$ can then be evaluated using the method of partial waves. Putting (cf. eqn (2.5.84))

$$f_{m,n} = -(\mu/2\pi\hbar^2)\langle \mathbf{k}_m, m_1^- | V_2 | \mathbf{k}_n, n_1^+ \rangle, \quad (4.61)$$

we can write

$$f_{m,n} = (k_m k_n)^{-\frac{1}{2}} \sum_l (2l+1)\exp[i(\delta_l^m + \delta_l^n)]\beta_l^{m,n} P_l(\cos\theta), \quad (4.62)$$

where we have put $\quad L_\mathbf{k}^n(\mathbf{R}) = \langle \mathbf{R}, n | \mathbf{k}, n_1^+ \rangle,$

and, from eqns (2.2.71) and (2.2.83),

$$L_\mathbf{k}^n(\mathbf{R}) = \sum_l i^l(2l+1)\frac{U_{kl}^n(R)}{R}\exp(i\delta_l^n)P_l(\cos\theta) \quad (4.63)$$

and $\quad \beta_l^{m,n} = (2\mu/\hbar^2)(k_m k_n)^{\frac{1}{2}} \int U_{kml}^m(R)V_{m,n}(R)U_{knl}^n(R)\,\mathrm{d}R \quad (4.64)$

is a function of k_m and k_n.

§ This approximation also implies that the angular dependence of the scattering amplitude is determined by $\hat{\mathbf{k}}_m \cdot \hat{\mathbf{k}}_n$ only (cf. the derivation of eqn (2.2.71)).

The differential cross-section for the transition $n \to m$ is given by

$$d\sigma_{m,n} = \frac{k_m}{k_n}|f_{m,n}|^2 \, d\omega, \qquad (4.65)$$

and the cross-section for the transition $n \to m$ is

$$\sigma_{m,n} = (\pi/k_n^2) \sum_l (2l+1) 4 |\beta_l^{m,n}(k_m, k_n)|^2. \qquad (4.66)$$

In the modified wave-number approximation (Takayanagi 1963), the summation over l is performed by the introduction of the variable k_n',

$$(k_n')^2 = k_n^2 - l(l+1)/R_M^2 \qquad (4.67)$$

and writing

$$\sigma_{m,n} \approx (2\pi R_M^2/k_n^2) \int_0^{k_n} |\beta_0^{m,n}(k_m', k_n')|^2 \, dk_n'. \qquad (4.68)$$

Using eqn (4.64), we only need to evaluate β for a single one-dimensional (s-wave) collision. This approximation has been used extensively in the theory of energy transfer (see Herzfeld and Litovitz 1959, Mies and Shuler 1962). In practice, one also often assumes that the potentials that determine the relative motion are the same in the initial and final states, $\langle m|V|m\rangle = \langle n|V|n\rangle$, independently of the internal state (see, however, Mies 1964), or that $V(R)$ is the average of $V_{m,m}(R)$ and $V_{n,n}(R)$ (see, for example, Marchi and Smith 1965).

We can also evaluate the summation over l in eqn (4.62) using the stationary phase method. (For full details see the discussion of the elastic case in Bernstein 1965, section V. Higher-order corrections are discussed by Smith, Mason, and Vanderslice 1965.) For large l we use the result

$$P_l(\cos\theta) \simeq [2/\pi(l+\tfrac{1}{2})\sin\theta]^{\frac{1}{2}} \sin[(l+\tfrac{1}{2})\theta + \tfrac{1}{4}\pi]$$

(valid for $\sin\theta \geqslant 1/l$) in the series (4.62) and convert the sum into an integral

$$f_{m,n} \simeq [-2/\pi k_m k_n \sin\theta]^{-\frac{1}{2}} \int_0^\infty dl \, (l+\tfrac{1}{2})^{\frac{1}{2}} [\exp(iB_+) - \exp(iB_-)] \beta_l^{m,n},$$

where

$$B_{\pm} = \delta_l^n + \delta_l^m \pm [(l+\tfrac{1}{2})\theta + \tfrac{1}{4}\pi].$$

Significant contribution to the integral over l will come from those values that make the Bs stationary,

$$\left.\frac{\partial B_{\pm}(l)}{\partial l}\right|_{l=L} = 0 \quad \text{or} \quad \Theta = \frac{\partial \delta_l^m}{\partial l} + \frac{\partial \delta_l^n}{\partial l}, \qquad (4.69)$$

where Θ is the deflexion function, $\theta = |\Theta|$. Equation (4.69) shows two contributions to the deflexion, one from the relative motion before the interaction ($\partial \delta_l^n/\partial l$) and one from the relative motion after the interaction ($\partial \delta_l^m/\partial l$). In the purely elastic case§ we have $\Theta = 2\partial \delta_l/\partial l$.

Let $B(L)$ be a stationary value of $B(l)$. We evaluate the integral by expanding $B(l)$ about L,

$$B(l) = B(L) + \frac{1}{2}\left(\frac{\partial^2 B}{\partial l^2}\right)_{l=L}(l-L)^2 + \ldots, \quad (4.70)$$

and by evaluating all quantities which depend weakly on l at $l = L$. Thus

$$f_{m,n} \simeq i \sum_L [(2L+1)/\pi k_m k_n \sin\theta]^{\frac{1}{2}} \exp[iB(L)]\beta_L^{m,n} J, \quad (4.71)$$

where

$$J = \int_0^\infty dl \exp\left[i\left(\frac{\partial^2 B}{\partial l^2}\right)_{l=L}(l-L)^2/2\right]$$

and the summation over L is a summation over all the values of L that make either of the Bs stationary. The integral over l is a Fresnel-type integral. We put

$$\Theta'_l = \frac{\partial^2 B(l)}{\partial l^2}$$

and find

$$J = (2\pi/|\Theta'_L|)^{\frac{1}{2}} \exp[i\pi\Theta'_L/4|\Theta'_L|].$$

Each term in eqn (4.71) can thus be written as

$$[(2L+1)/2k_m k_n \sin\theta |\Theta'_L|]^{\frac{1}{2}} 2i\beta_L^{m,n} \exp(i\gamma_L), \quad (4.72)$$

where $\gamma_L = \delta_L^n + \delta_L^m - (L+\tfrac{1}{2})\Theta_L + (\Theta'_L/|\Theta'_L| + \Theta_L/|\Theta_L|)(\pi/4)$.

The term in the square bracket can be interpreted as a cross-section. To see this we consider first the purely elastic case, where $k_m = k_n = k$, and $2i\beta_l = \exp(2i\delta_l) - 1$. The term -1 does not contribute to the integral (since it corresponds to scattering in the forward direction) and so $(2i\beta_L) \equiv \exp(2i\delta_L)$. Introducing the impact parameter b by $bk = L + \tfrac{1}{2}$ we get for the pure-elastic case

$$\frac{d\sigma_c}{d\omega} = |f(k,\theta)|^2 = b/\sin\theta \left|\frac{\partial\Theta}{\partial b}\right|_{bk=L+\frac{1}{2}}, \quad (4.73)$$

which is the correct classical result. In deriving this result we have intentionally failed to take account of the possibility that $f(k,\theta)$ is a sum of several terms. In this case we shall have the possibility of interference between these terms when we superpose the amplitudes (see Bernstein 1965). If we neglect the interference between different L values that lead to scattering into the same angle we can write

$$\frac{d\sigma_c}{d\omega} = \sum_{b(L)} b/\sin\theta \left|\frac{\partial\Theta}{\partial b}\right|_{b(L)}. \quad (4.74)$$

§ In the semi-classical region we can replace $\beta_k(b)$ by β_{kl} in eqn (4.46) and use the asymptotic value
$$J_0(x) \to (\pi/2x)^{\frac{1}{2}} \sin(x + \tfrac{1}{4}\pi).$$

To extend the discussion to the inelastic case we define the deflexion function§

$$\Theta^{m,n} = \tfrac{1}{2}(\Theta^n + \Theta^m) = \partial \delta_l^m/\partial l + \partial \delta_l^n/\partial l, \qquad (4.75)$$

which is the average of the deflexion functions for elastic scattering when the internal state m is occupied (Θ^m) and when the internal state n is occupied (Θ^n). We also introduce the (geometric) average impact parameter $b(k_m k_n)^{\frac{1}{2}} = L + \tfrac{1}{2}$, and put

$$\frac{\mathrm{d}\sigma_c^{m,n}}{\mathrm{d}\omega} = b/\sin\theta \left|\frac{\partial \Theta^{m,n}}{\partial b}\right| \qquad (4.76)$$

when we can write

$$f_{m,n} = \sum_{b(L)} \left|\frac{\mathrm{d}\sigma_c^{m,n}}{\mathrm{d}\omega} P_{m,n}(b)\right|^{\frac{1}{2}} \exp(i\gamma_b) \qquad (4.77)$$

where, from eqns (4.72) and (4.73), we can identify $2\beta_L^{m,n}$ with $P_{m,n}^{\frac{1}{2}}[b(L)]$. Neglecting interference between the amplitudes with different L values, we have in the classical unit

$$\frac{\mathrm{d}\sigma_{m,n}}{\mathrm{d}\omega} = \sum_{b(L)} \frac{\mathrm{d}\sigma_c^{m,n}}{\mathrm{d}\omega} P_{m,n}(b). \qquad (4.78)$$

Finally, if $E \gg E_m - E_n$ and we make the approximation

$$\langle n|V|n\rangle = \langle m|V|m\rangle$$

(which is correct for rigid spheres, for example), $\delta_l^m = \delta_l^n$ and so σ_c does not depend on the internal state.

In deriving eqn (4.78) we have gone beyond the semi-classical approximation, since we have superposed probabilities rather than amplitudes. Most of the available experimental evidence contradicts this approximation. (See, for example, the interference patterns reported by Dworetsky *et al.* 1967.) However, in truly reactive scattering it is very often the repulsive branch of the interaction that leads to the reaction. In such a case there may well be only one stationary L value. Within the semi-classical approximation we have truncated the expansion (4.70) after the first correction, which is not always sufficient (see Ford and Wheeler 1959). From a scattering point of view we have used a DWB approximation which is a weak-coupling approximation.

To avoid making the DWB approximation we can use the exact result for the component of the wavefunction that corresponds to inelastic scattering (cf. eqn (2.5.72)) $G_1^+ V_2 \psi^+$. G_1^+ is the outgoing Green's operator for the Hamiltonian $H_0 + V_1$ and since by assumption V_1 is a

---

§ Recall that by assumption the scattering is in a plane since l is conserved.

central potential we can perform a simple partial wave expansion. Thus, using eqns (2.5.73) and (4.63),

$$\lim_{R\to\infty} \langle \mathbf{R}, m | G_1^+ | m, \mathbf{R}' \rangle = \sum_l -(\mu/2\pi\hbar^2) R^{-1} \exp(ik_m R) \times$$
$$\times \frac{U_{k_m l}^m(R')}{R'} \exp(i\delta_l^m) P_l[\cos(\pi-\theta)], \quad (4.79)$$

where $E = E_m + \hbar^2 k_m^2/2\mu$ and $\cos\theta = \hat{\mathbf{R}}.\hat{\mathbf{R}}'$. Consequently we can always write down

$$f_{m,n} = (k_m k_n)^{-\frac{1}{2}} \sum_l (2l+1)\exp(i\delta_l^m)\beta_l^{m,n} P_l(\cos\theta), \quad (4.80)$$

where $\beta_l^{m,n}$ is no longer a real quantity and, since the phase of $\beta_l^{m,n}$ is not known from elastic scattering only, we have lost the ability to interpret

$$\Theta^{m,n} = \frac{\partial \delta_l^m}{\partial l} + \frac{\partial \arg \beta_l^{m,n}}{\partial l}$$

in a simple way. The reason is quite obvious. When we go beyond the DWB approximation the transition $n \to m$ can proceed via any number of intermediate stages, so that the observed deflection receives contributions from all these possible paths.

To gain a qualitative understanding of the observed differential cross-section for inelastic processes one can therefore try the form (see, for example, Marchi and Smith 1965, Green and Johnson 1966, Rosenfeld and Ross 1966),

$$f_{m,n} = \left(\frac{k_n}{k_m}\right)^{\frac{1}{2}} \sum_i \left(P_i \frac{d\sigma_i}{d\omega}\right)^{\frac{1}{2}} \exp(i\gamma_i), \quad (4.81)$$

where the summation is over the possible postulated paths for the transition.

Several 'direct' (section 3.2.3) reactions, for which a DWB treatment is expected to be valid, have been studied using crossed molecular-beam techniques. It is often observed that in the region where several stationary values of L can contribute to eqn (4.77) (namely at angles less than the rainbow (cf. Fig. 2.2) angle) the elastic differential cross-section has the appearance expected for a non-reactive system of similar characteristics as the reactants. However, where only one term in eqn (4.77) contributes§ the elastic differential cross-section shows a marked decline (cf. Fig. 3.5), which is interpreted as due to the occurrence of inelastic processes. If we interpret $1-P_{n,n} = P(b)$ as the total inelastic-opacity function we can write, using eqn (4.78) with one term,

$$P(b) = (d\sigma_c^{n,n} - d\sigma_{n,n})/d\sigma_c^{n,n}. \quad (4.82)$$

§ Namely, at angles above the rainbow or, in other words, for impact parameters below some critical value.

This relation has been used by Greene, Ross, and their co-workers for the analysis of reactive scattering. (See, for example, Beck, Greene, and Ross 1962, Greene, Moursund, and Ross 1965.) The 'experimental'

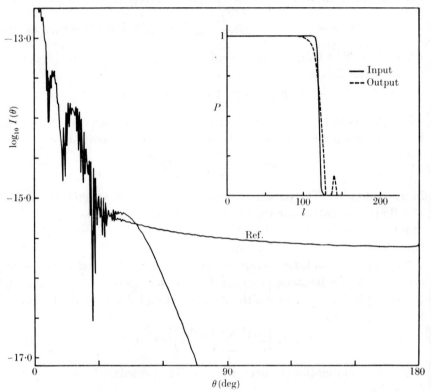

FIG. 3.5. An opacity function analysis of a differential elastic cross-section (Bernstein and Levine 1968). The reference case corresponds to $\eta_l = 0$ in eqn (4.27) (i.e. no inelastic scattering) and the phase shifts δ_l are the same as those used in Fig. 2.3. The second curve was computed using an opacity function given by the input graph in the insert (i.e. a slightly rounded step). The two curves were analysed using eqn (4.78). Only one term contributes in the present case, and the deflection angle is given in Fig. 2.2. The resulting opacity function is shown as the output graph in the insert.

points in Fig. 3.8 were obtained in this way. $d\sigma_C^{n,n}$ was computed by assuming a reasonable central potential between the reactants and adjusting the potential parameters to predict correctly $d\sigma_{n,n}$ for angles less than and about the rainbow angle.

A computational study of this technique was made by Bernstein and Levine (1968). An example is given in Fig. 3.5.

It should be remembered, however, that eqn (4.82) rests on several assumptions. Irrespective of the validity of its derivation, $P(b)$, as defined, is the opacity function for all inelastic processes, whether reactive or not. It is not the opacity function for reactive collisions only. Moreover, the simple result, (4.82), is only valid in the semi-classical limit, when β_l depends smoothly on l (so that eqn (4.71) holds), and when only one term contributes to eqn (4.78).

3.4.2. *The impact parameter method*

An alternative approach to obtain an opacity function is by the impact parameter method (Mott 1931, Bates 1962, Bates and Holt 1966). In this method the relative motion is treated classically, and in the simplest form any deviations from a straight-line trajectory are ignored. The mutual interaction $V(\mathbf{R})$ is considered as $V[\mathbf{R}(t)]$ with t as the independent variable, and the time dependence of the relative motion is obtained either from Newton's equations (neglecting the change in the kinetic energy due to the transitions) or from a straight-line approximation

$$\mathbf{R} = \mathbf{b} + \mathbf{v}t, \quad \mathbf{b} \cdot \mathbf{v} = 0, \qquad (4.83)$$

where b is the impact parameter and $\mathbf{v}$ the initial relative velocity. Scattering theory is then formulated from a time-dependent point of view, and time integrations are converted to coordinate integrations using eqn (4.83). In the lowest order in perturbation theory (cf. eqn (2.7.24)), the amplitude to find the internal state m occupied after the collision if the internal state n was initially occupied is

$$B(m, E) = (-i/\hbar v) \int_{-\infty}^{\infty} V_{m,n}(\mathbf{R}) \exp(i\omega_{m,n} Z/\hbar v)\, \mathrm{d}Z, \qquad (4.84)$$

where we have taken $\mathbf{v}$ along the Z axis and have put

$$\langle \mathbf{R}, m | V | n, \mathbf{R} \rangle = V_{m,n}(\mathbf{R}) \quad \text{and} \quad \omega_{m,n} = E_m - E_n.$$

The relation of this result to the conventional Born approximation (eqn (2.3.61)) is apparent, $\omega_{m,n}/v$ being the momentum transfer. The opacity function for the transition $n \to m$ is then given by

$$P_{m,n}(b) = |B(m, E)|^2. \qquad (4.85)$$

The present approximation, which can be written as

$$B(m, E) = \langle m|(-i/\hbar) \int_{-\infty}^{\infty} \exp(iH_0 t/\hbar) V \exp(-iH_0 t/\hbar)\, \mathrm{d}t | n \rangle = \langle m | 2iA | n \rangle \qquad (4.86)$$

(A is defined by the last expression), violates the optical theorem. To obtain a unitary approximation we may try to assume that eqn (4.86) is the lowest-order result for the series

$$B(m, E) = \langle m|\exp(2iA)|n\rangle. \qquad (4.87)$$

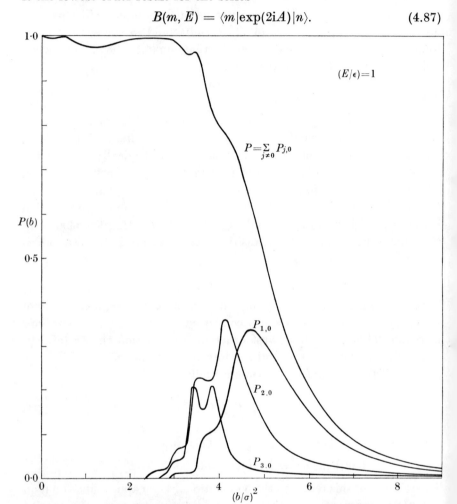

Fig. 3.6. The opacity functions $P_{j',j} = \sum_{m_j}\sum_{m_{j'}} |S_{j'm_{j'},jm_j}|^2$ in the sudden approximation, from the computation of Bernstein and Kramer (1966) for rotational excitation by an unisotropic potential. The quantum numbers j' and j refer to the rotational states. The area under each curve is $\sigma_{j',j}$ in units of $\pi\sigma^2$. Indicated also is the total inelastic opacity function. For $b > b^*$, $(b^*/\sigma)^2 \simeq 5$, the three transitions indicated explicitly, account for more than 95 per cent of the total, while for $b < b^*$, the results correspond to the statistical approximation (sections 3.5.1 and 3.5.2) where many states equally contribute.

(The approximation (4.87) was discussed by Takayanagi (1963) and Callaway and Bauer (1965).) In other words, if we define a matrix $\mathbf{V}(t)$ by

$$\hbar[\mathbf{V}(t)]_{m,n} = \langle m|V|n\rangle\exp(-i\omega_{n,m}t/\hbar) \qquad (4.88)$$

then an approximate but unitary **S** matrix (eqn (2.7.22)) is given by

$$\mathbf{S} = \exp\left[-i \int_{-\infty}^{\infty} \mathbf{V}(t)\,dt\right]. \tag{4.89}$$

If we make the further approximation of neglecting the changes in internal energies, $\omega_{n,m} \simeq 0$, we obtain the **S** matrix of Kramer and Bernstein (1964). A result of such computation is presented in Fig. 3.6.

To derive eqn (4.89) in the impact-parameter method we consider the total Hamiltonian as $h+V(t)$, where h is the Hamiltonian for the internal degrees of freedom

$$h\varphi_n = E_n \varphi_n$$

and $V(t) = V[\mathbf{R}(t)]$. Due to the (classical) relative motion the mutual interaction is a time-dependent interaction. We expand the total wavefunction as

$$\psi(t) = \sum_n b_n(t)\varphi_n \exp[-iE_n t/\hbar] \tag{4.90}$$

and substitute in Schrödinger's time-dependent equation,

$$i\hbar \frac{\partial \psi(t)}{\partial t} = (h+V(t))\psi(t),$$

premultiply by $\varphi_m^* \exp(iE_m t/\hbar)$, and integrate over internal coordinates to find

$$\frac{db_m(t)}{dt} = (-i/\hbar) \sum_n V_{m,n}(t)\exp(-i\omega_{n,m} t/\hbar)b_n(t) \tag{4.91}$$

or, in matrix form,

$$\frac{d\mathbf{B}(t)}{dt} = -i\mathbf{V}(t)\mathbf{B}(t), \tag{4.92}$$

where $\mathbf{V}(t)$ is defined by eqn (4.88). We can conjecture the solution

$$\mathbf{B}(t) = \exp\left[-i \int_{-\infty}^{t} \mathbf{V}(t')\,dt'\right]\mathbf{B}(-\infty), \tag{4.93}$$

and eqn (4.89) follows in the limit $t \to \infty$. To check that eqn (4.93) is a solution, we substitute it in the differential equation (4.92). To differentiate the exponential we have to expand it in a power series. Considering the third term, for example, we have

$$-\frac{d}{dt}\left|\int_{-\infty}^{t} \mathbf{V}(t')\,dt'\right|^2 = -\mathbf{V}(t)\int_{-\infty}^{t} \mathbf{V}(t')\,dt' - \int_{-\infty}^{t} \mathbf{V}(t')\,dt'\cdot\mathbf{V}(t),$$

since $\mathbf{V}(t)$ need not commute with $\int_{-\infty}^{t} \mathbf{V}(t')\,dt'$. If $\mathbf{V}(t)$ commutes with the integral, eqn (4.93) is an exact solution.

Our approximate solution is, in fact, the first approximation to the exact solution proposed by Magnus (1954; see also the extensive discussion by Pechukas and Light 1966a),

$$\mathbf{B}(t) = \exp\left[\sum_n \mathbf{A}_n(t)\right]\mathbf{B}(-\infty), \tag{4.94}$$

where the A_ns are anti-Hermitian operators, $A_n^\dagger = -A_n$, so that truncating the Magnus approximation to any order will always yield a unitary **S** matrix

$$\mathbf{A}_1(t) = -i \int_{-\infty}^{t} \mathbf{V}(t')\, \mathrm{d}t',$$

$$\mathbf{A}_2(t) = \tfrac{1}{2} \int_{-\infty}^{t} \mathrm{d}t' \int_{-\infty}^{t'} \mathrm{d}t''\, [\mathbf{V}(t''), \mathbf{V}(t')].$$

Higher-order terms involve higher-order commutators. All these terms would vanish in the absence of any time correlation between $\mathbf{V}(t'')$ and $\mathbf{V}(t')$, namely, in the limit of impulsive collisions. Also if we can put $\mathbf{V}(t) = \mathbf{V}f(t)$, where **V** is time-dependent and $f(t)$ is an arbitrary function of time, the approximation becomes the exact solution. The approximation requires $pa \gg \hbar$ for the use of a classical trajectory and a momentum transfer $\Delta p \ll p$ to ensure an impulsive limit. In the extreme case we have $\Delta pa \ll \hbar$ or

$$\tau_c \ll \hbar/\Delta E, \tag{4.95}$$

where $\tau_c = a/v$ is the duration of the collision and is required to be much shorter than the 'duration', $\hbar/\Delta E$, of the internal transition. Equation (4.95) is known as the 'sudden' limit in contrast to the other extreme, the adiabatic limit. In the sudden limit $(\omega_{m,n}\tau_c/\hbar) \simeq 0$ and eqn (4.89) reduces to the **S** matrix in the sudden approximation.

We can also formulate a distorted wave-Magnus approximation (Levine 1967c; see also Cross 1967) by analogy to the distortion correction of the usual Born approximation (section 2.5.2). If $\{|n\rangle\}$ is the set of internal states of the non-interacting molecules we resolve the mutual interaction $V = V_1 + V_2$ into a part V_1 which leads to distortion

$$V_1 = \sum_n |n\rangle V_{n,n} \langle n| \tag{4.96a}$$

and a part V_2, which can cause only internal transitions

$$V_2 = \sum_{n \neq m} |n\rangle V_{n,m} \langle m|. \tag{4.96b}$$

The zero-order Hamiltonian H_0' is taken as

$$H_0' = h + V_1[\mathbf{R}(t)] \tag{4.97}$$

with the eigenfunctions φ_n and the eigenvalues $E_n + V_{n,n}(t)$, and we use the expansion

$$\psi(t) = \sum_n d_n(t)\varphi_n \exp\left[-i\left(E_n t/\hbar + \int V_{n,n}(t')\,dt'/\hbar\right)\right], \quad (4.98)$$

so that eqn (4.92) is modified to

$$\frac{d\mathbf{D}(t)}{dt} = -i\mathbf{V}'(t)\mathbf{D}(t). \quad (4.99)$$

Here
$$\hbar[\mathbf{V}'(t)]_{m,n} = (1-\delta_{m,n})V_{m,n}(t)\exp[-i\omega_{n,m}(t)/\hbar], \quad (4.100)$$

with
$$\omega_{m,n}(t) = (E_m - E_n)t + \int^t (V_{m,m} - V_{n,n})\,dt'. \quad (4.101)$$

The S matrix in the first Magnus approximation is then given, in the H_0' representation, by

$$\mathbf{S}' = \exp\left[-i\int_{-\infty}^{\infty} \mathbf{V}'(t)\,dt\right]. \quad (4.102)$$

Comparing eqns (4.98) and (4.91),

$$d_n(t) = b_n(t)\exp\left[i\int^t V_{n,n}(t')\,dt'/\hbar\right].$$

We make the phase unique by requiring that $d_n(0) = b_n(0)$, when we can write
$$\mathbf{S} = \exp(i\mathbf{\Delta})\mathbf{S}'\exp(i\mathbf{\Delta}) \quad (4.103)$$

where, provided $V_{n,n}$ is a central potential,

$$\Delta_{n,m} = \frac{\delta_{n,m}}{2\hbar}\int_{-\infty}^{\infty} V_{n,m}(t)\,dt$$

is the semiclassical phase shift, due to the distortion.

The main contribution to the transition $m \leftarrow n$ is from those values of R where the phase of the matrix element $(\mathbf{V}')_{m,n}$ is stationary. Putting

$$\frac{\partial \omega_{m,n}(t)}{\partial t} = 0,$$

we find that this occurs when R is determined by

$$E_m - E_n = \Delta E = V_{n,n}(R) - V_{m,m}(R)$$

or
$$E_n + V_{n,n}(R) = E_m + V_{m,m}(R). \quad (4.104)$$

If eqn (4.104) can be satisfied for a particular value of $R = R_c$, we say that a crossing point exists. It may be that eqn (4.104) has roots when R is considered as a complex variable. In this case the integral

$$\int_{-\infty}^{\infty} \mathbf{V}'(t)\,dt$$

may still be evaluated by the method of saddle point integration (Morse and Feshbach 1953). When a real crossing point exists we can evaluate the integral by a stationary phase method. Let

$$\hbar\gamma_{m,n} = \int_{-\infty}^{\infty} V_{m,n}(t)\exp[-i\omega_{n,m}(t)/\hbar]\,dt \qquad (4.105)$$

and let $R = R_c$ be the crossing point. Then, recalling that $R = R(t)$,

$$\omega_{n,m}(t) = \omega_{n,m}(R_c) + \frac{1}{2}\left(\frac{\partial^2 \omega_{n,m}}{\partial t^2}\right)(t-t_c)^2 + \dots$$

where $R_c = R(t_c)$, and so, by taking stationary quantities outside the integral sign,
$$\hbar\gamma_{m,n} = V_{m,n}(R_c)\exp[-i\omega_{n,m}(R_c)]J,$$
where

$$J = \int_0^\infty \exp\left\{-i\frac{\partial}{\partial R}(V_{n,n}-V_{m,m})_{R=R_c}(R-R_c)^2/2\hbar v\right\}\frac{dR}{v}$$

and we have put $R = vt$. Regarding the velocity as constant (linear trajectory approximation), J is a Fresnel-type integral,

$$J = \frac{1}{v}\left(\frac{2\pi\hbar v}{|F|}\right)^{\frac{1}{2}}\exp(-i\pi F/4|F|),$$

where
$$F = \frac{\partial}{\partial R}(V_{n,n}-V_{m,m})_{R=R_c}$$

is the difference between the forces at the crossing point. We thus obtain

$$\hbar\gamma_{m,n} = \frac{1}{v}\left(\frac{2\pi\hbar v}{|F|}\right)^{\frac{1}{2}}V_{m,n}(R_c)\exp[i\theta(R_c)], \qquad (4.106)$$

where the last factor is the contribution of all the phase factors.

Expanding the exponential in eqn (4.102) we have, to the lowest order,

$$|S_{m,n}|^2 = |\gamma_{m,n}|^2 = \frac{2\pi|V_{m,n}(R_c)|^2}{\hbar v|(\partial/\partial R)(V_{n,n}-V_{m,m})_{R=R_c}|} \qquad (4.107)$$

and, for $m \neq n$, $\qquad \sigma_{m,n} = 2\pi\int b|S_{m,n}|^2\,db. \qquad (4.108)$

To compute the cross-section we express the relative velocity along the line of relative motion, $v(R)$ at $R = R_c$ as

$$v_n^2 = V_{n,n}(R_c) + v_n^2 b^2/R_c^2 + [v(R_c)]^2$$

(cf. eqn (4.4)), where v_n is the initial relative velocity. Changing the variable of integration to b/R_c we see that $\sigma_{m,n}$ is indeed proportional to πR_c^2.

In general, we can try to write

$$\sigma_{m,n} = \pi b^{*2},$$

where b^* is determined by $|S_{m,n}(b^*)|^2 = \tfrac{1}{2}$. In the simplest approximation we can equate b^* to b_m (eqn (4.5)), where b_m is determined not from the dynamics but from some reaction criterion. An alternative possibility for estimating a b^* for the total inelastic cross-section,

$$\sum_m \sigma_{m,n} = \pi b^{*2}$$

is to compute b^* from the Born approximation, eqn (4.86). The total inelastic cross-section estimated in this fashion is found to be very near the exact sudden result, computed by summing eqn (4.108) over $m \neq n$ (Bernstein and Kramer 1966). We can thus conclude that for $b > b^*$ most of the inelastic transitions are to final states that are allowed by first-order perturbation theory.

Our lowest-order approximation, eqn (4.107), is equivalent to the lowest-order approximations in the semi-classical analysis of a two-state problem (Stueckelberg 1932, Zener 1932, Landau and Lifshitz 1965). The distorted wave Magnus matrix element is essentially equivalent to the semi-classical matrix element when we use the W.K.B. approximation for the relative motion. The semi-classical method has also been used (Landau and Lifshitz 1965, Nikitin 1965a; this paper contains several references to previous Russian work on this problem), when a crossing point occurs only for a complex value of R.

The assumption that the phase factor in eqn (4.105) is the most rapidly varying part of the integrand need not be valid (Coulson and Zalewski 1962). In the sudden limit

$$\frac{\omega_{m,n} a}{v} < \hbar,$$

where a is the range of $V_{m,n}$ and the phase is nearly constant. Then

$$S_{m,n} \simeq \frac{1}{\hbar v} \int V_{m,n}(\mathbf{R}) \, \mathrm{d}R. \tag{4.109a}$$

When the approximation of constant velocity is not made one replaces the time integration by an integration along the classical trajectory under the influence of an average potential $V(R)$. From the first integral of Newton's second law

$$\frac{\mathrm{d}R}{\mathrm{d}t} = \left\{\frac{2}{\mu}[E - V(R) - l^2/2\mu R^2]\right\}^{\tfrac{1}{2}} \tag{4.110}$$

and if R_0 is the classical turning point, $(dR/dt)_{R=R_0} = 0$,
$$R_0^2[1-V(R_0)/E] = b^2,$$
we have
$$S_{m,n} \simeq \int_{R_0}^{\infty} V_{m,n}(\mathbf{R})(dR/dt)^{-1} \, dR. \tag{4.109b}$$

In practice one often neglects $V(R)$ in eqn (4.110), so that
$$\frac{dR}{dt} = v(1-b^2/R^2)^{\frac{1}{2}}$$
and $R_0 = b$. When this form is used in eqn (4.109b) it is clear that $S_{m,n}(b)$ tends to zero for high values of b.

Within the framework of the Magnus approximation, the two-state problem can be solved exactly. To see this we note that the diagonal elements of $\mathbf{S}$ can be multiplied by a constant phase without modifying the computed probabilities. We thus define
$$\mathbf{S}'' = \exp\left[-i\int_{-\infty}^{\infty} \mathbf{V}(t) \, dt - \frac{i}{N} \mathrm{Tr} \int_{-\infty}^{\infty} \mathbf{V}(t) \, dt \cdot \mathbf{I}\right]. \tag{4.111}$$

For a two-by-two problem we put $\mathbf{S}'' = \exp(-i\mathbf{A})$ so that
$$A_{11} = -A_{22} = \frac{1}{2\hbar} \int_{-\infty}^{\infty} [V_{11}(t) - V_{22}(t)] \, dt,$$
$$A_{12} = A_{21}^* = \int_{-\infty}^{\infty} [\mathbf{V}(t)]_{1,2} \, dt, \tag{4.112}$$

and $\mathbf{A}^2$ is proportional to the unit matrix. With the definition
$$\alpha^2 = A_{11}^2 + A_{12}^2$$
one can show that
$$S_{11} = \cos\alpha - iA_{11}(\sin\alpha)/\alpha \tag{4.113a}$$
and
$$S_{12} = -iA_{12}(\sin\alpha)/\alpha. \tag{4.113b}$$
The two other matrix elements are determined by unitarity. Thus
$$|S_{21}(b)|^2 = |A_{12}|^2 (\sin^2\alpha)/\alpha^2. \tag{4.114}$$

The transition is said to be a symmetric (or resonance) transition when $V_{11}(t) = V_{22}(t)$. (For example, excitation transfer between identical atoms or molecules, $B+B^* \to B^*+B$.) Then $A_{11} = 0$ and
$$|S_{21}(b)|^2 = \sin^2\left\{\int_{-\infty}^{\infty} [\mathbf{V}(t)]_{21} \, dt\right\}. \tag{4.115}$$

Equation (4.115) is identical to that derived by Rosen and Zener (1932; also Skinner 1961). For other applications see Gurnee and Magee (1957),

Mori, Watanabe, and Katsuura (1964). Nyeland and Bak (1965) have applied this model to reactive collisions of the type $A+BA' \to BA+A'$ where A and A' are the same species.

For sufficiently high energies we can regard the relative velocity as constant and write eqn (4.115) as

$$|S_{21}(b)|^2 = \sin^2\left[\frac{1}{\hbar v}\int_0^\infty V_{12}(\mathbf{R})\,\mathrm{d}R\right]. \qquad (4.116)$$

The opacity is thus an oscillating function of v^{-1}. The physical significance of this observation is clear from Massey's criterion. (See also Lichten 1963.) For an efficient excitation transfer the duration of the collision τ_c should equal $\hbar/\Delta E$, where ΔE is the energy difference, and eqn (4.116) can be rewritten

$$|S_{21}|^2 = \sin^2(\Delta E \tau_c/\hbar), \qquad (4.117)$$

where
$$\Delta E = \frac{1}{a}\int_0^\infty V_{12}(\mathbf{R})\,\mathrm{d}R.$$

In other words, we can consider the initial $(B+B^*)$ and final (B^*+B) states as two degenerate states that are interacting via V_{12}, and so are 'resonating' between the two structures. For an efficient transfer the collision time should be an odd multiple of the time $(\hbar/\Delta E)$ of transition between the two structures. Our analysis was based on a classical description of the nuclear motion, but one can show that in the quantal two-state approximation the opacity $|S_{21}|^2$, (eqn (4.22)), can be written as $\sin^2 \alpha_l$. However, eqn (4.116) is based on the sudden limit, (4.109a). When we use the more accurate form (4.109b), the oscillations are damped for high b values. Interesting experimental results were presented by Lockwood and Everhart (1962).

The impact approximation has been used extensively whenever a large number of partial waves contribute to a given transition and many elements of $\mathbf{S}$ are required (for example, Anderson 1949). In this case one can use the approximation of neglecting multiple transitions by putting $V_{m,m'} = 0$ if neither m nor m' is the initial-state index. The partitioning technique can then be applied very easily to solve the equations of motion.

3.4.3. *The adiabatic theory of reactive collisions*

In an adiabatic theory, the adiabatic approximation, in the sense of section 3.2.1, is employed with respect to some of the degrees of freedom of the colliding molecules. Common to all adiabatic theories is the assumption that the adiabatic approximation can be applied to the electronic motion (London 1929, Pelzer and Wigner 1932, Golden 1949).

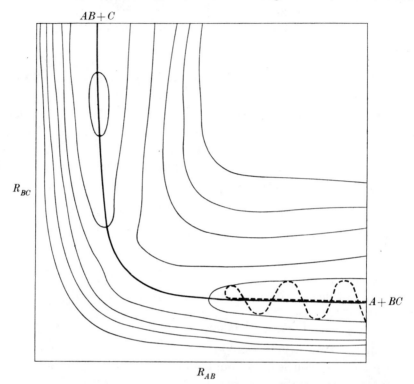

FIG. 3.7. A schematic potential energy surface for a linear collision of A and BC. The contours represent equipotentials. (The shallow basin in the product valley represents a mutual dispersion type attraction between the products.) The full line indicates the reaction coordinate (s) while the broken line indicates a vibrationally non-adiabatic trajectory that leads to internal vibrational excitation of the reactants. The trajectory analysis of the system $H+H_2$ (Karplus, Porter, and Sharma 1965) suggests that in that system such non-reactive trajectories are atypical. The reactive trajectories were more often vibrationally non-adiabatic. See also Marcus (1966b).

In this approximation we consider that the solution of the L.S. equation can be approximated by

$$\psi_i^+(\mathbf{X},\mathbf{r}) = \chi_i(\mathbf{r}|\mathbf{X})G_i^+(\mathbf{X}), \qquad (4.118)$$

where $\mathbf{X}$ is the set of nuclear coordinates, one of which is the relative separation of the reactants, and $\mathbf{r}$ is the set of electronic coordinates. The nature of the approximation is to replace the L.S. equation by

$$(K_\mathbf{r}+V)\chi_i(\mathbf{r}|\mathbf{X}) = E_i(\mathbf{X})\chi_i(\mathbf{r}|\mathbf{X})$$

and
$$[K_\mathbf{X}+E_i(\mathbf{X})]G_i^+(\mathbf{X}) = EG_i^+(\mathbf{X}) \qquad (4.119)$$

with suitable boundary conditions on $G_i^+(\mathbf{X})$. (The corrections to eqns (4.118) and (4.119) are discussed in section 3.2.1.)

The equivalent potential $E_i(\mathbf{X})$ is a function of $3N-3$ nuclear co-ordinates, where N is the number of nuclei, and can be thought of as defining a surface in a $(3N-3)$-dimensional space (see Fig. 3.7). It is commonly referred to as a potential energy surface, and can be computed variationally using the Rayleigh–Ritz method. (For some recent computations see Edmiston and Krauss 1965, Harris, Micha, and Pohl (1965).)

Such *a priori* computations to a 'chemical accuracy' (say, better than $\pm 0\cdot 1$ eV) are currently feasible only for very simple systems, although the situation is rapidly improving. In the absence of *a priori* computations one often uses semi-empirical methods (see Laidler and Polanyi 1965, Johnston 1966) or simply approximates the main features on the basis of chemical intuition. Two approximations that are often invoked are to replace $E_i(\mathbf{X})$ by either a sum of two-body terms, or a sum of a spherical part and angular terms. Thus for a collision $A+BC$ we can take

$$E_i = V(\mathbf{R}_{AB})+V(\mathbf{R}_{AC})+V(\mathbf{R}_{BC}) \qquad (4.120\,\mathrm{a})$$

or

$$E_i = V(R_i)+ \sum_n V_n(R_i, R_{BC})P_n(\hat{\mathbf{R}}_i \cdot \hat{\mathbf{R}}_{BC}), \qquad (4.120\,\mathrm{b})$$

where R_i is the coordinate of relative motion of A and BC. Such approximations are very natural for the purpose of computing vibrational or rotational excitation cross-sections. Moreover, at large relative separation the approximation (4.120 a) improves and asymptotically

$$E_i \to \begin{cases} V(\mathbf{R}_{BC}) & R_i \to \infty \\ V(\mathbf{R}_{AB}) & R_f \to \infty \end{cases}$$

where R_f is the coordinate of relative motion of AB and C. In the same fashion

$$\chi_i(\mathbf{r}|\mathbf{X}) \to \begin{cases} \varphi_{A(i)}\varphi_{BC(i)} & R_i \to \infty \\ \varphi_{AB(i)}\varphi_{C(i)} & R_f \to \infty. \end{cases} \qquad (4.121)$$

The 'chemical' nature of the function $\chi_i(\mathbf{r}|\mathbf{X})$ depends on the particular coordinate that we let tend to infinity, and one can construct correlation diagrams indicating the states of the reactants or products to which $\chi_i(\mathbf{r}|\mathbf{X})$ correlates as R_i or R_f increase. The important point (sometimes called the 'adiabatic hypothesis') is that in the adiabatic approximation transitions between the different functions $\chi_i(\mathbf{r}|\mathbf{X})$ cannot take place. Thus in terms of eqn (4.121) the transitions from A in the internal (electronic) state i to the internal (electronic) state, $j, j \neq i$, or to C in the internal (electronic) state j cannot take place in the adiabatic approximation. With respect to the (electronic) quantum number i the collision is elastic, although it may be a reactive collision from a chemical point of view since the process $A(i)+BC(i) \to AB(i)+C(i)$ can take place where the argument denotes the electronic state.

To determine the cross-section for reactive (or inelastic) processes we have to solve for the functions $G_i^+(\mathbf{X})$ which is, in itself, a major problem, as discussed in Part 2 (see, for example, Micha, 1965). We can try to obtain a qualitative understanding of the main features by considering the classical motion of the nuclei under the potential $E_i(\mathbf{X})$. (See the extensive discussion in Glasstone et al. 1941; also Hammond 1955, Polanyi 1959, and Kuntz et al. 1966.) Recently considerable work was done on obtaining exact numerical solutions for the classical equations of motion. (See the review of Bunker 1966; also Karplus, Porter, and Sharma 1965, Fig. 3.8, and Kuntz et al. 1966.) This procedure is referred to as a trajectory analysis. Classically, of course, a trajectory will either lead to products or not, and one has to average over several trajectories to obtain a reaction probability. The required averaging is over the initial phase of the mutual vibration (and rotation) of BC. In a quantum-mechanical theory this phase is random.§ Two trajectories with identical values for all initial quantum numbers, but with different initial phases can (and do) lead to different outcomes.

The conceptual and dynamical analysis of the collision on a potential $E_i(\mathbf{X})$ can be considerably simplified by the introduction of the 'reaction coordinate' s. Qualitatively the reaction coordinate is defined (Glasstone et al. 1941) as a coordinate of relative motion that changes continuously from the coordinate or relative motion of the reactants (when $s \to -\infty$, R_i large) to the coordinate of relative motion of the products ($s \to \infty$, R_f large). For the reaction $A+BC$ we expect that the physical motion for $s \sim 0$ corresponds to the asymmetric vibration of ABC. It is not at all clear how to construct such a coordinate in general. If, for example (Child 1966), B is infinitely heavy so that the motion is in a plane, we can define
$$s = -R_{BA}\cos\alpha + R_{BC}\sin\alpha, \qquad (4.122)$$
where α is the angle that the tangent to s makes with the R_{BA} axis (cf. Fig. 3.7). The general problem was discussed extensively by Marcus (1966b, 1968; see also Smith 1962b.)

We can now consider the expansion (Hofacker 1963, Child 1966)
$$G_i^+(\mathbf{X}) = \sum_m F_m^i(s)\eta_m(\mathbf{q}),$$

§ The phase is random since the initial number of BC molecules is well defined. See Levine (1966e).

where **q** is the remaining set of nuclear coordinates, and obtain a single set of coupled equations which describe the nuclear motion. The asymptotic behaviour of $F_m^i(s)$ determines whether the collision is reactive and also determines the internal state after the collision.

In the same fashion that we simplified the coupled equations for the electronic motion by the introduction of an adiabatic approximation we can simplify $G_i^+(\mathbf{X})$ by invoking adiabaticity with respect the nuclear motion, for example, by assuming that

$$G_i^+(\mathbf{X}) \simeq F^i(s)\eta_i(\mathbf{q}|s), \qquad (4.123)$$

where $\qquad [K_\mathbf{q}+E_i(\mathbf{X})]\eta_i(\mathbf{q}|s) = E_i(s)\eta_i(\mathbf{q}|s).$

Physically, eqn (4.123) assumes that the motion in the reaction coordinate is under the averaged influence of all other nuclear degrees of freedom (Hirschfelder and Wigner 1939).

Within the framework of the approximation (4.123) the transition state criteria for reaction can be formulated as: when $s > d$ the motion is such that s is monotonically increasing. In other words, once the system has passed the ('saddle') point $s = d$ it will proceed to form the products. Corrections to this picture have been discussed by Hirschfelder and Wigner (1939) who have also shown that when the density of internal states is stationary and the reactants are in equilibrium, the equilibrium is maintained for any value of s. The proof is based on eqn (2.4.80). By the equilibrium assumptions,

$$p_i(s) = p_i(s \to -\infty)\exp\{-\beta[E_i(s)-E_i(s \to -\infty)]\}$$

and

$$p_i(s \to -\infty) = p_j(s \to -\infty)\exp\{-\beta[E_i(s \to -\infty)-E_j(s \to -\infty)]\}.$$

Thus the second relation holds for all s. From the transition state criteria we can evaluate the reaction rate by evaluating the forward probability current at $s = d$, using the technique of section 2.6.4.

The construction of solutions to eqn (4.123) can be further simplified. We can, for example, invoke vibrational adiabaticity

$$\eta_i(\mathbf{q}|s) \simeq g_i(z|s)f_i(\omega|s), \qquad (4.124)$$

where for $s \to -\infty$, z is the vibration of BC; for $s \sim 0$, z is the symmetric vibration of ABC; and for $s \to \infty$, z is the vibration of AB (say $z = R_{BA}\sin\alpha + R_{BC}\cos\alpha$). ω refers to the rotation of ABC. In this approximation we can construct correlation diagrams for the vibrational states.

In the vibrational adiabatic theory non-reactive vibrational excitation cannot take place, and the products appear in a specified vibrational state, determined from the limit of $g_i(z|s)$, $s \to \infty$. Corrections for this approximation can be made in principle using the method of scattering by two potentials (Hofacker 1963, Marcus 1966b). In this approximation $E_i(s)$ is a sum of rotational and vibrational energies. In a transition-state theory the barrier to reaction is $E_i(d)$, while the total energy is

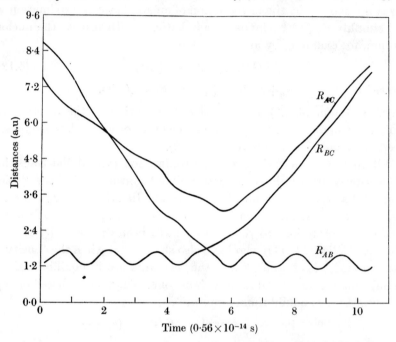

FIG. 3.8. A reactive trajectory for the collision $H + H_2$ computed (classically) by Karplus, et al. (1965). See also Wall, Hiller, and Mazur (1958) for the first trajectory analysis of this system. Initially H_2 is in the ground vibrational and fifth rotational state. The initial relative velocity is $1 \cdot 18 \times 10^6$ cm/s. The opacity function for reaction found by these authors (H_2 in ground vibrational and rotational states, relative velocity $1 \cdot 17 \times 10^6$ cm/s), does not have a strong decline over a small interval of b values, but was better represented by $P(b) \simeq 0 \cdot 39 \cos(\pi b/2 b_m) \theta(b_m - b)$, $b_m \simeq 1 \cdot 85$ a.u.

$e + E_i(s \to -\infty)$ where e is the initial translational energy. In a classical language we can therefore say that the energy $E_i(s \to -\infty) - E_i(d)$ is an additional kinetic energy that is available to surmount the barrier (see Marcus 1964, 1966a). We can then determine the maximum impact parameter that will lead to reaction, from $e + E_i(s \to -\infty) \geqslant E_i(d)$. (The index i depends on the orbital angular momentum.) In practice, the quantitative discussions of the adiabatic theory have been restricted to arguments of this type. Any attempt to remove the restrictive assumptions of vibrational (and rotational) adiabaticity necessarily requires a solution of a coupled-equations problem. The recent computational advances in handling this problem suggests that in the future this technique will replace the classical trajectory-analysis method. In the immediate future one expects that non-adiabatic corrections will be introduced via the Magnus and similar approximations.

3.5. Statistical theories

3.5.0. The physical idea that the rate of a given process can be expressed as the probability p_r of realizing a particular intermediate state times the rate k_r^f of the transition from the intermediate to the final state,

$$\text{rate} = p_r k_r^f, \tag{5.1}$$

is common to a large number of model theories of rate processes, and was part of chemical kinetics long before quantum mechanics. Most quantitative applications were based, however, on quantum-mechanical formulations.

The probability of realizing the required intermediate may depend on the initial state. From microscopic reversibility (for degeneracy averaged rates), $p_r^i k_r^f = p_r^f k_r^i$, or

$$\frac{p_r^i}{k_r^i} = C_r, \tag{5.2}$$

where C_r is a constant, which is a property of the intermediate only, irrespective of the initial state from which it was formed. We can then write

$$\text{rate} = C_r k_r^i k_r^f. \tag{5.3}$$

The rate of formation of the intermediate from the initial state is

$$\sum_f C_r k_r^i k_r^f$$

and the reaction rate can also be written

$$\text{rate} = \frac{k_r^f}{\sum_f k_r^f} \sum_f C_r k_r^i k_r^f. \tag{5.4}$$

The first factor is the fraction of intermediates that break down into the final state and the second factor is the rate of formation of the intermediate. In terms of cross-sections the second term is the cross-section for the formation of the complex while the first factor is essentially an opacity function.

In quantum mechanics we must recognize that there can be several collision intermediates, corresponding to different eigenvalues of constants of motion, and eqn (5.4) should be summed over r.

As an example, we consider the collision of an atom A with a rotor B to form an intermediate complex $[AB]_r$. To compute k_r^i, the rate of breakdown of the complex to the initial state, we use microscopic reversibility to equate the (degeneracy averaged) rates of formation and breakdown. Let the rotor be in the state j, let J be the (conserved) total angular momentum and σ_r^i be the cross-section for the formation of the complex. The rate of complex formation is then, at an energy of E to $E+\mathrm{d}E$ (cf. eqn (2.6.19))

$$(2j+1)(2\pi\hbar)^{-1}\pi^{-1}k^2\sigma_r^i, \tag{5.5}$$

while the rate of breakdown is

$$(2J+1)k_r^i \frac{\mathrm{d}N(E)}{\mathrm{d}E}, \tag{5.6}$$

where $\mathrm{d}N(E)$ is the number of collision complexes, of total angular momentum J, in the energy interval E to $E+\mathrm{d}E$, and we put $D = (\mathrm{d}N/\mathrm{d}E)^{-1}$, so that D is the energy spacing between the levels. (Note that the rates (5.5) and (5.6) are not degeneracy averaged, but include the degeneracy factors as is required by eqns (2.8.39) and (2.8.40).) Thus§

$$\sigma_r^i = \frac{\pi}{k^2} \cdot \frac{2J+1}{2j+1} \cdot 2\pi \frac{\hbar k_r^i}{D}. \tag{5.7}$$

We shall see below, eqn (5.38), that when every collision of A and B leads to a complex formation,

$$\sigma_r^i = \frac{\pi}{k^2} \cdot \frac{2J+1}{2j+1}$$

and so we conclude that

$$2\pi \frac{\hbar k_r^i}{D} \leqslant 1$$

or

$$k_r^i \leqslant (D/2\pi\hbar). \tag{5.8}$$

Replacing the inequality in eqns (5.8) by an equality we obtain the basic (implicit) assumption of all quasi-equilibrium theories of molecular collisions. (See in particular Rice 1961.) To correct this approximation it is conventional to introduce the 'transmission coefficient' P_r^i,

$$P_r^i = 2\pi \frac{\hbar k_r^i}{D} \leqslant 1 \tag{5.9}$$

so that

$$k_r^i = \frac{DP_r^i}{2\pi\hbar} \tag{5.10}$$

and

$$\sigma_r^i = \frac{\pi}{k^2} \cdot \frac{2J+1}{2j+1} P_r^i. \tag{5.11}$$

In practice, P_r^i is put equal to 1 (or occasionally $\frac{1}{2}$) and not further considered in simple applications. In this approximation k_r^i does not depend on the index i. Thus from eqn (5.2), the probability p_r^i does not depend on the initial channel that led to the formation of the complex. This is the so-called 'quasi-equilibrium' approximation.

§ The quantity $\hbar k_r^i/D$ is sometimes referred to as the 'strength' function.

From eqn (5.4),
$$\sigma_{fi} = \sigma_r^i \frac{k_r^f}{\sum_f k_r^f},$$
so that, putting
$$k_r^f = D P_r^f / 2\pi\hbar$$
and using eqn (5.11),
$$\sigma_{fi} = \frac{\pi}{k^2} \cdot \frac{2J+1}{2j+1} \cdot \frac{P_r^f P_r^i}{P_r}, \qquad P_r = \sum_f P_r^f. \tag{5.12}$$

In the quasi-equilibrium approximation $P_r^i = 1$ (or in general the number of initial channels) and P_r = total number of channels. Equation (5.12) with the quasi-equilibrium specification of P_r^i has recently been used extensively by Light and co-workers (see below).

To introduce the exit-state approximation it is convenient to consider eqn (5.8). D is the average energy interval between successive levels and so any measurement done to identify a single level must have an energy resolution better than D. We can then identify a lifetime τ_r^i of a particular level only if
$$\frac{D \tau_r^i}{2\pi} \geqslant \hbar.$$

Putting $\tau_r^i = (k_r^i)^{-1}$ we recover eqn (5.8). Another useful interpretation of D is to regard $2\pi\hbar/D$ as the recurrence time of the system, by which we mean that the quantum-mechanical probability of a given state is periodic with a period $2\pi\hbar/D$ (Blatt and Weisskopf 1952, section VIII F; Levine 1966e). To see this we expand a given state in the energy eigenstates of the system where, by assumption, $E_n = E_0 + nD$, or
$$|\psi(t+\tau)|^2 = \left| \sum_n \langle \psi | n \rangle \exp[-i(E_0 + nD)(t+\tau)/\hbar] \right|^2$$
$$= |\psi(t)|^2 \quad \text{if} \quad \tau = 2\pi\hbar/D.$$

Consider now eqn (5.1) applied to the rate of breakdown of the complex. In other words, we assume that in order for the complex to break, it must be in a particular state which we call the exit state. Then
$$k_r^i = p_r^s k_{rs}^i, \tag{5.13}$$
where p_r^s is the probability of realizing the exit state, given that the system is the collision complex. We can compute p_r^s as the ratio of the lifetime of the exit state to the recurrence time
$$p_r^s = \tau_{rs}/(2\pi\hbar/D), \tag{5.14}$$
so that
$$k_r^i = \frac{D}{2\pi\hbar}(k_{rs}^i/k_{rs}), \quad (k_{rs})^{-1} = \tau_{rs}. \tag{5.15}$$

Comparing eqns (5.15) and (5.10) we see that the transmission factor is simply the fraction of exit states that break into the initial channel (the rest revert to the complex).

Two particular exit states have often been employed. The first is the Rice–Ramsperger (1927)–Kassel (1932), RRK, model, where the exit state corresponds to energy in excess of E_0 in the degree of freedom that corresponds to the relative motion in the channel i. k_r^i is computed from eqn (5.13) where k_{rs}^i is not specified and p_r^s is determined by level counting for specific molecular models.

p_r^s is a conditional probability and so we can write

$$p_r^s = \frac{p_{rs}}{p_r}, \qquad (5.16)$$

where p_r is the probability of realizing the rth intermediate and p_{rs} is the probability of realizing both the rth intermediate and the exit states. For example, let the intermediate be considered as s degenerate oscillators of frequency ω. The rth intermediate has an energy of $(m+i)\hbar\omega$, and the exit state corresponds to m of these (phonons) localized in a particular oscillator ($E_0 = m\hbar\omega$). Equating (unnormalized) probability with the number of events that lead to the required outcome, we have that p_r is proportional to the number of ways of distributing $m+i$ identical objects in s cells. To compute p_{rs} we note that the number of ways of realizing an energy in excess of E_0 in a particular oscillator is the number of ways of distributing i identical objects in s cells (since m objects are definitely assigned to the exit oscillator).§ Thus

$$p_r^s = \frac{(i+s-1)!/i!(s-1)!}{(m+i+s-1)!/(m+i)!(s-1)!}. \qquad (5.17)$$

Classical statistics should hold for $i \gg s$ and we find∥

$$p_r^s = \left(\frac{E-E_0}{E}\right)^{s-1}, \quad k_r^i = p_r^s k_{rs}^i. \qquad (5.18)$$

A second type of exit state was introduced by Slater (1939, 1959) and requires that a degree of freedom of relative motion be extended to a critical distance. (In Slater's model of the harmonic oscillator this condition is equivalent to specifying a critical energy.) Slater then

§ Note that this is also the number of ways of putting any number up to i objects in $s-1$ cells, since

$$\sum_{k=0}^{i} \frac{(k+s-2)!}{k!(s-2)!} = \frac{(i+s-1)!}{i!(s-1)!}.$$

∥ Since

$$\frac{(a+b)!}{a!b!} \to \frac{a^b}{b!}, \quad a \gg b.$$

computes the (classical) recurrence frequency and uses eqn (5.10) with $P_r^i = 1$ so that k_r^i is the recurrence frequency.

To determine k_{rs}^i in the RRK model we must recognize the possibility that there may be several exit states corresponding to the internal degeneracy of energy levels. Let N_r be the number of the exit states associated with the rth complex; then

$$k_r^i \leqslant (2\pi\hbar)^{-1} D N_r. \tag{5.19}$$

Regarding eqn (5.19) with an equality sign,

$$k_r^i = (2\pi\hbar)^{-1} \frac{N_r}{\mathrm{d}N/\mathrm{d}E}, \tag{5.20}$$

where $(\mathrm{d}N/\mathrm{d}E)\mathrm{d}E$ is the number of levels of the collision complex in the energy interval E to $E+\mathrm{d}E$. Using classical statistics for the RRK model,

$$N(E) = \frac{(E/\hbar\omega)^s}{s!} \tag{5.21}$$

and (cf. footnote to eqn (5.17))

$$N_r(E) = \frac{((E-E_0)/\hbar\omega)^{s-1}}{(s-1)!}, \tag{5.22}$$

so that

$$k_r^i = \frac{\omega}{2\pi}\left(\frac{E-E_0}{E}\right)^{s-1} \tag{5.23}$$

and $k_{rs}^i = \omega/2\pi$. Equation (5.23) is also the result of Slater's theory for $2(s-1)$ degenerate oscillators. In general, if in the exit state the frequencies are modified from the collision complex,

$$k_{rs}^i = \prod_{i}^{s}\omega_i \Big/ 2\pi \prod_{i}^{s-1}\omega_i' \tag{5.24}$$

where the primes refer to the exit state. (To compare this result to the discussion in section 3.4.0 recall that the model used there corresponds to $s = 2$.)

Equation (5.20) of the RRK model can be written

$$(\mathrm{d}N/\mathrm{d}E)k_r^i = (2\pi\hbar)^{-1} N_r \tag{5.25}$$

where, in accordance with the quasi-equilibrium assumption, the density of levels of the intermediates is independent of the initial channel. N_r is the number of exit states associated with the rth intermediate.

We can compute N_r by the transition state method, where N_r is the number of internal levels of the transition state with energies less than $E-E_0$. Then

$$(\mathrm{d}N/\mathrm{d}E)k_r^i = (2\pi\hbar)^{-1} \int_0^{E-E_0} \rho_{\ddagger}(E-E_0-e)\,\mathrm{d}e. \tag{5.26}$$

This model is referred to as the RRK-M model, after Marcus and Rice (1951) and Marcus (1952). Essentially similar ideas were discussed by Rosenstock, Wallenstein, Wahrhaftig, and Eyring (1952) who assumed that, irrespective of the mode of formation, at a given energy E the intermediates are drawn from a microcanonical ensemble, and dN/dE is the number of states in this ensemble. We stress again (following Bohr (1936)) that this assumption should read 'a microcanonical ensemble, with a given value of all conserved quantum numbers'.

In actual applications these approaches require an assumption about the structure of the transition state and efficient techniques for level counting in a given structure. For recent reviews see Rabinovitch and Setser (1964) and Nikitin (1966).

From eqns (5.26), (5.6), and (5.5) we thus have the usual result of transition state theory,
$$Y_f^i(E) = N_r, \tag{5.27}$$
for the transition rate $i \to f$. If we do not make the quasi-equilibrium assumption, then
$$Y_f^i(E) = p_r^i(E) k_r^f. \tag{5.28}$$
Moreover, the result
$$k_r^f = (2\pi\hbar)^{-1} \frac{N_r}{(dN/dE)} P_r^f \tag{5.29}$$
is still valid as it is based only on microscopic reversibility.

To summarize, we have assumed that the transition $A+B \to C+D$ proceeds via the mechanism
$$A+B \to \underset{\text{'collision complex'}}{[AB]_r} \to C+D$$
and have obtained eqns (5.1)–(5.19). We have then made the quasi-equilibrium assumption, to the effect that the probability of forming the complex in a collision is independent of the channel in which the collision took place, and simplified matters even further by taking this probability to be unity.

At this point we analysed the formation and breakdown of the complex in the exit-state picture. That is, we have used the mechanism
$$\underset{\substack{\text{'collision}\\\text{complex'}}}{[AB]_r} \to \underset{\text{'exit state'}}{[AB]_{rs}^f} \to \underset{\text{'products'}}{C+D}$$
and similarly with respect to $A+B$, so that
$$A+B \to [AB]_{rs}^i \to [AB]_r \to [CD]_{rs}^f \to C+D.$$

We have labelled the exit states with channel indices since in principle they need not be the same. In the quasi-equilibrium approximation they are, however, identical. Using the RRK exit state we obtained eqn (5.20), which is equivalent to the transition state approximation.

The previous discussion is taken from the author's thesis (unpublished) and is based on the ideas of Rice and Ramsperger (1927), Rice (1930), Kassel (1932), Evans and Polanyi (1935), Eyring (1935), Bohr (1936), Landau (1936), Weisskopf (1937), Bethe (1937), Slater (1939), Bohr, Peierls, and Placzek (1939), Bohr and Wheeler (1939), Marcus and Rice (1951), Rosenstock, Wallenstein, Wahrhaftig, and Eyring (1952), Hauser and Feshbach (1952), Blatt and Weisskopf (1952), Keck (1958), Slater (1959), and Bernstein, Dalgarno, Massey, and Percival (1963). Some of these ideas have also been reviewed by Rice (1967, Chapters 18–20).

In the following three sections we examine the theoretical background of our physical discussion. In section 3.5.1 we try to avoid the *a priori* concept of the reaction mechanism and replace the physical picture by a more formal discussion based on the **S** matrix. The dynamical background for our assumptions is discussed in section 3.5.2, and a reformulation of our ideas using the partitioning technique is considered in section 3.5.3. Some further comments are given in sections on unimolecular breakdown, where the collision complex is not formed by the collision of $A+B$, rather the assumed mechanism is

$$A+B \to [A]_r + B'$$
$$\underset{\text{complex}}{[A]_r} \to \underset{\text{exit state}}{[A]_{rs}} \to \text{final states.}$$

3.5.1. *The statistical approximation*

Statistical theories intend to reproduce the average behaviour of the **S** matrix over an energy interval (Weisskopf 1937). Under certain conditions the average behaviour may be particularly simple, and under these conditions the statistical theory is particularly useful. As an example we consider (following Bernstein *et al.* 1963) the rotational excitation of a rigid rotor by an atom. When the off-diagonal matrix elements (with respect to the states of the rotor) of the potential are large, many internal states are coupled together, and we expect the elements of the **S** matrix to vary rapidly as functions of the energy. If we diagonalize the **S** matrix, the phase shifts will be large (in absolute value) and rapidly varying, thus the actual observables that are determined by the phase shift (up to a multiple of 2π) will be determined essentially by a random§ phase shift. We can then replace the energy average by an average over an ensemble of **S** matrices. The definition of

§ When we diagonalize the $\mathbf{S}(E)$ matrix, the eigenvalues, $\exp[2i\delta_j(E)]$, do not (barring additional constraints) cross when considered as functions of the parameter E (the energy). Thus one cannot regard the different phases as truly uncorrelated variables.

a suitable ensemble is by no means trivial,§ and instead of proceeding in this rigorous fashion we shall simply assume various reasonable properties about the averages computed using some (unspecified) ensemble.

In the region of strong coupling, the simplest physical assumption that we can make is that the average probability of transition from a given initial channel into any final channel is independent of the final channel, subject to conservation of constants of motion. The conserved quantum numbers are the total energy, the momentum of the total centre of mass, the total angular momentum $\mathbf{J}$ and its projection on a fixed axis M, and any other symmetries of a particular system. Time-reversal invariance and parity conservation imply that microscopic reversibility, eqn (2.8.61), holds.

For a given total energy and total angular momentum J, in the centre of mass system, let $S^J(\alpha, \gamma)$ be the element of the $\mathbf{S}$ matrix for transitions from the channel γ to α (section 2.3.1).

Representing the average over a distribution by $\langle \ \rangle$ our assumptions imply that
$$\langle S^J(\alpha, \gamma)\rangle = \langle S^J(\beta, \gamma)\rangle \tag{5.30}$$
and
$$\langle |S^J(\alpha, \gamma)|^2\rangle = \langle |S^J(\beta, \gamma)|^2\rangle \tag{5.31}$$
for any possible final channels α and β. Since the matrix $\mathbf{S}$ is unitary,
$$1 = \left\langle \sum_\alpha |S^J(\alpha, \gamma)|^2 \right\rangle = \sum_\alpha \langle |S^J(\alpha, \gamma)|^2\rangle = N_J \langle |S^J(\alpha, \gamma)|^2\rangle, \tag{5.32}$$
where N_J is the number of possible final channels consistent with the given total angular momentum. Thus∥
$$\langle |S^J(\alpha, \gamma)|^2\rangle = 1/N_J. \tag{5.33}$$

The mean, degeneracy averaged, transition rate $\gamma \to \alpha$ is obtained from eqn (2.7.67) as $(\gamma \neq \alpha)$
$$\left\langle \frac{\mathrm{d}\dot{N}^J(\alpha \leftarrow \gamma)}{\mathrm{d}E} \right\rangle = (2\pi\hbar)^{-1}\pi^{-1}k_j^2 \langle \sigma^J(\alpha \leftarrow \gamma)\rangle$$
$$= \frac{2J+1}{2j+1}(2\pi\hbar)^{-1}\langle |S^J(\alpha, \gamma)|^2\rangle$$
$$= \frac{2J+1}{2j+1}(2\pi\hbar)^{-1}N_J^{-1}, \tag{5.34}$$

§ Bernstein et al. (1963) have assumed an ensemble of symmetric unitary matrices, for which $\langle \mathbf{S}\rangle = \langle \mathbf{RS}\rangle$, where $\mathbf{R}$ is an arbitrary orthogonal matrix and the brackets denote the mean over the ensemble. We can consider such an ensemble as a random distribution of points on the unit, N-dimensional sphere. A given initial channel corresponds to a column of the $\mathbf{S}$ matrix, and each column is represented by a point on the sphere, with the matrix elements corresponding to the direction cosines of the point. The properties of such ensembles (mainly for the matrix representation of the Hamiltonian) are discussed in Porter (1965).

∥ See also Krieger (1967).

and the mean rate of all inelastic transitions is

$$(2\pi\hbar)^{-1}\pi^{-1}k_j^2 \sum_{\alpha \neq \gamma} \langle \sigma^J(\alpha \leftarrow \gamma)\rangle$$
$$= (2\pi\hbar)^{-1}\frac{2J+1}{2j+1}\langle(1-|S^J(\alpha,\alpha)|^2)\rangle = (2\pi\hbar)^{-1}\frac{2J+1}{2j+1}(1-N_J^{-1}). \quad (5.35)$$

Equation (5.35) determines the mean inelastic cross-section from the channel γ for a given J.

To evaluate the elastic transition rate we shall assume (eqn (5.64))

$$\langle S^J(\gamma,\gamma)\rangle = 0. \quad (5.36)$$

Thus, using eqn (2.7.64),

$$\left\langle \frac{\mathrm{d}\dot{N}^J(\gamma \leftarrow \gamma)}{\mathrm{d}E}\right\rangle = (2\pi\hbar)^{-1}\frac{2J+1}{2j+1}\langle|\mathbf{T}^J(\gamma,\gamma)|^2\rangle$$

$$= (2\pi\hbar)^{-1}\frac{2J+1}{2j+1}\langle|1-S^J(\gamma,\gamma)|^2\rangle$$

$$= (2\pi\hbar)^{-1}\frac{2J+1}{2j+1}\left(1+\langle|S^J(\gamma,\gamma)|^2\rangle-2\,\mathrm{re}\langle S^J(\gamma,\gamma)\rangle\right)$$

$$= (2\pi\hbar)^{-1}\frac{2J+1}{2j+1}(1+N_J^{-1}). \quad (5.37)$$

The first contribution in eqn (5.37) is the shadow scattering rate, while the second contribution is from counting the initial channel as any other channel. For this reason one sometimes (see section 3.5.2) refers to the second term as a 'fluctuation' or 'compound' contribution. The picture is that of a formation of a 'collision complex' due to the strong coupling. The shadow elastic scattering is that required by the optical theorem, for the formation of the complex. Subsequent to its formation the complex can break down to any possible channel, and the 'compound elastic' contribution represents the fragmentation of the complex into the initial channel. As we have seen, the shadow contribution is related to the effective range of the interaction region, and so the shadow contribution is sometimes called the 'shape elastic' contribution. It should be stressed, however, that this is just a convenient language to describe the result. We do not imply that a 'physical' complex is actually formed. Of course, if a complex is formed, our terminology does acquire a physical significance.

Adding the compound elastic cross-section to the inelastic cross-section we obtain

$$(\pi/k_j^2)\frac{2J+1}{2j+1}, \quad (5.38)$$

which we can regard as a cross-section for the formation of the collision complex. In the field of reactive molecular collisions this physical model was used extensively by Light (1964). See, however, the revised version, Pechukas, Light, and Ronkin (1966) and Nikitin (1965b).

The total cross-section is

$$\left\langle \sum_\alpha \sigma^J(\alpha \leftarrow \gamma) \right\rangle = (2\pi/k_j^2)\frac{2J+1}{2j+1} \qquad (5.39)$$

and is the sum of the shadow (or shape) elastic cross-section and the cross-section for the formation of the collision complex. Both these contributions are independent of the number of channels N_J into which the complex can break down.

The case $N_J = 1$ corresponds to the RPH approximation in a single-channel collision. To obtain analogous results to eqns (4.32)–(4.34) we sum eqn (5.39) over all J values from $|j-l|$ to $j+l$ to obtain the total cross-section from an initial state of given j and l. We find

$$(2\pi/k_j^2)(2l+1) \qquad (5.40)$$

for the total cross-section, in agreement with eqn (4.32). It is obvious that all the other relations hold and that

$$P_l^J = (N_J-1)/N_J. \qquad (5.41)$$

Microscopic reversibility is ensured since we have

$$\langle |S^J(\alpha,\gamma)|^2 \rangle = \langle |S^J(\gamma,\alpha)|^2 \rangle = 1/N_J. \qquad (5.42)$$

The extension of our discussion to reactive collisions is clear. Within the framework of the physical model of the collision complex we must assume that all the reactive events proceed via the formation and breakdown of the complex. The unitarity of the $\mathbf{S}^J$ matrix is now extended by summing over both the different reactive channels and over the internal states in each reactive channel. If we consider a collision of an atom and a diatomic

$$A+BC \rightarrow AB+C$$

then $\qquad \langle \sigma(\alpha_f, \gamma_i) \rangle = (\pi/k_n^2)(2j_i+1)^{-1} \sum_{|l_i-j_i| \leqslant J \leqslant l_i+j_i} \sum_M N_J^{-1}, \qquad (5.43)$

where N_J^{-1} is the total number of states into which, at a given energy and other good quantum numbers, the complex can break. We have also assumed that this number is independent of M (the projection of the total angular momentum on a fixed axis), as is consistent with rotational invariance.

We can now sum over various quantum numbers to obtain less detailed information. For a given initial internal state n, and initial orbital angular momentum l_i we can sum over all final states and over J to obtain the RPH value for the total cross-section, $(2\pi/k_n^2)(2l_i+1)$.

To obtain the cross-section between two internal states we sum over all initial and final orbital angular momenta, to get

$$\langle \sigma(m_f, n_i) \rangle = (\pi/k_n^2)(2j_i+1)^{-1} \sum_J (2J+1) \frac{N_J^f(m) N_J^i(n)}{N_J}, \quad (5.44)$$

where $N_J^i(n)$ is the number of states in the reactive channel i, of internal quantum numbers n that can be formed from the collision complex at a given total energy, and total angular momentum. We can interpret $N_J^f(m)/N_J$ as the probability, $P_J(m_f) < 1$, that the collision complex will break down into the internal state m in the reactive channel f, and write

$$\langle \sigma(m_f, n_i) \rangle = \sum_J \langle \sigma^J(n_i) \rangle P_J(m_f). \quad (5.45\,\text{a})$$

We can consider $\langle \sigma^J(n_i) \rangle$ as the cross-section for the formation of the complex from the initial state. Thus, subject to the statistical assumption, we can interpret the rate (or yield) of reactive collisions at a given energy E as a product of the rate (or yield) of formation of a collision complex from the initial state and the probability of its breakdown into the final state, summed over all possible complexes. We stress, however, that this is an interpretation of eqn (5.45) rather than an assumption.

From eqn (5.44) we can write

$$\langle \sigma^J(n_i) \rangle = (\pi/k_n^2) \sum_{l_i} (2l_i+1) \frac{2J+1}{(2j_i+1)(2l_i+1)}. \quad (5.46)$$

Within the context of the collision complex model we can interpret the last fraction in eqn (5.46) as the fraction of collisions with given l_i and j_i that couple to form a resultant J, $|l_i - j_i| \leqslant J \leqslant l_i + j_i$.

In the zeroth approximation we can take $P_J(m_f) = 1$ so that

$$\sum_J \langle \sigma^J(n_i) \rangle = (\pi/k_n^2) \sum_{l_i} (2l_i+1)$$

$$\simeq 2\pi \int b \, db, \quad (5.47)$$

in agreement with the simple collision theory where every collision is reactive. Combining eqns (5.46) and (5.45 a) and performing the summation over l_i at the last stage

$$\langle \sigma(m_f, n_i) \rangle = (\pi/k_n^2) \sum_{l_i} (2l_i+1) \sum_J \frac{2J+1}{(2j_i+1)(2l_i+1)} P_J(m_f). \quad (5.45\,\text{b})$$

Replacing the summation over l_i by an impact parameter integration we can write the reactive cross-section in terms of an opacity function. In particular, if, as is often the case, $j_i \ll l_i$ we can take as the first approximation $J = l_i$, $J = \mu v b$ and put $P_J(m_f) = P(m_f; b)$ and eqn (5.45 b) can be written in the approximate form

$$\langle \sigma(m_f, n_i) \rangle = \int P(m_f; b) \, \mathrm{d}\langle \sigma(n_i) \rangle = 2\pi \int b P(m_f; b) \, \mathrm{d}b. \quad (5.45\,\mathrm{c})$$

The summation over l_i in eqn (5.47) should be truncated at some maximum value l_m, since for high l values the coupling between the internal and relative motion is not likely to be strong and the statistical approximation is invalid. In the same fashion we have to restrict the summation over l_f that is involved in computing N_J. In other words, we need a prescription for the computation of $N_J^i(n)$ and $N_J^f(n)$ in eqn (5.44).

The statistical theory for the purely elastic case (Massey and Mohr 1934) determines l_m such that $|\delta_{l_m}| = \frac{1}{2}$ where δ_l is computed in the Born approximation. We can extend this prescription by introducing for each channel the diagonal elements of the potential (with respect to the internal states) and determine δ_l for each channel. However, if the coupling is strong, the Born approximation is not accurate, and we no longer have a known, simple analytic dependence of δ on l (cf. Bernstein et al. 1963). Within the framework of the collision complex model, Pechukas and Light (1965) have suggested that l_m be determined by analogy to the collision theory, namely

$$E = V(d) + \hbar^2 l_m(l_m+1)/2\mu d^2 + E_{\mathrm{int}},$$

where $V(R)$ is some specified potential. For reactions without activation energy they took

$$V(R) = -C_n/R^n,$$

namely, the lowest contribution to the dispersion force ($n = 4$ or 6 for charged-neutral or neutral-neutral collision respectively), where d is the position of the maximum of $V(R)$ plus the centrifugal term (cf. eqn (4.7)). For reactions with activation energy (Lin and Light 1966) $d = R_m$ is considered to be some distance of closest approach and

$$V(R_m) = E° - \lambda_v E_{\mathrm{vib}} - \lambda_r E_{\mathrm{rot}}, \quad (5.48)$$

where $E°$ is the activation energy from the ground state of the reactants, E_{vib} and E_{rot} are the vibrational and rotational energies of the reactants, and $0 < \lambda < 1$. An opacity function computed using this method is shown in Fig. 3.9.

The computed yield functions and rate constants are in good to fair agreement with experiment. More detailed information is predicted less accurately. This is to be expected, as the statistical approximation becomes more reliable the less detailed is the information required.

The statistical approximation can, in principle, be applied to compute differential cross-sections. The basic physical assumption is that the only correlation between the initial and final directions of relative motion

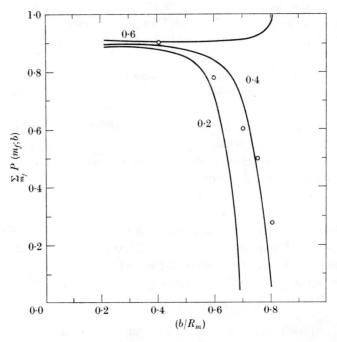

FIG. 3.9. The total reactive opacity function computed by Lin and Light (1966) for the reaction $K + HBr \to KBr + H$. The parameters are $E^\circ = 0.15$ kcal/mol; $\lambda_v = 0.9$; λ_r as indicated in the figure. HBr is initially in the ground vibrational and rotational level and the initial translational energy is 0.08 eV. The circles refer to the analysis of Beck *et al.* (1962). (See, however, Airey *et al.* 1967.)

is that determined by the conservation of good quantum numbers. (See Wolfenstein 1951.)

As a particular example, we consider the statistical average of the differential cross-section for an atom–rotor collision. In the notation of section 2.3.1, the basic statistical approximation is taken to be

$$\langle T^{J'}(j'l', jl) T^{J*}(j'l'_1, jl_1) \rangle = \delta_{J,J'} \delta_{l',l'_1} \delta_{l,l_1} \langle |T^J(j'l', jl)|^2 \rangle. \quad (5.49)$$

We are thus assuming the absence of interference between the different partial waves that contribute to a given J value as well as the absence of correlations between the different J values. The degeneracy averaged cross-section is obtained by summing over $m_{j'}$ and averaging over m_j.

Using eqn (2.3.55) we can average the (degeneracy averaged) differential cross-section ($j' \neq j$),

$$\langle 4k_j^2\, d\sigma_{j',j}/d\omega\rangle$$
$$= 4(2\pi)^2(2j+1)^{-1}\Big\langle \sum_{m_j}\sum_{m_{j'}}\Big|\sum_{JM}\sum_{ll'} S^J(j'l',jl)D^{Mm_{j'}}_{J\gamma'}(\hat{\mathbf{k}}_{j'})D^{Mm_j*}_{J\gamma}(\hat{\mathbf{k}}_j)\Big|^2\Big\rangle.$$

In view of the statistical approximation, eqn (5.49), and taking the incident direction, $\hat{\mathbf{k}}_j$, along the z axis, we obtain, using eqn (2.3.36),

$$4\pi \sum_{JM}\sum_{ll'}(2l+1)\langle|S^J(j'l',jl)|^2\rangle \sum_{m_j}\sum_{m_{j'}}\sum_{m_{l'}} \times$$
$$\times |Y^{m_{l'}}_{l'}(\hat{\mathbf{k}}_{j'})|^2 |\langle jm_j\,l0|jlJM\rangle|^2|\langle j'm_{j'}\,l'm_{l'}|j'l'JM\rangle|^2. \quad (5.50)$$

Integrating over the direction (θ,ϕ) of $\hat{\mathbf{k}}_{j'}$, and using the sum rule

$$\sum_{m_j}|\langle jm_j\,l0|jlJm_j\rangle|^2 = (2J+1)/(2l+1),$$

we find

$$\langle \pi^{-1}k_j^2\,\sigma_{j',j}\rangle = (2j+1)^{-1}\sum_J(2J+1)\sum_{ll'}\langle|S^J(j'l',jl)|^2\rangle,$$

as obtained previously.

The angular distribution predicted by eqn (5.50) is symmetric about $\theta = \tfrac{1}{2}\pi$, and this feature is often used to decide whether a complex is indeed formed in the reaction. (For a classical discussion, see Herschbach (1962).) For structureless particles, eqn (5.50) can be further simplified to give

$$\langle 4k^2(d\sigma/d\omega)\rangle = \sum_l (2l+1)^2\langle|S_l-1|^2\rangle P_l^2(\cos\theta). \quad (5.51)$$

Equation (5.51), which can also be derived directly (using eqn (5.49)) from, say, eqn (2.2.74), exhibits explicitly the symmetry about $\theta = \tfrac{1}{2}\pi$. (Recall that $P_l(\cos(\pi-\theta)) = (-1)^l P_l(\cos\theta)$.)

3.5.2. *The optical potential*

So far we have discussed the statistical theory in the absence of any detailed information about the dynamics of the reaction. The concept of a collision complex was introduced as an interpretation of the structure of the formal results rather than as a dynamical concept. It is clearly of some interest to try to derive the statistical approximation from a dynamical theory.

One possible approach is based on the theory of scattering by two potentials (section 2.5.1). Let $\mathbf{S}_1$ be the S matrix for the scattering by a potential V_1,

$$S_1 = \Omega_1^\dagger \Omega_1, \quad (5.52)$$

and let the actual potential V be the sum $V = V_1+V_2$, so that (eqns (2.5.30) and (2.5.65))

$$\begin{aligned}S &= I-2\pi i T \\ &= I-2\pi i(T_1+\Omega_1^\ddagger T_2\, \Omega_1) \\ &= S_1-2\pi i \Omega_1^\ddagger T_2\, \Omega_1 \\ &= \Omega_1^\ddagger(I-2\pi i T_2)\Omega_1 = \Omega_1^\ddagger S_2\, \Omega_1.\end{aligned} \qquad (5.53)$$

Introducing the eigenfunctions of T_2 (by analogy to eqn (2.5.54)) we can write

$$S = \sum_j \Omega_1^\ddagger |\kappa_j\rangle \exp(2i\Delta_j)\langle \kappa_j|\Omega_1. \qquad (5.54)$$

We now make the physical assumption that V_1 cannot induce transitions between different inelastic (or reactive) channels. Assuming that V_1 is a central potential we can perform a partial wave analysis of the scattering by V_1 and write, with a real $|i_1\rangle$,

$$\Omega_1|i\rangle = |i_1^+\rangle = \exp(i\delta_i)|i_1\rangle, \qquad (5.55)$$

so that
$$S_{fi} = \exp[i(\delta_i+\delta_f)] \sum_j \langle f_1|\kappa_j\rangle \exp(2i\Delta_j)\langle \kappa_j|i_1\rangle, \qquad (5.56)$$

or
$$\mathbf{S} = \mathbf{R}^T\mathbf{S}_2\, \mathbf{R}, \qquad (5.57)$$

where $R_{ji} = \langle \kappa_j|i_1^+\rangle = \langle i_1^+|\kappa_j\rangle^* = \langle i_1^-|\kappa_j\rangle = R_{ij}.$

We also note§ that
$$\mathbf{S}_1 = \mathbf{R}^T\mathbf{R}. \qquad (5.58)$$

In the RPH approximation‖
$$\mathbf{S}_1 = i\mathbf{I}.$$

Equation (5.57) is the starting point of the statistical approximation. Dynamical models can now be introduced by explicitly averaging eqn (5.57) over an energy interval, rather than by replacing the energy average by an ensemble average. (In the statistical approximation the ensemble we have used corresponds to Δ_j being uniformly distributed over 0 to 2π.) Consider an energy interval ΔE over which $\mathbf{R}$ can be assumed energy independent. Then

$$\langle \mathbf{S} \rangle = \mathbf{R}^T\langle \mathbf{S}_2\rangle \mathbf{R}, \qquad (5.59)$$

where the brackets denote an average over this interval, and we can now put
$$\mathbf{S} = \mathbf{R}^T\langle \mathbf{S}_2\rangle \mathbf{R}+\mathbf{R}^T[\mathbf{S}_2-\langle \mathbf{S}_2\rangle]\mathbf{R}. \qquad (5.60)$$

The physical significance of eqn (5.60) is based on the significance of an energy average of the $\mathbf{S}$ matrix (Friedman and Weisskopf 1955). $\langle \mathbf{S}\rangle$ represents those events that occur on a time scale $\hbar/\Delta E$. For a given ΔE

§ For example, by putting $\mathbf{S}_2 = \mathbf{I}$.
‖ By our physical assumption $\mathbf{S}_1$ is always diagonal and unitary, but in general $\mathbf{S}_1 = \exp(2i\boldsymbol{\delta})$ where $\boldsymbol{\delta}$ is diagonal with elements δ_i.

eqn (5.60) makes a distinction between the fast processes, represented by the first term, and the slow processes that occur on a time scale longer than $\hbar/\Delta E$. The rapid processes are labelled 'direct', and can include, in principle, both inelastic and reactive transitions. The slower processes described by the second term in eqn (5.60) are the 'compound' processes. The fundamental assumption of the statistical approximation in the strong-coupling region is that all inelastic (and reactive) transitions are compound processes, or that $\langle \mathbf{S} \rangle$ is a diagonal matrix. When direct inelastic transitions take place $\langle \mathbf{S} \rangle$ is no longer a diagonal matrix, and in this case it is convenient to modify the definition of V_1, so that V_1 is the portion of the interaction leading to the rapid, direct processes. The 'optical' potential is defined so that

$$\mathbf{S}_0 = \langle \mathbf{S} \rangle,$$

where $\mathbf{S}_0$ is the $\mathbf{S}$ matrix computed as if the optical potential is the exact potential. (See, for example, Feshbach 1960, Brown 1959 and references therein.)

Let ΔE be an energy interval such that $\hbar/\Delta E$ is shorter than the duration of the compound processes. On the time scale $\hbar/\Delta E$ the formation of a collision complex represents a transition out of the elastic channel and the optical potential will be complex whenever compound processes take place§ even in the absence of truly inelastic transitions, since the compound elastic scattering is an inelastic process when viewed on a time scale of $\hbar/\Delta E$. In other words, there is a time-lag between the shadow elastic and the compound elastic (and compound inelastic) processes and the averaging procedure that leads‖ to a distinction between direct and compound processes is sufficient to lead to a complex optical potential.

As an example, we return to the problem of rotational excitation. From the optical theorem (for example, eqn (2.4.60)) we know that the total transition rate is linear in $\mathbf{S}$. Thus

$$2\pi\hbar \sum_\alpha \frac{\mathrm{d}\dot{N}^J(\alpha \leftarrow \gamma)}{\mathrm{d}E} = \frac{2J+1}{2j+1}\left[1 - 2\,\mathrm{re}\, S^J(\gamma,\gamma) + \sum_\alpha |S^J(\alpha,\gamma)|^2\right]$$
$$= 2\frac{2J+1}{2j+1}[1 - \mathrm{re}\, S^J(\gamma,\gamma)]. \quad (5.61)$$

§ In contrast, the equivalent potential is complex whenever truly inelastic transitions take place.

‖ Our discussion assumed the existence of a suitable energy interval ΔE, over which a distinction between the direct and compound processes can be made. Strictly speaking, the average separates out the average energy-dependence from the variations about the average. It is not necessarily true that the variation in energy (particularly if these are slow variations, ΔE large) correspond to 'slower' processes.

Denoting an energy average by a bracket,

$$2\pi\hbar \left\langle \sum_\alpha \frac{\mathrm{d}\dot{N}^J(\alpha \leftarrow \gamma)}{\mathrm{d}E} \right\rangle = 2\frac{2J+1}{2j+1}[1-\mathrm{re}\langle S^J(\gamma,\gamma)\rangle]$$

$$= 2\frac{2J+1}{2j+1}[1-\mathrm{re}\,S_0^J(\gamma,\gamma)]. \quad (5.62)$$

The total average transition rate is thus predicted correctly by the optical potential. If the energy interval ΔE is sufficiently small $\Delta E \ll E$, we can regard k^2 as a constant and write

$$\sum_\alpha \langle \sigma^J(\alpha \leftarrow \gamma)\rangle = (2\pi/k^2)\frac{2J+1}{2j+1}[1-\mathrm{re}\,S_0^J(\gamma,\gamma)].$$

The rates (and cross-sections) for the elastic and inelastic events separately are not linear in **S**. For example,

$$2\pi\hbar\left\langle \frac{\mathrm{d}\dot{N}^J(\gamma \leftarrow \gamma)}{\mathrm{d}E}\right\rangle = \frac{2J+1}{2j+1}\langle |1-S^J(\gamma,\gamma)|^2\rangle$$

$$= \frac{2J+1}{2j+1}|1-\langle S^J(\gamma,\gamma)\rangle|^2 + \frac{2J+1}{2j+1}\{\langle |S^J(\gamma,\gamma)|^2\rangle - |\langle S^J(\gamma,\gamma)\rangle|^2\}. \quad (5.63)$$

The first contribution is that predicted by the optical potential and corresponds to direct elastic scattering, while the second term is the compound elastic scattering. We see clearly that the compound contribution corresponds to the fluctuation from the average. To see the equivalence of these results to those of section 3.5.1 we note that in section 3.5.1 we have assumed that $\langle \mathbf{S}^J\rangle = \langle \mathbf{RS}^J\rangle$. Taking $\mathbf{R} = -\mathbf{I}$ we see that

$$\langle S^J(\gamma,\gamma)\rangle = -\langle S^J(\gamma,\gamma)\rangle \quad (5.64)$$

for all γ, or $\langle S^J(\gamma,\gamma)\rangle = 0$ in the statistical approximation.

The total transition rate for inelastic events is given by the difference between the total rate and the elastic rate. The optical potential thus predicts a total inelastic rate as

$$(2\pi\hbar)^{-1}\frac{2J+1}{2j+1}\left\{\sum_\alpha \langle |S^J(\alpha,\gamma)|^2\rangle - |\langle S^J(\gamma,\gamma)\rangle|^2\right\}$$

$$= (2\pi\hbar)^{-1}\frac{2J+1}{2j+1}P_\gamma^J, \quad P_\gamma^J = 1-|\langle S^J(\gamma,\gamma)\rangle|^2, \quad (5.65)$$

where we have subtracted the first term of eqn (5.63) from eqn (5.62). Even if **S** is diagonal, this term does not vanish, but merely reduces to the compound elastic term (second term of eqn (5.63)). The inelastic rate computed using the optical potential is thus the rate of formation of the collision complex. The actual inelastic rate is obtained by subtracting-off the compound elastic rate. P_J^γ is the opacity function computed from the optical potential. If every collision in the channel γ with total angular momentum J led to formation of a complex the resulting rate would be (cf. eqn (5.46))

$$(2\pi\hbar)^{-1}(2l+1)\frac{2J+1}{(2j+1)(2l+1)},$$

so that P_J^γ has the physical significance as the fraction of such collisions that lead to a complex. For this reason it is sometimes known as the 'sticking' probability§ or the 'transmission coefficient'. In the statistical approximation $P_J^\gamma = 1$ (or in general $P_J^\gamma = N_J^\gamma$, where N_J^γ is the number of levels in the initial channel, γ, into which the complex can break).

To satisfy eqn (5.65) and the condition of microscopic reversibility we can write the rate of transitions $\gamma \to \alpha$ for a given J as

$$\left\langle \frac{\mathrm{d}\dot{N}^J(\alpha \leftarrow \gamma)}{\mathrm{d}E} \right\rangle = (2\pi\hbar)^{-1}\frac{2J+1}{2j+1}\cdot\frac{P_J^\alpha P_J^\gamma}{P_J}, \qquad (5.66)$$

where
$$P_J = \sum_\alpha P_J^\alpha.$$

When $\Delta E \ll E$ we can write eqn (5.66) in terms of the averaged cross-section. The result is known as the Hauser–Feshbach (1952) formula and provides a generalization of eqn (5.44) of the statistical approximation. We can sum over the initial and final orbital angular momenta to obtain

$$\left\langle \frac{\mathrm{d}\dot{N}(j' \leftarrow j)}{\mathrm{d}E} \right\rangle = (2\pi\hbar)^{-1}(2j+1)^{-1}\sum_J (2J+1)\frac{P_J^{j'} P_J^j}{P_J}, \qquad (5.67)$$

where $P_J^j = \sum_l P_J^\alpha$, $\alpha = (j, l)$, and the summation over l is restricted to satisfy $|j-l| \leqslant J \leqslant j+l$ and $l \leqslant l_\mathrm{m}$ as in eqn (5.40). An alternative summation chain can be used to write the results in terms of partial waves of given l. We then define

$$P(\alpha \leftarrow \gamma) = \sum_J \frac{2J+1}{(2j+1)(2l+1)}\frac{P_J^\alpha P_J^\gamma}{P_J} \qquad (5.68)$$

with $|j-l| \leqslant J \leqslant j+l$, sum $P(\alpha \leftarrow \gamma)$ over all possible values of l' (or alternatively of l) to obtain $P(j' \leftarrow \gamma)$ (or $P(\alpha \leftarrow j)$) and finally sum over

§ This label can be justified when we note that in the statistical approximation these probabilities are directly related to the coefficients $|\langle i_1^-|\kappa_j\rangle|^2$ of eqn (5.57).

l (or l') up to l_m (or l'_m),

$$\left\langle \frac{\mathrm{d}\dot{N}(j' \leftarrow j)}{\mathrm{d}E} \right\rangle = \sum_{l=0}^{l_m} (2l+1)P(j' \leftarrow \gamma)$$

or
$$= \sum_{l'=0}^{l'_m} (2l'+1)P(\alpha \leftarrow j). \qquad (5.69)$$

Like other opacity functions $P(j' \leftarrow \gamma)$ can be approximated on the basis of one-dimensional models. (In particular these methods find their formal expression in the R matrix theory. See Blatt and Weisskopf 1952, chapter 8.)

To conclude, we point out that eqns (5.61)–(5.63) have been obtained by performing an averaging of exact results and do not depend on any assumptions. To obtain eqn (5.65) we have computed the total inelastic rate that is predicted by $\langle \mathbf{S} \rangle$, the average $\mathbf{S}$ matrix. We have then proposed eqn (5.66) as a possible parametrization of the average inelastic rate, which satisfies eqn (5.65) and microscopic reversibility. The average elastic rate is given by eqn (5.63) which has two contributions; the first is the average elastic rate required by the optical theorem to support the average total inelastic rate (5.65), and the second contribution is the elastic term in eqn (5.66). The statistical approximation can now be used by replacing the energy average by an average over an ensemble of $\mathbf{S}$ matrices.

In other words, our derivation rests on the division of the average elastic rate into two contributions: one predicted by $\mathbf{S}_0 = \langle \mathbf{S} \rangle$ and one determined by the deviations of $\mathbf{S}$ from its average value. When we compute the average total inelastic rate predicted by $\langle \mathbf{S} \rangle$, the second contribution to the elastic rate is counted as an inelastic contribution. The average inelastic rate as computed by $\langle \mathbf{S} \rangle$ can then be parametrized as in eqn (5.66). It should be noted, however, that there may be other parametrizations that will satisfy eqn (5.65) and microscopic reversibility.

Serauskas and Schlag (1966, 1965; see also Hoare 1961) have discussed a statistical theory for energy transfer and have suggested that the range of final states that are accessible for a given collision be restricted by the duration of the collision in accordance with the Massey criteria. That is, a change in internal energy from E_n to E_m can only occur if the collision duration τ_c is shorter than $\hbar/(E_n-E_m)$. An average over τ_c is performed to obtain the average transition probability.

3.5.3. *Statistical theory for overlapping resonances*

When the rapid energy variations in the **S** matrix are due to collision intermediates that can be introduced by the partitioning technique (sections 3.2.0, 3.2.2) we know an exact parametrization of the **S** matrix, and an explicit energy average can be carried out by averaging over the parameters. In this situation the compound contribution can be given a well-defined physical meaning.

From eqn (2.110)

$$S_{fi} = (S_{fi})_d - 2\pi i \sum_s \frac{A^s_{f,i}}{E-\lambda_s}, \tag{5.70}$$

where (cf eqn (2.113))

$$\pi \sum_s A^s_{f,i} = \pi \sum_n \langle f_1^-|H|n\rangle\langle n|H|i_1^+\rangle \tag{5.71}$$

and $(S_{fi})_d$ is the S matrix for those processes that do not proceed via the resonance intermediates. By assumption $(S_{fi})_d$ is a slowly varying function of energy.

If there are no direct inelastic transitions and the initial and final channels are labelled by an orbital angular momentum we can introduce explicitly the phase shifts for direct scattering and put

$$\langle f_1^-| = \exp(2i\delta_f)\langle f_1^+|.$$

Averaging eqn (5.70) over an energy interval $(f \neq i)$,

$$\langle S_{fi}\rangle = \langle S_{fi}\rangle_d - 2\pi i \sum_s \frac{1}{\Delta E} \int_E^{E+\Delta E} \frac{A^s_{f,i}}{E'-\lambda_s}\,dE'$$

$$= (S_{fi})_d - 2\pi i \left[-\frac{2\pi i}{\Delta E}\sum_s \langle A^s_{f,i}\rangle\right]. \tag{5.72}$$

In the statistical approximation $\langle S_{fi}\rangle = 0$. The first term in eqn (5.72) vanishes since in the statistical approximation we assume the absence of direct inelastic transitions. From eqn (5.72) we see that we must also assume

$$\langle A^s_{f,i}\rangle = 0 \quad (f \neq i) \tag{5.73}$$

to ensure the statistical approximation. Since we can regard $A^s_{f,i}$ as a transition amplitude from channel i to channel f $(f \neq i)$, we can interpret eqn (5.73) as the assumption of no correlation between the formation $(i \to s)$ and breakdown $(s \to f)$ amplitudes of the resonance. For the diagonal element we have

$$\langle S_{ii}\rangle = (S_{ii})_d\left[1 - 2\pi\left(\frac{2\pi}{\Delta E}\sum_n |\langle n|H|i_1^+\rangle|^2\right)\right],$$

where we have used eqn (5.71) and assumed the absence of direct inelastic processes. In the statistical approximation $\langle S_{ii} \rangle = 0$. If there are N resonances in the interval ΔE, and D is the average spacing between the resonances $\Delta E = ND$ (we assume $\Delta E > D$), an average width $\langle \Gamma_i \rangle$ can be defined by

$$\langle \Gamma_i \rangle = \frac{\pi}{N} \sum_n |\langle n|H|i_1^+\rangle|^2 \tag{5.74}$$

so that
$$\langle S_{ii} \rangle = (S_{ii})_d \left[1 - 4\pi \frac{\langle \Gamma_i \rangle}{D}\right]. \tag{5.75}$$

In the statistical approximation

$$2\pi\hbar \frac{2\langle \Gamma_i \rangle}{\hbar D} = 1. \tag{5.76}$$

As we have already indicated, $2\Gamma_{n,i}/\hbar$ can be regarded as the (primary) rate of decay of the nth resonance to the initial channel. Equation (5.76) is thus in accord with the quasi-equilibrium result (eqn 5.8). The width of a resonance can be considered as the interval of energy, about its energy, over which the resonance makes a significant contribution (cf. eqns (2.85), (2.86)). Equation (5.76) can thus be interpreted (following Rice 1961) to mean that on the average there is no region in energy to which a resonance does not contribute.

To compute rates for particular transitions we need $\langle |S_{f,i}|^2 \rangle$. Using eqn (5.70),

$$\langle |S_{fi}|^2 \rangle = |(S_{fi})_d|^2 + 4\pi^2 \sum_{s,s'} \frac{1}{\Delta E} \int \frac{A_{f,i}^{s*} A_{f,i}^{s'}}{(E-\lambda_s^*)(E-\lambda_{s'})} \, dE. \tag{5.77}$$

There is no interference term between the direct and compound processes when we use eqn (5.73). Neglecting the correlation between different resonances and using eqns (5.73) and (5.71) we can write the second term above as

$$4\pi^2 \sum_s \frac{\pi \langle A_{i,i}^s \rangle \langle A_{f,f}^s \rangle}{D \langle \Gamma \rangle} = 4\pi \frac{\langle \Gamma_i \rangle \langle \Gamma_f \rangle}{D \langle \Gamma \rangle}, \tag{5.78}$$

where $\langle \Gamma \rangle = \sum_f \langle \Gamma_f \rangle$. Moreover,

$$\sum_f \langle |S_{f,i}|^2 \rangle = 4\pi \frac{\langle \Gamma_i \rangle}{D} + |(S_{i,i})_d|^2$$

$$= 1. \tag{5.79}$$

We can introduce the transmission coefficients

$$P^i = 4\pi \frac{\langle \Gamma_i \rangle}{D} \tag{5.80}$$

so that, with $P = \sum_f P^f$,
$$\langle |S_{fi}|^2 \rangle = P^f P^i / P \tag{5.81}$$
in agreement with eqns (5.66) and (5.44). In the statistical approximation $P_f = P_i = 1$ and $P = \sum_f 1 = N$, where N is the number of channels.

In section 3.5.1 we have defined the compound processes essentially by a distance criterion, in that we applied the statistical approximation only to those collisions where the orbital angular momentum was less than some maximum l_m. In the semi-classical approximation this corresponds to defining a maximum impact parameter b_m. One can then try to define in general a compound collision as those events where during the collision the molecules enter a specified volume§ about the system centre of mass. This is the point of view of the R matrix theory (Eu and Ross 1966, Lane and Thomas 1958). Equation (5.70) holds also in the R matrix, and A_{fi}^s has the same significance. One can also derive this model by the partitioning technique, when we let Q (which projects onto the compound part of the wavefunction) be a (physical) space projection on the region internal to the collision complex volume (Feshbach 1962).

§ The radius of the collision complex volume may be different for different channels.

3.6. Unimolecular reactions

3.6.0. The theory of unimolecular reactions is concerned with reactions where the observed chemical change can be represented as

$$A \to \text{products}, \qquad (\text{I})$$

and on a macroscopic time-scale the rate of depletion of A equals the rate of formation of products. Even when the reaction (I) is exothermic it is normally not spontaneous. An isolated molecule A is stable. The reaction is assumed to proceed via the mechanism (see Johnston 1966, Slater 1959, Lindemann 1922),

$$A+M \to A^*+M', \qquad (\text{II a})$$

$$A^* \to \text{products}. \qquad (\text{II b})$$

We shall refer to the second stage as a unimolecular breakdown. In the statistical theory it is usually assumed that the molecule A^* need be characterized by a set of good quantum numbers only, so that the subsequent breakdown is independent of the method of formation. The assumed mechanism requires that the duration of the collision (II a) be much shorter than the lifetime of A^* in order that the excitation and breakdown stages be distinct. We can therefore introduce a time interval Δt, which is longer than the time required for the formation of an intermediate A^* but shorter compared to the time required for its subsequent breakdown. As discussed in sections 3.5.0. and 3.5.2 one can introduce statistical concepts for the computation of averaged rates, on physical grounds (the required averaging is clearly on an energy interval of about $\hbar/\Delta t$), since the fragmentation act is distinct from the excitation. The discussion of the rate of breakdown of intermediate states given in section 3.5.0 also applies to the present problem.

If the excitation process (II a) occurs by collisions with other molecules in the gas phase, one should consider also the competing process

$$A^*+M \to A+M'. \qquad (\text{II c})$$

However, one cannot in general assume that the molecule A that results from the collision (II c) will not undergo a unimolecular breakdown prior to any further excitation. (Primarily, since there can be several different states of A^* and process (II c) can induce transition between these states.)

We shall consider first the process of unimolecular breakdown (section 3.6.1), elaborate on our discussion of rate processes in ensembles (section 3.6.2), and finally include the excitation stage (section 3.6.3). Other discussions of unimolecular reactions have been given by Rice (1967, 1961, 1930, 1929), Mies and Krauss (1966), Thiele (1966), Nikitin (1966), Hofacker (1965), Slater (1959), and Rosen (1933).

3.6.1. *Unimolecular breakdown*

To study the process $A^* \to$ products, we begin by assuming that at the time $t = 0$ the system is in a particular state that corresponds to the species A^* and that no products are present. To avoid considering the excitation process that led to the species A^* we shall assume that the products of the breakdown, at $t \to +\infty$, are known and try to construct a total wavefunction for the system that leads at $t \to +\infty$ to the known products, while at $t = 0$ it corresponds to a particular state that characterizes A^*. Our discussion is based on the ideas introduced in section 2.7.2 and Hall and Levine (1966).

Let $\phi(t')$, $t' \to +\infty$, represent the products of the breakdown

$$\phi(t) = \sum_{\mathbf{n}} \int dE\, B(E, \mathbf{n}; t)|E, \mathbf{n}\rangle, \tag{6.1}$$

where $|E, \mathbf{n}\rangle$ is an eigenstate of the Hamiltonian H_0 of the non-interacting products, with a total (continuous) energy E, and other quantum numbers $\mathbf{n}$. Then

$$\langle E', \mathbf{m}|E, \mathbf{n}\rangle = \delta(E-E')\delta_{\mathbf{n},\mathbf{m}} \tag{6.2}$$

and
$$B(E, \mathbf{n}; t) = \exp(-iEt/\hbar)B(E, \mathbf{n}). \tag{6.3}$$

From eqns (6.2) and (6.1) we can regard $|B(E, \mathbf{n})|^2$ as the probability to observe the products with a total energy E and other quantum numbers $\mathbf{n}$, provided we take $\phi(t)$ to be normalized,

$$\langle \phi(t)|\phi(t)\rangle = 1 = \sum_{\mathbf{n}} \int dE\, |B(E, \mathbf{n})|^2. \tag{6.4}$$

Let $\psi(t)$ be the wavefunction of the system, which evolves under the full Hamiltonian H. We require that as $t \to \infty$, $\psi(t)$ will tend to $\phi(t)$. The solution of this boundary value problem was given in section 2.7.2 (in particular, eqn (2.7.34)) as

$$\psi(t) = \sum_{\mathbf{n}} \int dE\, B(E, \mathbf{n}; t)|E, \mathbf{n}^-\rangle, \tag{6.5}$$

where $|E, \mathbf{n}^-\rangle$ is the solution of the incoming§ L.S. equation

$$|E, \mathbf{n}^-\rangle = |E, \mathbf{n}\rangle + (E^- - H_0)^{-1}V|E, \mathbf{n}^-\rangle. \tag{6.6}$$

§ We are using the substitution $\Omega^- = \sum_{\mathbf{n}} \int dE\, |E, \mathbf{n}^-\rangle\langle E, \mathbf{n}|$. Using eqn (2.7.35) one can directly confirm that $\psi(t) \to \phi(t)$ as $t \to \infty$. (See also the discussion of eqn (6.21).)

Here $V = H - H_0$ is the interaction between the products and $E^- = E - i\epsilon$ where $\epsilon \to +0$ only after taking the matrix elements of eqn (6.6).

At $t = 0$ we require that only A^* and no products be present. If $|s\rangle$ is a state vector that characterizes A^*,

$$|\psi_s(t = 0)\rangle = |s\rangle. \tag{6.7}$$

Equations (6.7) and (6.5) are compatible when

$$B(E, \mathbf{n}) = \langle E, \mathbf{n}^-|s\rangle, \tag{6.8}$$

so that (cf. eqn (6.12))

$$|\psi_s(t = 0)\rangle = \sum_\mathbf{n} \int dE |E, \mathbf{n}^-\rangle\langle E, \mathbf{n}^-|s\rangle = |s\rangle. \tag{6.9}$$

Thus the probability to observe the state $|E, \mathbf{m}\rangle$ in the products, given that at $t = 0$ the system was in the state $|\mathbf{s}\rangle$, is

$$|\langle E, \mathbf{m}^-|s\rangle|^2, \tag{6.10}$$

while the probability $A_s(E)$ to observe the products with energy E, is

$$A_s(E) = \sum_\mathbf{n} |\langle E, \mathbf{n}^-|s\rangle|^2 = \sum_\mathbf{n} \langle s|E, \mathbf{n}^-\rangle\langle E, \mathbf{n}^-|s\rangle. \tag{6.11}$$

Introducing the relation (cf. (2.2.7))

$$\sum_\mathbf{n} |E, \mathbf{n}^-\rangle\langle E, \mathbf{n}^-| = \delta(E - H), \tag{6.12}$$

we can write

$$A_s(E) = \langle s|\delta(E - H)|s\rangle = (2\pi i)^{-1}\langle s|G(E^-) - G(E^+)|s\rangle, \tag{6.13}$$

where in the second line we have used eqn (2.1.9), and

$$G(E^\pm) = (E - H \pm i\epsilon)^{-1}$$

with $\epsilon \to +0$ after the evaluation of the matrix element.

The wavefunction for the system can now be written as

$$|\psi_s(t)\rangle = \sum_\mathbf{n} \int dE \, |E, \mathbf{n}^-\rangle\langle E, \mathbf{n}^-|s\rangle \exp(-iEt/\hbar) \tag{6.14a}$$

$$= \int dE \, \delta(E - H)|s\rangle\exp(-iEt/\hbar) \tag{6.14b}$$

$$= (2\pi i)^{-1} \int dE \, [G(E^-) - G(E^+)]|s\rangle\exp(-iEt/\hbar). \tag{6.14c}$$

While t in this equation is not restricted to $t > 0$, it is not physically realistic to consider the region $t < 0$ since this equation was set up to describe the breakdown ($t > 0$) stage only.

We leave it to the reader to show that alternative forms of eqn (6.14) are

$$|\psi_s(t)\rangle = \sum_\mathbf{n} \int dE \, |E, \mathbf{n}^+\rangle\langle E, \mathbf{n}^+|s\rangle\exp(-iEt/\hbar) \tag{6.15a}$$

and

$$= (2\pi i)^{-1} \int_\mathrm{sp} d\lambda \, G(\lambda)|s\rangle\exp(-i\lambda t/\hbar) = \exp(-iHt/\hbar)|s\rangle. \tag{6.15b}$$

Equation (6.15 b) is to be contrasted with the function

$$\theta(t)\exp(-iHt/\hbar)|s\rangle.$$

In this latter case we avoid the problem of 'preparation' by 'switching on' the interaction instantaneously at $t = 0$. While this concept is perhaps appealing, it leads in the author's opinion to unphysical difficulties associated with the lack of apparent 'source' of the state $|s\rangle$.

The normalization of a wavefunction of the form (6.14 a) is conserved in time (see also Appendix 2.D) since

$$\langle E', \mathbf{m}^-|E, \mathbf{n}^-\rangle = \delta(E-E')\delta_{\mathbf{n},\mathbf{m}}, \tag{6.16}$$

so that
$$\langle \psi(t)|\psi(t)\rangle = \sum_{\mathbf{n}} \int dE\, |B(E, \mathbf{n})|^2 = \langle \phi(t)|\phi(t)\rangle. \tag{6.17}$$

In particular, using equation (6.14 a) for $\psi_s(t)$ at $t = 0$ and at t,

$$\langle \psi_s(t=0)|\psi_s(t=0)\rangle = \langle s|s\rangle = \langle \psi_s(t)|\psi_s(t)\rangle$$

$$= \sum_{\mathbf{n}} \int dE\, |\langle E, \mathbf{n}^-|s\rangle|^2$$

$$= \int dE\, A_s(E). \tag{6.18}$$

The conservation of norm is clearly consistent with eqn (6.11).

The probability amplitudes for different processes can now be obtained from eqn (6.14 a). For example, the probability to observe the state $|E', \mathbf{m}\rangle$ at time t, $|D(\mathbf{m}, E'; t)|^2$, can be evaluated as

$$D(\mathbf{m}, E'; t) = \exp(iE't/\hbar)\langle E', \mathbf{m}|\psi_s(t)\rangle$$
$$= \sum_{\mathbf{n}} \int dE\, \langle E', \mathbf{m}|E, \mathbf{n}^-\rangle\langle E, \mathbf{n}^-|s\rangle\exp[-i(E-E')t/\hbar]. \tag{6.19}$$

Using eqn (2.7.35) and the result (obtained from eqn (6.6), $E^- = E - i\epsilon$),

$$\langle E', \mathbf{m}|E, \mathbf{n}^-\rangle = \delta(E'-E)\delta_{\mathbf{m},\mathbf{n}} + (E^- - E')^{-1}\langle E', \mathbf{m}|V|E, \mathbf{n}^-\rangle, \tag{6.20}$$

we find that as $t \to \infty$ the second term in eqn (6.20) will not contribute to eqn (6.19), and so

$$\lim_{t \to \infty} D(\mathbf{m}, E', t) = \langle E', \mathbf{m}^-|s\rangle = B(\mathbf{m}, E'), \tag{6.21}$$

as expected, in view of eqn (6.10).

In the same fashion the probability $A_s(t)$ to observe the state $|s\rangle$ at time t, $t > 0$ is determined by the amplitude

$$\alpha_s(t) = \langle s|\psi_s(t)\rangle = \sum_{\mathbf{n}} \int dE\, \exp(-iEt/\hbar)|\langle E, \mathbf{n}^-|s\rangle|^2$$

$$= \int dE\, \exp(-iEt/\hbar) A_s(E), \tag{6.22}$$

and $A_s(t) = |\alpha_s(t)|^2$. To evaluate this amplitude it is convenient to introduce the projection operator P, $P = I - |s\rangle\langle s|$, where I is the identity operator ($P^2 = P$ since $\langle s|s\rangle = 1$). From eqn (6.13) we can write
$$A_s(E) = -\pi^{-1}\mathrm{im}\langle s|G(E^+)|s\rangle. \tag{6.23}$$
The matrix element of $G(E^+)$ can be evaluated, using eqn (3.13), so that
$$\langle s|G(E^+)|s\rangle^{-1} = E - \langle s|H_0 + \Gamma(E^+)|s\rangle. \tag{6.24}$$
If $PH|s\rangle$ is non-vanishing (so that $|s\rangle$ is not an eigenstate of H) and the Hamiltonian H_0 has a continuous spectrum, $\langle s|\Gamma(E^+)|s\rangle$ will possess a non-positive imaginary part (cf. eqn (2.64)). Putting
$$E_s = \langle s|H_0|s\rangle + \mathrm{re}\langle s|\Gamma(E^+)|s\rangle \tag{6.25 a}$$
and
$$\Gamma_s = -\mathrm{im}\langle s|\Gamma(E^+)|s\rangle \tag{6.25 b}$$
we obtain (cf. eqn (2.86))
$$A_s(E) = \pi^{-1}\Gamma_s/[(E-E_s)^2 + \Gamma_s^2]. \tag{6.26}$$
To evaluate $\alpha_s(t)$ we arrange that the threshold of the continuous spectrum of H_0 begins at zero energy. Then, provided $E_s > \Gamma_s$, we can extend the range of integration over E to $-\infty$. Regarding E_s and Γ_s as energy independent (which is not necessarily the case) we find for $t > 0$
$$\alpha_s(t) = \exp(-iE_s t/\hbar)\exp(-\Gamma_s t/\hbar), \tag{6.27}$$
using the method of residues for the integration.

The result
$$A_s(t) = |\langle s|\psi_s(t)\rangle|^2 = \exp(-2\Gamma_s t/\hbar), \quad t > 0 \tag{6.28}$$
is known as the exponential decay law. The present derivation holds in general only if the continuous spectrum of H_0 extends from minus to plus infinity (for example, the Stark effect). If the continuous spectrum has a finite threshold there are additional contributions to $\alpha_s(t)$ which can be interpreted as due to a simultaneous specification of a finite initial time and an exact finite threshold (cf. Levine 1966b).

In general, the conclusion that
$$\lim_{t\to\infty} \langle s|\psi_s(t)\rangle \to 0 \tag{6.29}$$
depends on the condition that $\int |A_s(E)|\, dE$ exist, and, by the Riemann–Lebesgue lemma (Lighthill 1958) applied to eqn (6.22), eqn (6.29) follows. From the boundary condition $\psi_s(t) \to \phi(t)$ as $t \to \infty$, one can conclude that§
$$\lim_{t\to\infty} \langle s|\phi(t)\rangle \to 0 \tag{6.30}$$

§ For a rigorous proof one can use the analogue of eqns (D. 7) and (D. 8) of Appendix 2.D, since $\langle s|s\rangle$ is finite.

or, with $P = I - |s\rangle\langle s|$,
$$\lim_{t \to \infty} P\psi_s(t) \to \phi(t). \tag{6.31}$$

Equation (6.7) also implies
$$\lim_{t \to 0} P\psi_s(t) \to 0. \tag{6.32}$$

Equations (6.31) and (6.32) suggest that we interpret $P\psi_s(t)$ as the portion of the wavefunction representing the evolution of the products. Moreover, the conservation of the norm of $\psi_s(t)$ implies that

$$\langle P\psi_s(t)|P\psi_s(t)\rangle = \langle \psi_s(t)|P|\psi_s(t)\rangle = 1 - |\langle \psi_s(t)|s\rangle|^2 \tag{6.33}$$

and that
$$-\frac{d}{dt}A_s(t) = \frac{d}{dt}\langle \psi_s(t)|P|\psi_s(t)\rangle. \tag{6.34}$$

We have thus shown that the total wavefunction representing the system can be partitioned into two orthogonal components: one, representing the products, is the only component that remains as $t \to \infty$ but is absent as $t \to 0$ and the other, representing the intermediate state A^*, is the only component present at $t = 0$ but does not contribute at $t = \infty$. Both components are normalized and the rate of depletion of one component equals the rate of build up of the other.

The explicit construction of the two components of $\psi_s(t)$ was discussed by Hall and Levine (1966). In practice and for the purpose of gaining physical insight, it is often easier, once the validity of the description is established in principle, to construct the two components by an *a priori* specification of two projection operators, P and Q,

$$P + Q = I, \quad P^2 = P \quad \text{so that} \quad PQ = 0 \tag{6.35}$$

such that $P\psi(t)$ represents the products. The construction of such operators was discussed in Chapter 3.1. Only in special circumstances (namely when there are no other intermediates with energies near E_s) can Q be regarded as one-dimensional, $Q = |s\rangle\langle s|$.

Having selected some preliminary representation for P and Q one can represent the total Hamiltonian as

$$H = \begin{pmatrix} PHP & PHQ \\ QHP & QHQ \end{pmatrix} \tag{6.36}$$

and can introduce a model zero order 'decoupled' Hamiltonian

$$H' = \begin{pmatrix} PHP & 0 \\ 0 & QHQ \end{pmatrix}. \tag{6.37}$$

The Hamiltonian H' does not take into account the coupling between the P and Q components, and thus does not allow any breakdown. The

Hamiltonian PHP has a continuous spectrum that represents the unbounded relative motion of the products, while the spectrum of QHQ is, by construction, discrete. It simplifies the algebra (see Mies and Krauss 1966, Fano 1961) if the basis selected for P and Q leads to a diagonal form for PHP and QHQ, but this is not essential.

It is of interest to obtain eqn (6.31) by computing $\langle \psi_s(t)|P|\psi_s(t)\rangle$ directly. From eqn (2.2.7) one can write

$$\langle \psi_s(t)|P|\psi_s(t)\rangle = \sum_{\mathbf{m}} \int dE\, \langle \psi_s(t)|P|E,\mathbf{m}\rangle\langle E,\mathbf{m}|P|\psi_s(t)\rangle. \quad (6.38)$$

Using eqn (6.15 b) the amplitude $D(\mathbf{m}, E'; t)$ to observe the state $|E', \mathbf{m}\rangle$ in the product component can be written

$$D(\mathbf{m}, E'; t) = \exp(iE't/\hbar)\langle \mathbf{m}, E'|P\psi_s(t)\rangle$$

$$= (2\pi i)^{-1} \int_{sp} \exp[-i(\lambda-E')t/\hbar]\langle \mathbf{m}, E'|G(\lambda)|s\rangle\, d\lambda. \quad (6.39)$$

The required matrix element of the Green's operator was obtained in eqn (3.12),

$$\langle \mathbf{m}, E'|PG(\lambda)|s\rangle = \langle \mathbf{m}, E'|PW(\lambda)|s\rangle\langle s|G(\lambda)|s\rangle \quad (6.40)$$

and, from eqn (3.28 b),

$$\langle \mathbf{m}, E'|PW(\lambda)|s\rangle = (\lambda-E')^{-1}\langle \mathbf{m}, E'|\Gamma(\lambda)|s\rangle. \quad (6.41)$$

The integration path in eqn (6.39) can be deformed (as in eqn (3.34)) to a counter-clockwise path around the real axis. The only contribution to the integral is thus due to the discontinuity of the integrand on crossing the real axis, which is non-vanishing when E is in the continuous spectrum of H_0. There are two contributions. One from the discontinuity of the wave operator (eqn (3.1)) and one from the Green's function $\langle s|G(\lambda)|s\rangle$.

$$\langle \mathbf{m}, E'|[PG(E^-)-PG(E^+)]|s\rangle$$
$$= \langle \mathbf{m}, E'|PW(E^+)|s\rangle\langle s|[G(E^-)-G(E^+)]|s\rangle -$$
$$- \langle \mathbf{m}, E'|[PW(E^+)-PW(E^-)]|s\rangle\langle s|G(E^-)|s\rangle. \quad (6.42)$$

The discontinuity in the wave operator is evaluated from eqn (3.28 b) when we recall that E' is in the continuous spectrum of H_0 so that, using eqn (2.89),

$$-\lim_{\epsilon \to +0} \langle \mathbf{m}, E'|[PW(E^+)-PW(E^-)]|s\rangle = 2\pi i \delta(E'-E)\langle \mathbf{m}, E'|\Gamma(E)|s\rangle. \quad (6.43)$$

Using eqn (6.24) to evaluate the contribution of the first term in eqn (6.42), we obtain for the integral

$$D(\mathbf{m}, E'; t) = \langle \mathbf{m}, E'|PW(\lambda_s)|s\rangle\exp[-i(\lambda_s-E')t/\hbar]+$$
$$+\langle \mathbf{m}, E'|\Gamma(E')|s\rangle\langle s|G(E')|s\rangle, \quad (6.44)$$

where $\lambda_s = E_s - i\Gamma_s$ (eqn 6.25). In the limit $t \to \infty$ only the second term contributes and

$$\lim_{t\to\infty} D(\mathbf{m}, E'; t) = \langle \mathbf{m}, E'|\Gamma(E')|s\rangle\langle s|G(E')|s\rangle$$
$$= \langle \mathbf{m}, E'^-|s\rangle, \qquad (6.45)$$

where we have used the Hermitian conjugate of eqn (3.8) to obtain the last line. Equation (6.45) is in agreement with eqn (6.21). The finite lifetime $(\hbar/2\Gamma_s)$ of the state $|s\rangle$ implies that the energy of the products is not sharp but has a distribution with a width Γ_s. Using eqns (6.11) and (6.13),

$$\sum_{\mathbf{m}} |\langle \mathbf{m}, E^-|s\rangle|^2 = \pi^{-1}\Gamma_s/[(E-E_s)^2+\Gamma_s^2] = A_s(E). \qquad (6.46)$$

In the limit $\Gamma_s \to 0$,

$$A_s(E) = \delta(E-E_s). \qquad (6.47)$$

For finite times, if we approximate

$$\langle \mathbf{m}, E'|PW(\lambda_s)|s\rangle = (\lambda_s - E')^{-1}\langle \mathbf{m}, E'|\Gamma(\lambda_s)|s\rangle$$
$$\simeq -(E'-\lambda_s)^{-1}\langle \mathbf{m}, E'|\Gamma(E')|s\rangle, \qquad (6.48)$$

we can write, using eqns (6.44) and (6.48),

$$D(\mathbf{m}, E'; t) \simeq \frac{\langle \mathbf{m}, E'|\Gamma(E')|s\rangle}{E'-E_s+i\Gamma_s} \times$$
$$\times \{1-\exp(-\Gamma_s t/\hbar)\exp[-i(E_s-E')t/\hbar]\} \qquad (6.49)$$

and

$$|\langle E', \mathbf{m}|P\psi_s(t)\rangle|^2 \simeq \frac{|\langle \mathbf{m}, E'|\Gamma(E')|s\rangle|^2}{(E'-E_s)^2+\Gamma_s^2} \times$$
$$\times \{1-2\exp(-\Gamma_s t/\hbar)\cos[(E'-E_s)t/\hbar]+\exp(-2\Gamma_s t/\hbar)\}. \qquad (6.50)$$

The oscillatory time-dependence of the second term is normally not observed, since it is averaged out in an experiment with a poor resolution (cf. eqn (6.54) below). It has, however, been detected in refined measurements of decay of nuclear excited states by Mossbauer spectroscopy.

To sum eqn (6.50) over $\mathbf{m}$ we note (eqn (2.118)) that

$$\Gamma_s(E') = \pi \sum_{\mathbf{m}} |\langle \mathbf{m}, E'|\Gamma(E')|s\rangle|^2. \qquad (6.51)$$

Recognizing (section 2.3.0) that $\mathbf{m}$ is at least a pair of indices $\mathbf{m} = (\hat{\mathbf{k}}_m, m)$ where $\hat{\mathbf{k}}_m$ denotes the direction of relative motion of the products and m the internal state, we can write

$$\Gamma_s(E) = \sum_m \Gamma_s^m(E), \qquad (6.52)$$

$$\Gamma_s^m(E) = \pi \int d\hat{\mathbf{k}}_m |\langle \hat{\mathbf{k}}_m, m, E|\Gamma(E)|s\rangle|^2$$
$$= \pi \int d\hat{\mathbf{k}}_m \rho(k_m)|\langle \mathbf{k}_m, m|\Gamma(E)|s\rangle|^2, \qquad (6.53)$$

where in the last line we have changed the normalization of the wavefunction for relative motion of the products, and $\rho(k_m)$ is the density of translational states (eqn (2.2.10)).

Using the result

$$\int [(E-E_s)^2+\Gamma_s^2]^{-1} \cos[(E-E_s)t/\hbar]\, \mathrm{d}E = (\pi/\Gamma_s)\exp(-\Gamma_s t/\hbar) \quad (6.54)$$

and eqn (6.53) one obtains

$$\int \mathrm{d}\hat{\mathbf{k}}_m \int \mathrm{d}E |\langle \mathbf{m}, E'|P\psi_s(t)\rangle|^2 = (\Gamma_s^m/\Gamma_s)[1-\exp(-2\Gamma_s t/\hbar)]. \quad (6.55)$$

Summing this result over m confirms eqn (6.34).

The rate of formation of products in a particular internal state is given by the time derivative of eqn (6.50). In connection with the statistical theories (section 3.5.0) it is of interest to examine the rate of formation of products at a given total energy, given by

$$\frac{\mathrm{d}}{\mathrm{d}t} \sum_m |\langle \mathbf{m}, E'|P\psi_s(t)\rangle|^2 \leqslant (2\pi\hbar)^{-1}, \quad (6.56)$$

where we have used eqns (6.52) and (6.46) to obtain the upper bound.

One can extend the discussion to the case when at $t = 0$ the state of the system is represented by a density operator $\rho(t = 0)$ rather than by a pure state. (Equation (6.7) corresponds to $\rho(t = 0) = |s\rangle\langle s|$.) Such a description corresponds to the physical situation where we prepare (by an unspecified mechanism) an initial ensemble at $t = 0$. At some time $t > 0$ the density operator is

$$\rho(t) = \exp(-iHt/\hbar)\rho(0)\exp(iHt/\hbar) \quad (6.57)$$

and the probability to observe the products at some time t in the state $|E', \mathbf{m}\rangle$ is given by

$$\langle \mathbf{m}, E'|P\rho(t)P|E', \mathbf{m}\rangle. \quad (6.58)$$

In terms of a basis that diagonalizes $\rho(0)$, say§

$$\rho(0) = \sum_r p_r |r\rangle\langle r|, \quad \sum_r p_r = 1, \quad Q\rho(0)Q = \rho(0), \quad (6.59)$$

the rate of appearance of products in the state $|E', \mathbf{m}\rangle$ is given by

$$\frac{\mathrm{d}}{\mathrm{d}t}\langle \mathbf{m}, E'|P\rho(t)P|E', \mathbf{m}\rangle = \sum_r p_r \frac{\mathrm{d}}{\mathrm{d}t} |\langle \mathbf{m}, E'|P\psi_r(t)\rangle|^2. \quad (6.60)$$

Equation (6.60) has the form of a relaxation equation (section 2.4.0) and one can use the partitioning technique to obtain a Van Hove (1957)-type master equation for $\rho(t)$ given that $\rho(t = 0)$ is in the form of eqn (6.59). The details are given in Levine (1966c).

§ In general, we expect that such a basis is a bi-orthogonal set (see eqns (2.97)–(2.101)), so that a more general form of eqn (6.59) is $\rho(0) = \sum_r p_r |r\rangle\langle r^*|$.

In the quasi-equilibrium theory (Rosenstock et al. 1952, see also section 3.5.0) one makes the assumption that the probabilities p_r are independent of the mechanism of excitation that generated the ensemble at $t = 0$ and depend only on the total energy and other good quantum numbers.

To describe the bound states of A and the states A^* by the same Hamiltonian we introduce the projection operator C_i such that $C_i H C_i$ has a completely discrete spectrum, where H is the Hamiltonian of A. The eigenfunctions of $C_i H C_i$ with eigenvalues below the threshold for fragmentation represent the ordinary bound states of A, and the eigenfunctions with higher eigenvalues represent the states of A^*. We can now represent an equilibrium density matrix for A (or A^*) at the energy E by

$$\rho(E) = C_i \delta(E-H) C_i. \quad (6.61)$$

In a canonical ensemble

$$\rho_c = \int dE \exp(-\beta E) C_i \delta(E-H) C_i / Z_A, \quad (6.62)$$

where Z_A is the partition function for the undissociated molecule A,

$$Z_A = \int dE \exp(-\beta E) \text{Tr}\{C_i \delta(E-H)\}. \quad (6.63)$$

If the distribution of the total energy is not canonical the factor $\exp(-\beta E)/Z_A$ in eqn (6.62) is replaced by the appropriate density function.

If we regard ρ_c as the density matrix at $t = 0$ then

$$p_r(E) = \text{Tr}[|r\rangle\langle r|\rho(E)] = \langle r|C_i \delta(E-H) C_i |r\rangle \quad (6.64)$$

so that eqn (6.62) can be written

$$\rho_c = \int dE \exp(-\beta E) \sum_r p_r(E) |r\rangle\langle r|/Z_A \quad (6.65)$$

and the rate of formation of products is (from eqn (6.60))

$$\int dE \exp(-\beta E) \sum_r p_r(E) \frac{d}{dt} |\langle \mathbf{m}, E'|P\psi_r(t)\rangle|^2 / Z_A. \quad (6.66)$$

It is important to note that if E_r is below the threshold for fragmentation of A the state $|r\rangle$ is an eigenstate of H so that $HC_i|r\rangle = C_i E_r |r\rangle$ and $PHC_i|r\rangle = 0$. Thus the states of A do not contribute to the summation over r in eqn (6.66) (since $P\psi_r(t) = 0$ for these states). The projection Q on the activated states A^* is contained in C_i, $C_i Q = Q$. Only the portion $Q\rho(t=0)Q$ will contribute to the breakdown.

The results for the evolution of $\rho(t)$ can be expressed in a simple form using the bi-orthogonal set of states that diagonalizes $QG(\lambda)Q$, eqn (2.107).

In terms of these states

$$Q \exp(-iHt/\hbar)Q = (2\pi i)^{-1} \int_{sp} \exp(-i\lambda t/\hbar) Q G(\lambda) Q \, d\lambda$$

$$= (2\pi i)^{-1} \int_{sp} \exp(-i\lambda t/\hbar) \sum_s \frac{|s\rangle\langle s^*|}{\lambda - \lambda_s} \, d\lambda$$

$$= \sum_s |s\rangle\langle s^*| \exp(-iHt/\hbar) |s\rangle\langle s^*| \qquad (6.67)$$

since $\langle s^*|s'\rangle = \delta_{s,s'}$. There are no cross terms in the last equation so that the evolution of the activated molecule can be described by a set of non-interacting states. In contrast a basis that only diagonalizes QHQ will not diagonalize the operator $Q \exp(-iHt/\hbar)Q$. The reasons are apparent when we iterate $QG(\lambda)Q$, since such a basis will not diagonalize $Q\Gamma Q$ which describes the breakdown of A^* into products and the subsequent healing of the products to form A^* (cf. eqn (2.96)).

The rate of depletion of A^* is given by

$$\frac{d}{dt} \mathrm{Tr}\, Q\rho(t) = (i\hbar)^{-1} \mathrm{Tr}\, Q[\rho(t), H] = \sum_s p_s \frac{d}{dt} |\langle s^*|\psi_s(t)\rangle|^2$$

$$= -\frac{d}{dt} \mathrm{Tr}\, P\rho(t). \qquad (6.68)$$

In the canonical ensemble we can write this rate as

$$\sum_s \frac{d}{dt} |\langle s^*|\psi_s(t)\rangle|^2 \int dE \exp(-\beta E) p_s(E)/Z_A, \qquad (6.69)$$

where the integral is just the probability of occupation of $|s\rangle$ in the canonical ensemble. The rates (6.60) and (6.69) are time dependent due to the depletion of the initial ensemble. In a stationary experiment the initial ensemble is continuously regenerated.

For thermal bimolecular reactions where the initial ensemble is maintained by collisions of very short duration we can obtain a simple description of the system by the introduction of an 'averaged' density matrix (following Karplus and Schwinger (1948)). The averaging is over the time interval between successive collisions. If the collisions are random, with an average frequency ω, the probability of a time interval τ to $\tau + d\tau$ between successive collisions is $\exp(-\omega\tau)\omega\, d\tau$. An average, stationary, density operator $\bar\rho$ is defined,

$$\bar\rho = \omega \int_0^\infty \rho(\tau) \exp(-\omega\tau) \, d\tau \qquad (6.70)$$

so that $\dot{N}$, the rate of formation of products, given by (see also eqn (6.68))
$$\dot{N} = (i\hbar)^{-1}\text{Tr}\{Q[\bar{\rho}, H]\},$$
can be written
$$\dot{N} = \omega \int_0^\infty (i\hbar)^{-1}\text{Tr}\{Q[\rho(\tau), H]\}\exp(-\omega\tau)\,\mathrm{d}\tau \qquad (6.71)$$
$$= \omega \int_0^\infty \exp(-\omega\tau)\frac{\mathrm{d}}{\mathrm{d}\tau}\text{Tr}\{Q\rho(\tau)\}\,\mathrm{d}\tau.$$

In the limit $\omega \to \infty$ one can show by integration by parts that
$$\lim_{\omega\to\infty}\dot{N} = \lim_{\tau\to 0}\frac{\mathrm{d}}{\mathrm{d}\tau}\text{Tr}\{Q\rho(\tau)\}. \qquad (6.72\,\text{a})$$

The limit in eqn (6.72 a) is known as the 'high pressure' limit since the collision frequency is proportional to the pressure in the ensemble when only binary collisions can take place. The alternative limit $\omega \to 0$ can also be obtained in the same fashion, by first differentiating eqn (6.71) with respect to ω, when we obtain
$$\dot{N} \xrightarrow[\omega\to 0]{} \omega\lim_{\tau\to 0}\text{Tr}\{Q\rho(\tau)\}. \qquad (6.72\,\text{b})$$

Eqns (6.71) and (6.72) are the quantum phenomenological analogues of Slater's 'new approach to rate theory' (1959). See also Levine (1967c) and Thiele (1966).

A detailed discussion of excitation processes in ensembles is given in sections 2.6.2 and 2.6.3.

To extend the discussion to include the details of the excitation process we consider the association process
$$B+C \to A^* \to B+C$$
where $B+C$ are the products of the unimolecular breakdown. An initial state at $t \to -\infty$ that leads to the specified form of $\psi_s(t)$ (say eqn (6.15 a)) can be defined by (cf. eqn (2.7.46))
$$\psi_s(t) = \Omega^+\phi'(t) \qquad (6.73)$$
or since
$$\Omega^+|E, \mathbf{n}\rangle = |E, \mathbf{n}^+\rangle,$$
$$\phi'(t) = \sum_\mathbf{n} \int \mathrm{d}E\,|E, \mathbf{n}\rangle\langle E, \mathbf{n}^+|s\rangle\exp(-iEt/\hbar). \qquad (6.74)$$

Occupation probabilities can now be evaluated for all t. For example,
$$A_s(t) = |\langle s|\psi_s(t)\rangle|^2 = \exp(-2\Gamma_s|t|/\hbar). \qquad (6.75)$$
Equation (6.75) demonstrates the 'build up' of the activated state during the excitation period (when t is negative and increasing). The elements

of the **S** matrix are obtained immediately on comparing the expansion of the initial (eqn (6.74)) and final (eqn (6.1)) states, using eqn (2.7.22),

$$\mathbf{B} = \mathbf{SA}, \quad (6.76\,\mathrm{a})$$

or
$$\langle E, \mathbf{n}^-|s\rangle = \sum_{\mathbf{m}} S_{\mathbf{n},\mathbf{m}}\langle E, \mathbf{m}^+|s\rangle. \quad (6.76\,\mathrm{b})$$

Thus $S_{\mathbf{n},\mathbf{m}} = \langle E, \mathbf{n}^-|E, \mathbf{m}^+\rangle$ (as expected, cf. eqn (2.5.30)).

To relate the wavefunction of the system to the 'duration' of the excitation process we write for $t > 0$

$$|\psi(t)\rangle = \exp(-iHt/\hbar)\int_{-\infty}^{0} dt'\, f(t')\exp(iHt'/\hbar) \times$$
$$\times [\exp(-iE't'/\hbar)|E', \mathbf{n}\rangle], \quad (6.77)$$

and refer to $f(t')$ as the 'switching function'. The interpretation of eqn (6.77) is based on regarding the term in the square bracket as an initial state at the time t'. From the time t' onwards this state evolves under the full Hamiltonian H. (To see this one has to introduce the first factor in eqn (6.77) inside the integral to obtain $\exp[-iH(t-t')/\hbar]$.) The switching function $f(t')$ determines the distribution of time t' at which the interaction between B and C begins to operate. In a stationary experiment (equation (2.D.11)) $f(t') = \epsilon\exp(\epsilon t')$ and $\epsilon \to +0$ after the t' integration is performed. If the excitation period is short $f(t')$ is peaked near $t' = 0$.

Putting
$$|\psi(t)\rangle = \sum_{\mathbf{n}}\int dE\,|E, \mathbf{n}^+\rangle A(\mathbf{n}, E)\exp(-iEt/\hbar), \quad (6.78)$$

one finds

$$A(\mathbf{m}, E) = \langle E, \mathbf{m}^+|\psi(t=0)\rangle$$
$$= \int_{-\infty}^{0} dt\, f(t)\langle E, \mathbf{m}^+|\exp[i(H-E')t/\hbar]|E', \mathbf{n}\rangle. \quad (6.79)$$

In the stationary limit $f(t) = \epsilon\exp(\epsilon t)$ and using eqn (2.D.13) we see that in the limit $\epsilon \to +0$

$$A(\mathbf{m}, E) = \langle E, \mathbf{m}^+|E', \mathbf{n}^+\rangle = \delta(E-E')\delta_{\mathbf{m},\mathbf{n}}.$$

Under stationary conditions the activated state is formed at a constant rate and we expect that the probability to observe the activated state will be a stationary time-independent quantity. (This is an example of the so-called 'steady state' assumption.) Substituting the stationary value of $A(\mathbf{m}, E)$ in eqn (6.78) we obtain

$$A_s(t) = |\langle s|\psi(t)\rangle|^2 = |\langle s|E', \mathbf{n}^+\rangle|^2. \quad (6.80)$$

In a stationary experiment with well-defined energies one does not observe the time evolution of the activated state and its presence is manifested in the energy dependence of the scattering amplitude (cf. section 3.2.2). In general the transform relation (6.79) implies that the range of energies for which $A(\mathbf{m}, E)$ assumes significant values is inversely proportional to the range in time of the switching function.

In general for association processes one can determine the excitation mechanism by the partitioning technique, as is discussed in section 3.2.2. The solution of the L.S. equation can be written as (cf. eqn (1.26))

$$|E, \mathbf{m}^+\rangle = |E, \mathbf{m}_\mathrm{I}^+\rangle + G(E^+)|s\rangle\langle s|HP|E, \mathbf{m}_\mathrm{I}^+\rangle, \qquad (6.81)$$

where $P = I - |s\rangle\langle s|$. $|E, \mathbf{m}_\mathrm{I}^+\rangle$ is the component that evolves under the Hamiltonian PHP. It is not coupled to the activated state, and represents those collisions that do not proceed via the activated state. The second term in eqn (6.81) represents the collisions that do proceed via the activated state. In terms of time ordering (section 2.7.1) the activated state is first formed by the transition amplitude $\langle s|HP|E, \mathbf{m}_\mathrm{I}^+\rangle$ explicitly indicated in eqn (6.81). We can thus define

$$|z\rangle = |s\rangle\langle s|HP|E, \mathbf{m}_\mathrm{I}^+\rangle$$

and regard $|z\rangle$ as the initial state for the events subsequent to the first formation of the activated state.

Using eqn (6.81) we can write

$$\langle s|E, \mathbf{m}^+\rangle = \langle s|G(E^+)|z\rangle$$

$$= \sum_{n=0}^{\infty} (E^+ - \langle s|H_0|s\rangle)^{-1} [\langle s|\Gamma(E^+)|s\rangle \times$$

$$\times (E^+ - \langle s|H_0|s\rangle)^{-1}]^n \langle s|\Gamma(E^+)|E, \mathbf{m}\rangle. \qquad (6.82)$$

In section 3.2.2 (eqns (2.47)–(2.61)) we have discussed the significance of such 'time-ordered' iterations. Thus in the nth term in this expansion, subsequent to the first formation of the activated state, it undergoes n healing processes, in which the activated state breaks into products, which start to separate but form again the activated state. The description of the breakdown and the subsequent formation is contained in the rate operator Γ. When the activated state breaks into products not all the products heal, but a fraction escapes as products. Using the iteration (6.82) in eqn (6.78), we find with $A(\mathbf{m}, E) = \langle E, \mathbf{m}^+|s\rangle$,

$$\exp(\mathrm{i}\langle s|H_0|s\rangle t/\hbar)\langle s|\psi(t)\rangle = \sum_{n=0}^{\infty} [-\mathrm{i}\langle s|\Gamma|s\rangle t/\hbar]^n/n!$$

$$= \exp(-\mathrm{i}\langle s|\Gamma|s\rangle t/\hbar), \qquad (6.83)$$

where we regarded $\langle s|\Gamma|s\rangle$ as energy-independent. The exponential decay law corresponds to a coherent superposition of the different healing stages. (For a worked out example see Levine 1966c. The decay law found there is not exponential, however, since, for the model Hamiltonian used, the continuous spectrum had a finite range.)

In general, however, the projection Q will not be one dimensional, and instead of eqn (6.81) we have

$$|E, \mathbf{m}^+\rangle = |E, \mathbf{m}_1^+\rangle + G(E^+)QHP|E, \mathbf{m}_1^+\rangle$$
$$= |E, \mathbf{m}_1^+\rangle + WQG(E^+)QHP|E, \mathbf{m}_1^+\rangle. \qquad (6.84)$$

While in principle the Green's operator $QG(E^+)Q$ can be diagonalized, it is sometimes more convenient to introduce the concept of an exit state (section 3.5.0). In the present context we can define an exit state as a state that is coupled directly to $|E, \mathbf{m}_1^+\rangle$. The motivation for this definition stems from the fact that one can usually introduce a basis for the Q subspace, such that some states are not directly coupled to $|E, \mathbf{m}_1^+\rangle$ and can only be excited indirectly. As an example we can think of a collision of an atom with a rotor, in its ground state, with a total energy that is insufficient for rotational excitation. For the mutual interaction we assume $V = V_1(R) + aV_2(R)P_2(\hat{\mathbf{R}}.\hat{\mathbf{r}})$, where $\hat{\mathbf{r}}$ is the direction of the rotor axis and V_1 and V_2 are central potentials. The rotor can be internally excited during the collision thus leading to a temporary formation of bound states, as discussed in section 3.2.2 (see also Levine, Johnson, Muckerman, and Bernstein 1968). Only the second ($j = 2$) state of the rotor can be directly excited from the unbound relative motion since

$$\int \mathrm{d}\hat{\mathbf{r}}\ Y_0^{0*}(\hat{\mathbf{r}})P_2(\hat{\mathbf{R}}.\hat{\mathbf{r}})Y_j^{m_j}(\hat{\mathbf{r}}) = 0, \quad j > 2.$$

All higher (even j) states of the rotor can be internally excited, but only indirectly, that is by excitation transfer from the $j = 2$ state. The formation and breakdown of the activated states necessarily proceeds via the $j = 2$ state of the rotor.

If $|r\rangle$ is an activated state that diagonalizes $Q\mathscr{H}Q$ we can then write

$$\langle r|QHP|E, \mathbf{m}_1^+\rangle = \sum_s \langle r|rs\rangle\langle rs|QHP|E, \mathbf{m}_1^+\rangle, \qquad (6.85)$$

where (in conforming to the notation of section 3.5.0) we have denoted the exit state for the breakdown of the rth activated state by $|rs\rangle$. Thus from eqn (2.111),

$$\Gamma_r^m = \sum_s (|\langle r|rs\rangle|^2/\langle r|r\rangle)\Gamma_{rs}^m,$$
$$\Gamma_{rs}^m = \pi \int \mathrm{d}\hat{\mathbf{k}}_m\, \rho(k_m)|\langle rs|QHP|\mathbf{k}, m_1^+\rangle|^2.$$

In a similar fashion one can analyse the transition operator.

3.6.2. *Collision theory in ensembles*

The discussion of thermal unimolecular (and bimolecular) reactions can be based on the equation of motion of the density operator for the ensemble. In this way the problem of relaxation to a chemical equilibrium becomes a particular example of a relaxation process (see Chapter 3.7, in particular section 3.7.3). Our discussion follows Levine (1966e, 1966f), and refers explicitly to the sequence of reactions

$$A+M \to A^*+M',$$
$$A^* \to B+C.$$

These reactions can be observed in two types of experiments. In thermal unimolecular reactions we have a small concentration of A molecules in a host gas of molecules M. In this case deactivating collisions $A^*+M \to A+M$ can also take place. Moreover, our description implies the use of the binary collision approximation (eqns (2.5.117), (2.5.118)) since we neglect triple and higher collisions. In other situations M is a projectile (electron, photon, etc.); deactivating collisions are then excluded; for practical projectile intensities the binary collision approximation is exact and one observes the effects of a single collision only. Quite often the initial ensemble (before the collision) is subject to experimental control.

In the binary collision approximation the density operator for a pair of A and M satisfies the equation of motion (2.5.117),

$$i\hbar \frac{d}{dt} \rho_{AM} = [H_A + H_M + U, \rho_{AM}], \tag{6.86}$$

where H_A and H_M are the Hamiltonians of A and M respectively and U is their mutual interaction. We define a density operator for A by

$$\rho_A = \text{Tr}_M \rho_{AM}, \tag{6.87}$$

where the trace is over the coordinates of the molecule M. The average value of any observable that refers to A only, say Q_A, can be written $\text{Tr}\,\rho_A Q_A$, where the trace is over the coordinates of A.

The equation of motion for ρ_A in the binary approximation is obtained from eqns (6.86) and (6.87) as

$$i\hbar \frac{d}{dt} \rho_A = [H_A, \rho_A] + \text{Tr}_M\{[U, \rho_{AM}]\}. \tag{6.88}$$

The term $\qquad (i\hbar)^{-1} \text{Tr}_M\{[U, \rho_{AM}]\}$

represents the changes in ρ_A due to collisions of A with the host gas molecule M. In section 2.4.2 this term was evaluated for the case where

energy transfer is the only process taking place. The term
$$(i\hbar)^{-1}[H_A, \rho_A]$$
represents the changes in ρ_A due to the Hamiltonian H_A. If A^* is a stable molecule and ρ_A is diagonal in the states of A, this term does not contribute. In the present case we have already seen (eqn (6.68)) that this term represents the breakdown of A^* into the products.

For the analysis of macroscopic experiments one is often not interested in the detailed changes in ρ_A over time intervals of the order of the averaged duration of a collision. The quantity of interest is

$$\frac{\Delta \rho_A}{\Delta t}, \tag{6.89}$$

where $\Delta \rho_A = \rho_A(t+\Delta t) - \rho_A(t)$, and Δt is longer than the duration of a collision,
$$\Delta t > \tau_c \tag{6.90}$$
but is sufficiently small so that the change in ρ_A is linear in Δt. Under these conditions we can replace eqn (6.89) by the time derivative of ρ_A. It is understood, however, that the resulting equation of motion only provides the behaviour of ρ_A 'between collisions' and is not meant to represent ρ_A at all microscopic times. In other words, we are seeking an equation of motion for ρ_A where the time variable is the macroscopic time. On the macroscopic time scale the average collision appears to be instantaneous. If $\phi(t)$ is the initial state of the system before the collision, then after a single collision

$$\phi(t+\Delta t) = \exp(-iH_0 \Delta t/\hbar) S\phi(t), \tag{6.91}$$

where $\Delta t > \tau_c$, H_0 is the Hamiltonian in the absence of the interaction that leads to the collision and S is the S matrix, or

$$\phi_f(s \to +\infty) = S\phi_i(s \to -\infty),$$

where $\phi_i(s)$ is the initial state before the collision and s is the microscopic time. The collision act which is of very long duration on the microscopic scale s is instantaneous on the macroscopic scale t. In yet another way we can say that the collisions are sufficiently rapid, and the time between collisions is sufficiently long, that whenever the system is under observation the great majority of pairs A and M are non-interacting (that is, they are either before or after but not during a collision). We shall refer to this approximation as the 'Markoffian' approximation (see also section 3.7.3).

From a formal point of view one can analyse eqn (6.91) using the exact equation (2.7.6) for the forward evolution operator. We put

$$\phi(t+\Delta t) = U^+(\Delta t)\phi(t) \tag{6.92}$$

where the superscript $+$ is a reminder that U vanishes for negative values of the argument or, writing $\theta(t)$ for the step function in t,

$$U^+(\Delta t) = \theta(\Delta t)\exp(-iH\Delta t/\hbar) = U^+(t+\Delta t - t)$$

$$= U_0^+(\Delta t) - (i/\hbar) \int_{-\infty}^{\infty} dt'\, U_0^+(t+\Delta t-t')VU^+(t'-t), \quad (6.93)$$

and U_0^+ is the forward evolution operator for $H_0 = H - V$.

As discussed in section 2.7.2, the time t' in eqn (6.93) is the last time at which the molecules interact. From the time t' onwards the collision is over and the molecules evolve under H_0. In general, the actual range of t' is restricted by the time ordering (as manifested in the step function in eqn (6.93)) to $t' > t$ and $t+\Delta t > t'$, so that t' can assume any value in the interval Δt. If, however, we know that the collision is over by some time $t'' < t+\Delta t$ the range of integration can be further restricted to $t' < t''$, so that

$$U^+(\Delta t) = U_0^+(\Delta t) + U_0^+(t+\Delta t-t'') \times$$

$$\times (-i/\hbar) \int_{-\infty}^{\infty} U_0^+(t''-t')VU^+(t'-t)\, dt'. \quad (6.94)$$

If the collision is 'instantaneous' on the macroscopic t scale, we can put $t'' = t$ outside the integral, while $t'' \to \infty$ inside the integral (where the time scale is microscopic) so that

$$U^+(\Delta t) \simeq U_0^+(\Delta t)S \quad (6.95)$$

where S is the S operator (eqn (2.7.33)). The energy of the initial and final states is well defined to within $(\hbar/\Delta t)$, and a better energy resolution is not of operational importance (since time is defined to within Δt only).

The condition that the duration of the collision be much shorter compared with the time between collisions can also be formulated as a distance criteria. The time between collisions of an A molecule in a gas of $n = N/V$, M molecules per unit volume is l/v, where $l^{-1} = nv\sigma$ is the mean free path. The duration of a rapid collision is v/a, where a is the range of the interaction. Thus

$$a\sigma \ll V/N. \quad (6.96)$$

$a\sigma$ is the volume of the pair interaction region (recall that σ is the cross-section of the interaction region), while V/N is the volume per A–M pair. We can also interpret eqn (6.96) in the language of time-independent collision theory. Our assumption that one does not observe the molecules during a collision is equivalent to replacing the solutions of the L.S. equation by their asymptotic forms, valid for large relative separations of A and M. This approximation is valid outside the collision volume $a\sigma$. The inequality (6.96) ensures the validity of approximation for the volume V/N. In general $a \geqslant \lambdabar = k^{-1}$, where $\hbar k = p$ is the momentum of the relative motion since one can only define the relative separation to within k^{-1}. Thus $\tau_c \geqslant (\lambdabar/v) = \hbar/pv$. In thermal equilibrium $\tau_c \geqslant \hbar/kT \approx 10^{-11}/T$ s.

The present discussion is not required for the solution of eqn (6.88), but it does lead to considerable simplifications in the physical interpretation of the resulting solution. For example, if Q_n is the projection on a particular state of A, we have interpreted in section 2.4.2

$$\frac{d}{dt} \operatorname{Tr} Q_n \rho_A$$

as the rate of change of the probability to find the state n occupied. This interpretation is clearly subject to the assumption that ρ_A describes an isolated A molecule. In the same fashion our interpretation of Ehrenfest's theorem (section 2.4.1) is subject to the above approximation. Of course, in an isolated binary collision event our approximation becomes trivially exact, but in ensembles one cannot interpret Ehrenfest's theorem (or its approximation by the Golden rule) in the way we have used them without ensuring that initial and final non-interacting states can be defined. One should add perhaps that the definition of 'non-interacting' is an operational definition. One can, for example, compute transition rates between states that are interacting by a distortion potential that cannot lead to transitions (section 2.5.2). Another case in point is the activated states. Provided the lifetime of an activated state of A^* is longer than the duration of the activating collision, $A+M \to A^*+M'$, it becomes meaningful to consider A^* as the final state in a collision. One thus defines a unimolecular reaction by the condition

$$\tau_{A^*} > \tau_c. \tag{6.97}$$

In a collision of duration τ_c the energy is defined to within $\hbar/\tau_c = \Delta E_c$ only. Since (eqn (6.28)) $\tau_{A^*} = \hbar/2\Gamma_{A^*}$ we can require $\Delta E_c > 2\Gamma_{A^*}$ (for an isolated state). Equation (6.97) provides an alternative point of view to an identical conclusion found in the previous section. It is meaningful to consider an activated state as a distinct initial (or final) state in a collision process only if eqn (6.97) holds. One can also write eqn (6.97) as a distance criteria (Levine 1966f). If v is the relative velocity of A^* and M we require that

$$v\tau_{A^*} > a.$$

$v\tau_A$ is the average distance that A^* covers before breakdown. If A^* is to be observed after the collision, eqn (6.97) should hold as otherwise A^* will breakdown before A^* and M have terminated their mutual interaction.

At a risk of excessively labouring our point we re-derive Ehrenfest's theorem for ensembles from the Markoffian approximation. If B is measurable then the change in B due to a single collision is given by eqn (2.7.37),
$$\Delta B = S^\dagger BS - B. \tag{6.98}$$
We evaluate $\Delta B/\Delta s$, where Δs is a microscopically large energy interval, such that
$$\Delta s < \Delta t \quad \text{or} \quad (\hbar/\Delta s) < \Delta E \tag{6.99}$$
where $\Delta E = \hbar/\Delta t$ is the energy resolution. In view of eqn (6.99) we can use eqn (2.7.40) for Δs,
$$\Delta s \simeq \int_{-\infty}^{\infty} ds = 2\pi\hbar\delta(E_i - E_f),$$
where E_i and E_f are the initial and final energies (defined to within ΔE only). Using eqns (2.7.41) and (6.98),
$$\frac{\Delta\langle B\rangle}{\Delta s} = \sum_f (B_f - B_i) R_{f,i}, \tag{6.100}$$
where f sums over the final states, $B_f = \langle f|B|f\rangle$ is the expectation value of B in the state f, and $R_{f,i}$ is the rate of transitions $i \to f$. Equation (6.100) is Ehrenfest's theorem (cf. eqn (2.4.10)) for $d\langle B\rangle/dt$. Since B is a measurable, $[B, H_0] = 0$. Otherwise eqn (6.100) would contain an additional term $\langle[B, H_0]\rangle$. The results for the density matrix are a particular case of this derivation.

The question naturally arises of whether one can obtain a solution for $d\rho/ds$ (or dB/ds) where s is a microscopic time. By a solution we mean an algorithm for computing $d\rho(s)/ds$ given as a boundary condition $\rho_0(s)$, where $\rho_0(s)$ is the density operator for the non-interacting molecules, and starting from the exact equation of motion,
$$i\hbar \frac{d}{ds}\rho(s) = [H, \rho(s)].$$
The work of Zwanzig (1960, 1964) and Fano (1963) has provided an affirmative answer in principle. However the resulting expressions, when written in terms of the S or T operators, are at the moment unmanageable. In the lowest approximation these expressions reduce to a form equivalent to Ehrenfest's theorem (Fano 1963, Ben Reuven 1966, and Cooper 1967). Other interesting work on this problem was done by the

Belgian school (Resibois 1961, Prigogine and Resibois 1961, and by Lax 1964 and Argyres 1966). These studies show (see also section 3.7.3) that in general the time derivative of $\mathrm{Tr}\, B\rho(s)$ depends on the values of $\rho(s')$ for all $s' < s$ (as in eqn (6.93)). In the Markoffian approximation this dependence is replaced by a dependence on $\rho(s)$ only.

3.6.3. *Excitation processes*

The equation of motion for ρ_A,

$$\frac{\mathrm{d}}{\mathrm{d}t}\rho_A = (i\hbar)^{-1}[H_A, \rho_A] + (i\hbar)^{-1}\mathrm{Tr}_M\{[U, \rho_{AM}]\}, \qquad (6.88)$$

can now be analysed in the Markoffian approximation. Subject to the assumption that the average duration of a collision is much shorter than the lifetime of A^* we define, as in section 3.6.1, a projection operator C_i such that $C_i \rho_A C_i$ is the density operator for an undissociated A molecule, whether activated or not. In a stationary experiment we expect that

$$\frac{\mathrm{d}}{\mathrm{d}t}\mathrm{Tr}_A Q\rho_A(t) = 0 \qquad (6.101)$$

and, using eqn (6.88), interpret that to mean that the rate of breakdown into products is equal to the net rate of activation by collisions (Lindemann 1922). The rate of breakdown can be written as

$$(i\hbar)^{-1}\mathrm{Tr}_A\{Q[H_A, \rho_A(t)]\},$$

where Q is the projection on the activated states, so that $C_i Q = Q$ where C_i is the projection on all the state of A whether activated or not and we put $C_i = Q + C^i$, where C^i is the projection on the stable states of A.

In general, the state of an undissociated molecule at a given time depends on the outcome of the previous collisions. In the simplest case one can assume that the state of a molecule depends on the last collision only and, in the Markoffian approximation, argue as follows. Let $\rho_A(t, t_0)$ be the density operator at the time t when the last collision took place at t_0, $t_0 < t$. For $t > t_0$, $\rho_A(t, t_0)$ satisfies the equation of motion,

$$\frac{\mathrm{d}}{\mathrm{d}t}\rho_A = (i\hbar)^{-1}[H_A, \rho_A]. \qquad (6.102)$$

The effect of the collisions is incorporated by specifying

$$\lim_{t \to t_0 + 0} \rho_A(t, t_0),$$

i.e. the density operator immediately after the collision, which, by assumption, does not depend on the previous history of the system.

When, as a result of a collision at a time t_0 a molecule is activated, it can undergo a unimolecular breakdown. Thus $\text{Tr}\{Q\rho_A(t,t_0)\}$ is a decreasing function of t. At a given time t the above trace receives contributions from a whole range of t_0 values. Some molecules have just been activated, while others have undergone the last collision a long time in the past. To compute a unimolecular rate constant one has to average over the distribution of times t_0. If the collisions between A and M occur at random, with a mean frequency ω, one can define an average density operator $\bar{\rho}_A(t)$ (Karplus and Schwinger 1948) by

$$\bar{\rho}_A(t) = \omega \int_0^\infty \rho_A(t, t-\tau)\exp(-\omega\tau)\,d\tau, \qquad (6.103)$$

where $\omega \exp(-\omega\tau)\,d\tau$ is the probability that the last collision took place between $t-\tau$ and $t-(\tau+d\tau)$. Then

$$\frac{d}{dt}\bar{\rho}_A(t) = \omega \int_0^\infty \frac{d}{dt}\rho_A(t,t_0)|_{t_0=t-\tau}\exp(-\omega\tau)\,d\tau -$$

$$-\omega\int_0^\infty \frac{d}{d\tau}\rho_A(t,t-\tau)\exp(-\omega\tau)\,d\tau$$

$$= (i\hbar)^{-1}[H_A, \bar{\rho}_A(t)] - \omega[\bar{\rho}_A(t) - \rho_A(t,t)], \qquad (6.104)$$

where we have used eqn (6.102) in the first term and integrated the second term by parts. For stationary ensembles we regain eqn (6.70).

If we make the conventional assumption that every activated molecule is deactivated by collisions then

$$\text{Tr}\{Q\rho_A(t,t)\} = [A(t)], \qquad (6.105)$$

since any activated molecule present immediately after the collision is due to an activation of a stable molecule and, writing $\text{Tr}\{Q\bar{\rho}_A(t)\}=[A^*(t)]$, we obtain from eqn (6.104)

$$\frac{d}{dt}[A^*(t)] = -k[A^*(t)] - \omega[A^*(t)] + \omega[A(t)], \qquad (6.106)$$

where k is the average 'rate constant' for unimolecular breakdown

$$k = [A^*(t)]^{-1}\text{Tr}\{Q(i\hbar)^{-1}[H_A, \bar{\rho}_A]\}. \qquad (6.107)$$

If we furthermore assume that, as a result of these collisions, the density matrix immediately after a collision is an equilibrium density matrix, say

$$\rho_A(t,t) \propto \exp(-\beta H_A), \qquad (6.108)$$

then the collisions are referred to as 'strong collisions'. Further discussion of the Lindemann scheme is given in Levine (1967c). Similar

ideas can also be applied to the following mechanism:
$$A+A \to A_2^*$$
$$A_2^*+M \to A_2+M'$$
$$A_2^* \to A+A$$
of termolecular recombination, where A_2 is a stable molecule, and stabilization occurs on every collision (Levine 1967a). In the Markoffian approximation, and subject to the unimolecular criteria, eqn (6.97), we can regard the excitation process as an ordinary bimolecular reaction. The fact that the energy of an activated state is only defined to within Γ does not affect this conclusion since the total energy is only defined to within $(\hbar/\tau_c) > \Gamma$. The density matrix between collisions can then be written as $\rho_A^0 \rho_M^0$, where ρ_A^0 refers to an isolated A molecule, and the activation rate computed using a yield function. (See section 2.6.3, in particular eqn (2.6.44).)

The excitation rate can be written explicitly in terms of a transition operator $T_i(E^+)$ for the excitation process
$$T_i(E^+) = U+UG_{0,i}(E^+)T_i(E^+) = U\Omega_i(E^+) \qquad (6.109)$$
where, putting $H_A+H_M = H_{0,i}$,
$$G_{0,i}(E^+) = (E^+ - H_{0,i})^{-1}$$
and (cf. eqn (2.5.6))
$$\Omega_i = \int \delta(E-H_{0,i})\Omega_i(E^+) \, dE.$$
Using the identity
$$\delta(E-H_{0,i})\delta(E'-H_{0,i}) = \delta(E-E')\delta(E-H_{0,i}),$$
we obtain
$$\Omega_i^\dagger[Q,U]\Omega_i$$
$$= \int dE \, \{\delta(E-H_{0,i})QT_i(E^+)+\delta(E-H_{0,i})T_i^\dagger(E^+)G_{0,i}(E^-)QT_i(E^+)-$$
$$-T_i^\dagger(E^+)Q\delta(E-H_{0,i})-T_i^\dagger(E^+)QG_{0,i}(E^+)T_i(E^+)\delta(E-H_{0,i})\}, \qquad (6.110)$$
where we have used eqn (6.109) and its Hermitian conjugate and noted that
$$G_{0,i}^\dagger(E^+) = G_{0,i}(E^-).$$
In principle, Q is not a final state and Q and $G_{0,i}(E^+)$ do not commute since $[Q,H_A] \neq 0$. However, when we can regard the excitation process as an ordinary bimolecular process, Q can be regarded as measurable, $[Q,H_{0,i}] \simeq 0$. In terms of the identity
$$G_{0,i}(E^-)-G_{0,i}(E^+) = 2\pi i \delta(E-H_{0,i}),$$
eqn (6.110) is seen to be equivalent to Ehrenfest's theorem for the

measurable operator Q. The first and third terms correspond to the rate of depletion of $\langle Q \rangle$ and can be rewritten using the optical theorem (cf. eqns (2.4.6)–(2.4.8)). The second and fourth terms correspond to the rate of increase of $\langle Q \rangle$. Only these two terms will contribute to the rate when the initial state is an unactivated A molecule, $\mathrm{Tr}\, Q\rho_A^0 = 0$. Thus the activation rate is

$$\frac{2\pi}{\hbar}\mathrm{Tr}\left\{\rho_A^0 \rho_M^0 \int \mathrm{d}E\; T_i^\dagger(E^+) Q \delta(E-H_A-H_M) Q T_i(E^+)\right\}. \quad (6.111)$$

Equation (6.111) is the expected general result for a measurable operator Q in a bimolecular collision when the expectation value of Q vanishes in the initial state. (To compare eqn (6.110)) with, say, eqn (2.4.51) we have to expand the δ-function in a complete set of states for $H_{0,i}$.) We observe that, as expected, the excitation rate is time-independent only when ρ_A^0 and ρ_M^0 are stationary. An alternative derivation of eqn (6.111) was given in Levine (1966d).

When deactivating collisions can take place, eqn (6.111) is modified by the inclusion of a term representing the rate of collisions when A is activated before the collision (first and third terms in eqn (6.109)). Using the optical theorem (2.5.29), we obtain for the operator that determines this rate

$$(i\hbar)^{-1}Q[T_i^\dagger(E^+) - T_i(E^+)]Q = (2\pi/\hbar) Q T_i^\dagger(E^+) \delta(E-H_{0,i}) T_i(E^+) Q. \quad (6.112)$$

The nature of possible final states in the collision A^*+M is determined by the δ-function, $\delta(E-H_{0,i})$. If we assume that A^* does not undergo collision-induced dissociation, we can put

$$\delta(E-H_{0,i}) = C_i \delta(E-H_{0,i}) C_i = Q\delta(E-H_{0,i})Q + C^i \delta(E-H_{0,i}) C^i, \quad (6.113)$$

where $C_i = Q+C^i$, and we regard Q as measurable so that there are no cross terms. Putting eqn (6.113) into eqn (6.112) the first term is the rate of non-deactivating collisions,

$$A^* + M \to A^{*\prime} + M',$$

and the second term is the rate of deactivating collisions. The total rate of collisions of activated molecules is

$$\frac{2\pi}{\hbar}\mathrm{Tr}\left\{\rho_A^0 \rho_M^0 \int \mathrm{d}E\; \delta(E-H_{0,i}) Q T_i^\dagger(E^+) \delta(E-H_{0,i}) T_i(E^+) Q\right\}. \quad (6.114)$$

To conclude, when the collisions with the exciting species M can be regarded as bimolecular we can describe the collision rates by means of a yield operator (cf. section 2.6.3). Thus the rate of change of $\text{Tr}\,Q\rho_A$ due to collisions can be written

$$\left(\frac{\mathrm{d}}{\mathrm{d}t}\text{Tr}\,Q\rho_A\right)_{\text{collision}} = \text{Tr}_A\{\rho_A^0 Y_Q\} \qquad (6.115)$$

where $\qquad Y_Q = (i\hbar)^{-1}\text{Tr}_M\{\rho_M^0\Omega_i^\dagger[Q,U]\Omega_i\}.$

We can distinguish three contributions to Y_Q: (*a*) depletion by deactivating collisions where, from eqns (6.112)–(6.114),

$$Y_Q^a = \frac{2\pi}{\hbar}Q\,\text{Tr}_M\Big\{\rho_M^0\int \mathrm{d}E\; T_i^\dagger(E^+)C^i\delta(E-H_{0,i})C^iT_i(E^+)\Big\}; \quad (6.116\,\text{a})$$

(*b*) excitation of an initially unexcited A molecule where Y_Q^b is determined from eqn (6.111) as

$$Y_Q^b = \frac{2\pi}{\hbar}C^i\text{Tr}_M\Big\{\rho_M^0\int \mathrm{d}E\; T_i^\dagger(E^+)Q\delta(E-H_{0,i})QT_i(E^+)\Big\}; \quad (6.116\,\text{b})$$

and (*c*) change of state of excitation of an activated molecule, where Y_Q^c is determined by analogy to Y_Q^b (Q replaces C_i in front of the trace in eqn (6.116 b)).

Equation (6.115) can be written as a relaxation equation as in section 2.4.2. The resulting equations have been extensively discussed by Schlag and co-workers (Serauskas and Schlag 1965, Valance and Schlag 1966). In mass spectrometry and similar applications there are no deactivating collisions, so that every molecule that is activated will break down into products. The total rate of formation of products equals the rate of activation. The rate of formation of particular products can then be obtained by regarding $Q\delta(E-H_{0,i})Q$ in eqn (6.116 b) as an initial state for the breakdown (see Levine 1966*d*). It should be stressed, however, that the description of the evolution of activated molecules by a relaxation equation is subject to the approximations discussed in section 3.6.2. In particular, the states that diagonalize the evolution operator (eqn (6.67)) do not necessarily have an obvious chemical significance. Neither are these states identical to the eigenstates of QHQ, which describe the isolated molecule when the coupling that leads to breakdown has been 'turned off'.

3.7. The time-correlation method

3.7.0. THE time-correlation method is an alternative formulation of the problem of computing the rate of change of observables. It has a natural extension to ensembles, and in principle one can use it to write closed expressions for the rates in ensembles without invoking the Markoffian approximation (cf. section 3.6.2).

Schematically, the basic idea is as follows. Let $X(T)$ be such that the rate associated with some event can be written as

$$R = \lim_{T \to \infty} \frac{1}{T} |X(T) - X(0)|^2. \tag{7.1}$$

For example, $|X(T)-X(0)|^2$ can be the number of events observed in the time interval T. Putting

$$X(T) = X(0) + \int_0^T \dot{X}(t) \, dt, \tag{7.2}$$

one can write
$$R = \lim_{T \to \infty} \frac{1}{T} \int_0^T dt_1 \int_0^T dt_2 \, \dot{X}(t_1)\dot{X}(t_2). \tag{7.3}$$

Assuming that $\dot{X}(t_1)\dot{X}(t_2)$ depends only on t_2-t_1, we can perform the double integration by first holding t_2-t_1 constant, $t_2-t_1 = \tau$. Then

$$R = \lim_{T \to \infty} \frac{2}{T} \int_0^T (T-\tau) \dot{X}(0)\dot{X}(\tau) \, d\tau \tag{7.4}$$

and, provided
$$\lim_{\tau \to \infty} \dot{X}(0)\dot{X}(\tau) \to 0,$$

$$R = 2 \int_0^\infty \dot{X}(0)\dot{X}(\tau) \, d\tau. \tag{7.5}$$

It is now a matter of mathematical details to generalize eqn (7.5). In application to any particular process the first problem is to relate the actual observable under discussion to a time-correlation function. In the case where the Hamiltonian can be divided in a natural fashion into a zero-order Hamiltonian and a perturbation this can be done via the response theory (section 3.7.2), when the method is sometimes known as the Kubo (1957, also Kubo and Tomita 1954) method. Earlier work was done by Rice (1945), Kirkwood (1946), Callen and Welton (1951), Green (1952), Van Hove (1954), Anderson (1954), and Zemach and

Glauber (1956). Reviews, extensions, and applications were presented by Kaplan (1958), Bernard and Callen (1959), Kubo (1959), Zwanzig (1960, 1961a, 1964, 1965), Zubarev (1960), Glarum (1960), Yamamoto (1960), Abragam (1961), Alexander (1962), Fano (1963), Powles and Strange (1964), Shimizu (1965), Gordon (1965, 1966), Fujita (1966), Peterson (1967), Birnbaum (1967), and Bloom and Oppenheim (1967).

In the following section we introduce the concept of the time-correlation function and relate it to the rate of change of observables as discussed in Chapter 2.4. The linear response theory is outlined in section 3.7.2, while the evaluation of time-correlation functions is considered in section 3.7.3, using the Liouville-operator method.

3.7.1. *Time-correlation functions*

The time-correlation function of two operators A and B, that are not explicitly time-dependent, is defined by

$$F_{AB}(t, t') = \text{Tr}\{\rho_0 B(t')A(t)\}. \tag{7.6}$$

ρ_0 is the density operator of the system and $A(t)$ stands for the Heisenberg solution,
$$A(t) = \exp(iHt/\hbar) A \exp(-iHt/\hbar) \tag{7.7}$$

of the equation of motion of A,

$$i\hbar \frac{d}{dt} A(t) = [A(t), H] \tag{7.8}$$

and the boundary condition $A(0) = A$. H is the Hamiltonian of the system. For a stationary system $[\rho_0, H] = 0$ and, using the cyclic property of the trace and eqn (7.7), one finds that in a stationary system $F_{AB}(t, t')$ depends on $t-t'$ only,

$$F_{AB}(t-t') = \text{Tr}\{\rho_0 B(0)A(t-t')\}. \tag{7.9}$$

When ρ_0 is a canonical density operator,

$$\rho_0 = \exp(-\beta H)/Z, \qquad Z = \text{Tr}\{\exp(-\beta H)\},$$

we shall use the notation

$$F_{AB}(\tau) = \langle B(0)A(\tau) \rangle. \tag{7.10}$$

In general, $A(\tau)$ and B need not commute and one must distinguish $F_{AB}(\tau)$ from

$$F_{BA}(\tau) = \langle A(\tau)B(0) \rangle = \langle B(0)A(\tau+i\hbar\beta) \rangle. \tag{7.11}$$

To establish the second formula we have used the cyclic property of the trace and noted that (cf. eqn (7.7))

$$A(\tau+i\hbar\beta) = \exp(-\beta H)A(\tau)\exp(\beta H). \tag{7.12}$$

In applications, eqn (7.11) often plays the role of the statement of microscopic reversibility in the canonical ensemble.

Introducing a basis that diagonalizes the Hamiltonian H of the system
$$H|n\rangle = E_n|n\rangle, \tag{7.13}$$
one finds, for a stationary system,
$$F_{AB}(\tau) = \sum_{n,m} \rho_n B_{n,m}[A(\tau)]_{m,n}, \tag{7.14}$$
where $B_{n,m} = \langle n|B|m\rangle$ and $\rho_n = \langle n|\rho_0|n\rangle$ is diagonal in the basis that diagonalizes H. Since $\exp(-iH\tau/\hbar)|n\rangle = \exp(-iE_n\tau/\hbar)|n\rangle$,
$$F_{AB}(\tau) = \sum_{n,m} \rho_n B_{n,m} A_{m,n} \exp[-i(E_n-E_m)\tau/\hbar] \tag{7.15}$$
or $F_{AA}(\tau) = F_{AA}^*(-\tau)$. Introducing the identity
$$\exp[-i(E_n-E_m)\tau/\hbar] = \int_{-\infty}^{\infty} \exp(-i\omega\tau)\delta(E_n-E_m-\hbar\omega)\,d\omega, \tag{7.16}$$
we can write
$$F_{AB}(\tau) = \int_{-\infty}^{\infty} J_{AB}(\omega)\exp(-i\omega\tau)\,d\omega. \tag{7.17}$$
Here $J_{AB}(\omega)$ is the (real for $B = A^\dagger$) quantity
$$\begin{aligned}J_{AB}(\omega) &= \sum_{n,m} \rho_n B_{n,m} \delta(E_n-E_m-\hbar\omega) A_{m,n} \\ &= \sum_n \rho_n \langle n|B\Big\{\sum_m |m\rangle\delta(E_n-E_m-\hbar\omega)\langle m|\Big\}A|n\rangle \\ &= \sum_n \rho_n \langle n|B\delta(E_n-H-\hbar\omega)A|n\rangle.\end{aligned} \tag{7.18}$$
If the set of states $\{|m\rangle\}$ do not form a complete set, then with C_m as the projection on the relevant subspace
$$J_{AB}(\omega) = \sum_n \rho_n \langle n|BC_m\delta(E_n-H-\hbar\omega)A|n\rangle \tag{7.19}$$
and
$$F_{AB}(\tau) = \text{Tr}\{\rho_0 B(0)C_m A(\tau)\}. \tag{7.20}$$

If $J_{BA}(\omega)$ is the spectral density of $\langle A(\tau)B(0)\rangle$ we obtain by analogy to eqn (7.18),
$$J_{BA}(\omega) = \sum_{n,m} \rho_m B_{n,m} \delta(E_n-E_m-\hbar\omega) A_{m,n}. \tag{7.21}$$
In the canonical ensemble, $\rho_m = \exp(-\beta E_m)/Z$, and so, using the operational property
$$\begin{aligned}\rho(E_m)\delta(E_n-E_m-\hbar\omega) &= \rho(E_n-\hbar\omega)\delta(E_n-E_m-\hbar\omega) \\ &= \exp(\beta\hbar\omega)\rho(E_n)\delta(E_n-E_m-\hbar\omega),\end{aligned} \tag{7.22}$$

we obtain, on comparing with eqn (7.18),
$$J_{BA}(\omega) = \exp(\beta\hbar\omega)J_{AB}(\omega). \tag{7.23}$$
Equation (7.23) is the spectral density version of eqn (7.11).

As is perhaps already evident, eqn (7.18) is closely related (via the optical theorem, eqns (2.5.29) and (2.4.7)) to the rate of depletion (or depopulation) of the level n. As we demonstrate below, the reciprocity theorem can be invoked (essentially by considering $F_{BA}(\tau)$) to account for excitation (or population) rates as well. To see these points in detail, recall (eqn (2.4.51)) that the rate of change of an observable A can be written as
$$\frac{d\langle A\rangle}{dt} = (2\pi/\hbar)\sum_i p_i \sum_f (A_f - A_i)|\langle f|T|i\rangle|^2\delta(E_f - E_i), \tag{7.24 a}$$
where p_i is the occupation probability of the initial state i and A is measurable, $A|f\rangle = A_f|f\rangle$.

As was discussed in section 2.4.0 the first contribution to eqn (7.24) can be interpreted as due to transitions into the levels i and the second term is due to transitions out of these levels. To see this explicitly we note that since the indices i and f are dummy summation variables we can rewrite eqn (7.24 a) as
$$\frac{d\langle A\rangle}{dt} = \sum_i A_i\left\{\sum_f p_f R_{i,f} - \sum_f p_i R_{f,i}\right\} = \sum_i A_i \frac{dP_i}{dt}, \tag{7.24 b}$$
where dP_i/dt is the rate of change of the population of the ith level and
$$R_{f,i} = (2\pi/\hbar)|\langle f|T|i\rangle|^2\delta(E_i - E_f). \tag{7.25}$$
Each term in eqns (7.24 a) or (7.24 b) can be written as a spectral density function. In particular, when microscopic reversibility can be written as
$$R_{f,i} = R_{i,f} \tag{7.26}$$
(cf. section 2.8.1, eqns (2.8.35) and (2.8.45)) we can write eqn (7.24 b) as
$$\frac{d\langle A\rangle}{dt} = \sum_{f,i} (p_f - p_i) A_i R_{f,i}$$
$$= (2\pi/\hbar) \sum_{f,i} (p_f - p_i) A_i \langle f|T|i\rangle \delta(E_f - E_i)\langle i|T|f\rangle. \tag{7.24 c}$$

We recall that eqn (7.26) is always valid in the first Born approximation, so that the rates computed by the 'Golden rule' can always be written in the form of eqn (7.24 c). The connection with the physical problem is established when we divide the problem into an unperturbed 'system' and a perturbation. When the perturbation is an externally applied disturbance the division can be done in an obvious manner. ω then plays the role of the change in the energy of the unperturbed system which is brought about by the external perturbation. As an example, consider the rate of absorption of energy when an ensemble is placed in an electromagnetic field of frequency ω. ρ_0 is the density operator for the ensemble in the absence of the field and the indices m and n refer to the levels of the ensemble. T is the transition operator for the transitions induced by the field. Thus we should distinguish between the two possibilities

$$E_n = E_m \pm \omega,$$

where the top/bottom sign refer to induced absorption/emission.

$\dot{P}$, the rate of change of the internal energy of the ensemble, can be computed from eqn (7.24 c), with A being the internal energy. Using the operational property $x\delta(x-a) = a\delta(x-a)$ we obtain, after some rearrangement,

$$\dot{P} = (2\pi/\hbar)\hbar\omega \sum_{m,n} (\rho_m - \rho_n)|\langle n|T|m\rangle|^2 \delta(E_n - E_m - \hbar\omega), \qquad (7.27)$$

the first contribution corresponding to population and the second to depopulation of levels. Thus

$$\dot{P} = 2\pi\omega[\exp(\beta\hbar\omega) - 1]J_{TT}(\omega) \xrightarrow[\beta \to 0]{} 2\pi\omega\beta\hbar\omega J_{TT}(\omega). \qquad (7.28)$$

Up to this point the discussion has been rather general, and could have easily been applied to other situations. We now invoke the assumption that $R_{f,i}$ can be approximation by the 'Golden rule'. In other words, the transition operator in eqn (7.24) can be replaced simply by the applied perturbation Hamiltonian. It will become evident that this approximation is not essential to obtain the formal results. It does, however, provide an interpretation of the results which is useful, in that the external perturbation can be regarded as a probe to study the features of the unperturbed ensemble. In other words, in this approximation, we shall be able to relate $\dot{P}$ to properties of the unperturbed ensemble. No such factorization is, in general, possible when we keep higher order terms in $\dot{P}$. The formal results are still useful, however, in the general case as well. In general, external perturbations are several orders of magnitude smaller than intermolecular interactions at small separations and the Golden rule approximation is expected to be valid. As is well known it is valid for our particular example (excluding Lasers).

The external perturbation Hamiltonian H' which describes the interaction of the molecular dipoles with the external field $\mathbf{E}(t)$, is given by

$$H' = \mathbf{d} \cdot \mathbf{E}(t) \qquad (7.29)$$

where $\mathbf{d}$ is the dipole moment of the ensemble,

$$\mathbf{d} = \sum_j \mathbf{d}_j. \qquad (7.30)$$

Taking the external field in the z direction we obtain, for a field with a frequency ω in the Golden rule approximation,

$$R_{n,m} = (2\pi/\hbar)|E_z/2|^2|\langle n|d_z|m\rangle|^2\delta(E_n-E_m-\hbar\omega) \qquad (7.31)$$

where, for a field with a frequency ω,

$$E_z(t) = E_z \cos \omega t = (E_z/2)[\exp(i\omega t)+\exp(-i\omega t)].$$

Thus
$$\dot{P} = \tfrac{1}{2}\omega|E_z|^2\pi[\exp(\beta\hbar\omega)-1]J_{zz}(\omega), \qquad (7.32)$$

where $J_{zz}(\omega)$ is a property of the ensemble in the absence of the external perturbation, namely,

$$J_{zz}(\omega) = (2\pi)^{-1}\int_{-\infty}^{\infty} \mathrm{Tr}\{\rho_0 d_z d_z(t)\}\exp(i\omega t)\,\mathrm{d}t. \qquad (7.33)$$

The dipole correlation function for CO obtained by Gordon (1965) is shown in Fig. 3.10. In the classical limit the trace is replaced by integration over phase space, but the significance of $J_{zz}(\omega)$ as the spectral density of the correlation function remains. This is in accordance with the classical picture of radiation due to the motion of charges. The rate of emission (or absorption) at a given frequency is proportional to the square of the amplitude of the motion at that frequency.

It should be borne in mind that this analysis corresponds to a macroscopic description. $E_z(t)$ is the external electro-magnetic field and ρ_0 is the density operator of the ensemble. When the molecules of the ensemble are interacting, as is the case, for example, in pressure broadening (Anderson 1949, Fano 1963), the evaluation of a correlation function is far from simple. In general, only when we can invoke the binary collision approximation for the description of the collisions, and the Markoffian approximation for the evolution of ρ_0, can the correlation function be simply related to a molecular correlation function.

Thus the description of a dissipative process in an ensemble by means of the relaxation (7.24) can be transformed to a description by the correlation function
$$\mathrm{Tr}\{\rho_0[B, A(t)]\} \qquad (7.34)$$

and in a stationary ensemble this correlation function can be related to the ordinary correlation function $\mathrm{Tr}\{\rho_0 BA(t)\}$. Mathematically,

$$(2\pi)^{-1}\int_{-\infty}^{\infty} \mathrm{Tr}\{\rho_0[B, A(t)]\}\exp(i\omega t)\,\mathrm{d}t = J_{AB}(\omega)-J_{BA}(\omega) \qquad (7.35\,\mathrm{a})$$

and, in the canonical ensemble,

$$= J_{AB}(\omega)[1-\exp(\beta\hbar\omega)] \underset{\beta\to 0}{\to} -\beta\hbar\omega J_{AB}(\omega). \qquad (7.35\,\mathrm{b})$$

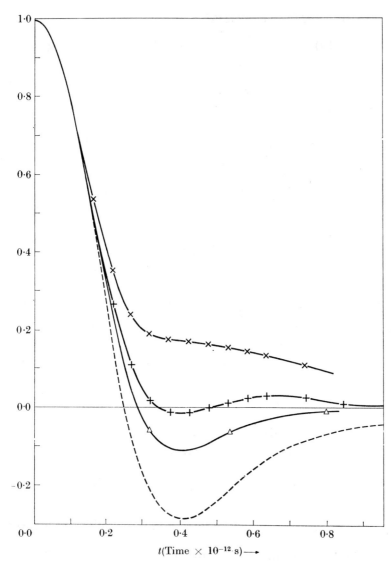

Fig. 3.10. CO dipole correlation function in various environments obtained by Gordon (1965) from experimental spectra ($\times$ in liquid $CHCl_3$; $+$ in liquid n-C_7H_{16}; Δ in Ar gas at 510 amagat) and computed for free CO (bottom line).

The relation between the correlation function for dissipative behaviour and the ordinary correlation function describing the fluctuation about equilibrium is known as the fluctuation–dissipation theorem (Callen and Welton 1951, Kubo 1957).

In a canonical ensemble we can write the correlation function (7.34) as $\langle[B, A(t)]\rangle$ and, from eqn (7.35 b),

$$(i\hbar)^{-1}\langle[B, A(t)]\rangle \xrightarrow[\beta\to 0]{} -\beta \int (-i\omega)\exp(-i\omega t) J_{AB}(\omega)\, d\omega$$

$$= -\beta \frac{\partial}{\partial t}\langle BA(t)\rangle = -\beta\langle B\dot{A}(t)\rangle = -\beta\langle \dot{B}A(t)\rangle, \quad (7.36)$$

$\dot{B} = (i\hbar)^{-1}[B, H]$ and the last line follows from the cyclic property of the trace. The correlation function (7.36) is sometimes known as the 'canonical' correlation function and denoted as $\beta\langle \dot{B}; A(t)\rangle$.

When the perturbation is not external, one can still express the rate as a correlation function, under some assumptions. As a simple example we can consider a polyatomic molecule in a monoatomic host gas. Let r and s be two internal states of the polyatomic with energy difference $\hbar\omega$. When a transition $r \to s$ is induced by a collision with a host gas atom $E_f - E_i = e_s - e_r - \hbar\omega$, where $e_r = \hbar^2 k_r^2/2\mu$ is the relative kinetic energy before the collision. In the Markoffian approximation the transition rate is given by

$$\dot{N}(s \leftarrow r) = (2\pi/\hbar) \sum_{\mathbf{k}_r} p(k_r) \sum_{\mathbf{k}_s} |\langle \mathbf{k}_s, s|T|r, \mathbf{k}_r\rangle|^2 \delta(e_s - e_r - \hbar\omega), \quad (7.37\,\text{a})$$

where $\hbar\mathbf{k}_r$ is the initial relative momentum and $p(k_r)$ is the distribution function for the relative momenta.

In the Golden rule approximation we can replace T by V to obtain (cf. eqn (7.18))

$$\dot{N}(s \leftarrow r) = (2\pi/\hbar) \sum_{\mathbf{k}_r} p(k_r) \langle \mathbf{k}_r|V_{rs}\delta(K - e_r - \hbar\omega)V_{sr}|\mathbf{k}_r\rangle, \quad (7.37\,\text{b})$$

where $V_{rs} = \langle r|V|s\rangle$ and K is the Hamiltonian for the relative motion, $K|\mathbf{k}_s\rangle = e_s|\mathbf{k}_s\rangle$. By analogy to eqn (7.17)

$$\dot{N}(s \leftarrow r) = \hbar^{-2} \sum_{\mathbf{k}_r} p(k_r) \int_{-\infty}^{\infty} \langle \mathbf{k}_r|V_{rs}(t)V_{sr}|\mathbf{k}_r\rangle \exp(i\omega t)\, dt$$

$$= \hbar^{-2} \int_{-\infty}^{\infty} \langle V_{rs}(t)V_{sr}\rangle \exp(i\omega t)\, dt. \quad (7.37\,\text{c})$$

Equation (7.37 c) provides yet another manifestation of the Zener–Massey adiabatic criteria (see also Zwanzig (1961b)). The transition rate is significant when the correlation function for the mutual interaction attains significant values for $t \sim \omega^{-1} = \hbar/\Delta E$, where ΔE is the energy transferred during the collision.

3.7.2. *Linear response theory*

From a mathematical point of view this section presents a particular example of the general theory of linear systems as outlined in Appendix

2.B. In particular, we want to derive eqns (2.B.13) and (2.B.18). The present discussion (following Kubo 1957) is much more specific in the description of the system.

Consider the change in the expectation value of an observable A, when the Hamiltonian, H, of the system is modified by switching on an external perturbation $H'(t)$,

$$H'(t) = -f(t)B. \tag{7.38}$$

$f(t)$ is the switching function that vanishes before the beginning of the experiment, say at $t \to -\infty$. We assume that the initial system is drawn from some (stationary) ensemble at $t \to -\infty$, and so is characterized by a density operator ρ_0. $\rho(t)$, the density operator at time t, satisfies the equation of motion

$$\frac{d}{dt}\rho(t) = (i\hbar)^{-1}[H+H'(t), \rho(t)] \tag{7.39}$$

and the boundary condition

$$\rho(t) \xrightarrow[t \to -\infty]{} \rho_0, \tag{7.40}$$

which can be combined into an integral equation

$$\rho(t) = \rho_0 + (i\hbar)^{-1} \int_{-\infty}^{t} \exp[iH(t'-t)/\hbar] \times$$
$$\times [H'(t'), \rho(t')] \exp[-iH(t'-t)/\hbar] \, dt', \tag{7.41}$$

as can be verified by differentiation. Equation (7.41) can be solved by iteration. The lowest order, which is linear in the perturbation, is obtained by replacing $\rho(t')$ by ρ_0 in the right-hand side. Thus, to the lowest order,

$$\langle A(t) \rangle = \text{Tr}\{\rho(t)A\} = \text{Tr}\{\rho_0 A\} -$$
$$- (i\hbar)^{-1} \text{Tr}\left\{ \int_{-\infty}^{t} \exp[iH(t'-t)/\hbar][B, \rho_0]\exp[-iH(t'-t)/\hbar]A \right\} f(t') \, dt' \tag{7.42}$$

or, using the cyclic property of the trace,

$$\langle A(t) \rangle = \text{Tr}\{\rho_0 A\} - (i\hbar)^{-1} \text{Tr}\left\{ \int_{-\infty}^{t} [B, \rho_0] A(t-t') \right\} f(t') \, dt'. \tag{7.43}$$

The (linear) response function $\phi_{AB}(t)$ is defined by writing

$$\Delta \langle A(t) \rangle = \text{Tr}\{[\rho(t)-\rho_0]A\} = \int_{-\infty}^{t} \phi_{AB}(t-t') f(t') \, dt'. \tag{7.44}$$

Comparing the two equations

$$\phi_{AB}(t) = -(i\hbar)^{-1} \text{Tr}\{[B, \rho_0]A(t)\} = (i\hbar)^{-1} \text{Tr}\{\rho_0[B, A(t)]\} \tag{7.45a}$$

and in the canonical ensemble (cf. eqns (7.35) and (7.36))

$$= (i\hbar)^{-1}\langle [B, A(t)] \rangle = \beta \langle \dot{B}; A(t) \rangle \xrightarrow[\beta \to 0]{} -\beta \langle \dot{B} A(t) \rangle. \quad (7.45\,\text{b})$$

f(t') is the 'switching on' function (eqn (7.38)) and is a measure of the magnitude of the external perturbation at the time t'. The range of values of t' that can contribute to $\Delta \langle A(t) \rangle$ is, from eqn (7.44), $t' < t$. In this sense we can refer to $\phi_{AB}(t-t')$ as the 'after-effect' function and interpret it as the change in $\langle A(t) \rangle$ due to a unit impulse at t'. Equation (7.44) is then a superposition of the responses due to the distribution of times t'. To see this explicitly we put f(t') = $\delta(t')$ and obtain

$$\Delta \langle A(t) \rangle = \begin{cases} 0, & t < 0 \\ (i/\hbar)\text{Tr}\{[B, \rho_0]A(t)\}, & t > 0. \end{cases} \quad (7.46)$$

Equation (7.44) then follows from the identity

$$\text{f}(t) = \int \text{f}(t')\,\delta(t-t')\,\mathrm{d}t',$$

which can be interpreted as representing a function by a superposition of impulses, weighed by their amplitudes.

We can extend the upper limit of integration in eqn (7.44) to $+\infty$ by introducing the Green's function,

$$G_{AB}(t-t') = -(i/\hbar)\theta(t-t')\text{Tr}\{\rho_0[A(t), B(t')]\}. \quad (7.47)$$

Using the cyclic property of the trace, eqn (7.44) can be written as

$$\Delta \langle A(t) \rangle = -\int_{-\infty}^{\infty} G_{AB}(t-t')\text{f}(t')\,\mathrm{d}t'. \quad (7.48)$$

$G_{AB}(t)$ is the Green's function for our (linear) inhomogeneous problem. It is a retarded Green's function of the type discussed by Zubarev (1960). Its physical interpretation is clear from eqn (7.46). It is the change in $\langle A(t) \rangle$ due to a unit impulse. One could also define the non-linear response of the system to the perturbation. In this case, however, the resulting response (or Green's) functions are no longer properties of the unperturbed system.

Particular choices of the switching on function f(t) can be made to represent different physical situations. The choice f(t) = f$\exp(\epsilon t)\theta(-t)$ with $\epsilon \to +0$ at the final stage can represent a continuous application of the external perturbation from $t = -\infty$ to $t = 0$. The response for

$t > 0$ is given by

$$\Delta \langle A(t) \rangle = \lim_{\epsilon \to +0} \int_{-\infty}^{0} \phi_{AB}(t-t') \mathrm{f}.\exp(\epsilon t')\, dt'$$

$$= \mathrm{f}. \lim_{\epsilon \to +0} \int_{t}^{\infty} \phi_{AB}(\tau) \exp(-\epsilon \tau)\, d\tau = \mathrm{f}.\Phi_{AB}(t), \qquad (7.49)$$

where the last line defines the relaxation function $\Phi_{AB}(t)$ and in obtaining the second line we have put $\tau = t-t'$. From eqn (7.49), for the canonical ensemble

$$\Phi_{AB}(t) = \int_{t}^{\infty} \phi_{AB}(t')\, dt' = \beta \int_{t}^{\infty} \langle B; \dot{A}(t') \rangle\, dt'$$

$$= -\beta \langle B; A(t) \rangle \xrightarrow[\beta \to 0]{} \beta \langle BA(t) \rangle, \qquad (7.50)$$

where in the second line we have used eqn (7.36).

In applying the theory we should bear in mind two points. We have not constrained our system to be in contact with a heat bath during the time at which the perturbation was operating. Also, the operators A and B refer to the whole ensemble, and provide a macroscopic description. The traditional example is the perturbation induced by an external field

$$H'(t) = -B_z E_z(t), \qquad (7.51)$$

where the field is in the z direction and B_z is the z component of the total dipole of the system,

$$\mathbf{B} = \sum_j e_j \mathbf{R}_j. \qquad (7.52)$$

e_j and $\mathbf{R}_j$ are the charge and position of the jth particle. The experiment consists in measuring the change in the current operator $\mathbf{j}$ ($\equiv A$). When a velocity can be defined

$$\mathbf{j} = \sum_j e_j \dot{\mathbf{R}}_j = \dot{\mathbf{B}}. \qquad (7.53)$$

Let $\phi_{xz}(t)$ be the response function for the change in the current in the x direction,

$$\phi_{xz}(t) = \beta \langle \dot{B}_z; \dot{B}_x(t) \rangle, \qquad (7.54)$$

where we have used eqns (7.53) and (7.36).

From eqns (7.44) and (7.51),

$$\Delta \langle j_x(t) \rangle = \int_{-\infty}^{t} \phi_{xz}(t-t') E_z(t')\, dt' \qquad (7.55)$$

to first order in the field. Analysing the field into its frequency components

$$E_z(t') = (2\pi)^{-1} \int_{-\infty}^{\infty} E_z(\omega) \exp[(-i\omega)t']\, d\omega \qquad (7.56)$$

we can write, following the derivation of convolution theorem (2.B.4),

$$\Delta \langle j_x(t) \rangle = (2\pi)^{-1} \int_{-\infty}^{\infty} \exp[(-i\omega)t] E_z(\omega) \times$$

$$\times \int_{-\infty}^{t} \phi_{xz}(t-t') \exp[(i\omega-\epsilon)(t-t')] \, dt' d\omega$$

$$= (2\pi)^{-1} \int_{-\infty}^{\infty} \exp[(-i\omega)t] E_z(\omega) \chi_{xz}(\omega) \, d\omega \quad (7.57)$$

where, using the substitution $\tau = t - t'$,

$$\chi_{xz}(\omega) = \int_{0}^{\infty} \exp[(i\omega-\epsilon)t] \phi_{xz}(t) \, dt \quad (7.58)$$

and $\epsilon \to +0$ in the final stage. In deriving eqn (7.57) we have restricted the time integration in eqn (7.58) to positive values only. In the long-time limit, $t \to \infty$, we obtain (cf. (2.B.8))

$$\Delta \langle j_x(t) \rangle \xrightarrow[t \to \infty]{} E_z(0) \chi_{xz}(0), \quad (7.59)$$

provided the limit exists. Equation (7.59) is then interpreted as Ohm's law. $E_z(\omega = 0)$ is the static electric field and $\chi_{xz}(0)$ is the static conductivity,

$$\chi_{xz}(0) = \lim_{\epsilon \to +0} \int_{0}^{\infty} \exp(-\epsilon t) \phi_{xz}(t) \, dt = \Phi_{xz}(0). \quad (7.60)$$

The previous example can clearly be generalized by introducing the concept of the admittance,

$$\chi_{AB}(\omega) = \lim_{\epsilon \to +0} \int_{0}^{\infty} \exp[(i\omega-\epsilon)t] \phi_{AB}(t) \, dt$$

$$= \Phi_{AB}(0) + i\omega \int_{0}^{\infty} \Phi_{AB}(t) \exp[(i\omega-\epsilon)t] \, dt, \quad (7.61)$$

where we have used eqn (7.50) to integrate the first line by parts. In terms of the admittance

$$\Delta \langle A(t) \rangle = (2\pi)^{-1} \int_{-\infty}^{\infty} \exp(-i\omega t) \chi_{AB}(\omega) f(\omega) \, d\omega, \quad (7.62)$$

where $f(\omega)$ is the Fourier frequency transform of the switching on function $f(t)$.

The admittance $\chi_{AB}(\omega)$ can be considered as the Fourier transform of the function $\theta(t) \phi_{AB}(t)$,

$$\chi_{AB}(\omega) = \lim_{\epsilon \to +0} \int_{-\infty}^{\infty} \exp[(i\omega-\epsilon)t] \theta(t) \phi_{AB}(t) \, dt, \quad (7.63)$$

where
$$\theta(t)\phi_{AB}(t) = \begin{cases} 0, & t < 0 \\ \phi_{AB}(t), & t > 0 \end{cases} = -G_{AB}(t). \quad (7.64)$$

Thus Titchmarsh's theorem (Appendix 2.B) can be invoked for the admittance. In particular, eqns (2.B.22) and (2.B.23) hold with respect to $\chi(\omega) - \chi(\infty)$ and are usually referred to, in the present context, as the Kramers–Kronig relations.

In practice, we expect that the response be integrable (in order that $\Phi(0)$ exist). On the other hand, by the Riemann–Lebesgue lemma (Lighthill 1958), if $\phi_{AB}(t)$ is absolutely integrable, $\chi_{AB}(\omega \to \infty) \to 0$. Thus a non-vanishing $\chi_{AB}(\omega \to \infty)$ can only be due to a possible $\delta(t)$ behaviour of $\phi_{AB}(t)$ near $t = 0$.

Equation (7.63) relates the admittance to the spectral density of the response function, via the convolution theorem. Thus from eqn (7.35), when ρ_0 is canonical,

$$(2\pi)^{-1} \int_{-\infty}^{\infty} \phi_{AB}(t)\exp(i\omega t)\, dt = (2\pi)^{-1} \int_{-\infty}^{\infty} (i\hbar)^{-1}\langle [B, A(t)]\rangle \exp(i\omega t)\, dt$$
$$= (i\hbar)^{-1}[J_{AB}(\omega) - J_{BA}(\omega)]. \quad (7.65)$$

And the convolution theorem and eqns (7.63) and (7.65) imply (see also eqn (2.B.21))

$$\chi_{AB}(\omega) = \int_{-\infty}^{\infty} \theta(t)\phi_{AB}(t)\exp(i\omega t)\, dt$$
$$= \hbar^{-1} \int_{-\infty}^{\infty} [J_{BA}(\omega') - J_{AB}(\omega')]\frac{d\omega'}{\omega^+ - \omega'}. \quad (7.66)$$

Then, with a real $J_{AB}(\omega)$,

$$\operatorname{im} \chi_{AB}(\omega) = \hbar^{-1} \int_{-\infty}^{\infty} [J_{BA}(\omega') - J_{AB}(\omega)]\tfrac{1}{2}[(\omega^+ - \omega')^{-1} - (\omega^- - \omega')^{-1}]\, d\omega'$$
$$= -(\pi/\hbar) \int_{-\infty}^{\infty} [J_{BA}(\omega') - J_{AB}(\omega')]\delta(\omega - \omega')\, d\omega'$$
$$= (\pi/\hbar)[J_{AB}(\omega) - J_{BA}(\omega)] \xrightarrow[\beta \to 0]{} -\pi\beta\omega J_{AB}(\omega). \quad (7.67)$$

For example, $\dot{P}$, the rate of absorption of power from an electromagnetic field, can be written (cf. eqns (7.32) and (7.67)) as

$$-\dot{P} = (\hbar\omega/2)|E_z|^2 \operatorname{im} \chi_{zz}(\omega). \quad (7.68)$$

Also, since $\phi_{AB}(t) = -d\Phi_{AB}(t)/dt$, it follows from the definition of the spectral density (eqn (7.65)) that the spectral density of $\Phi_{AB}(t)$ is given by $(i\omega)^{-1}$ times the spectral density of $\phi_{AB}(t)$. Using eqns (7.65) and (7.67),

$$(2\pi)^{-1} \int_{-\infty}^{\infty} \Phi_{AB}(t)\exp(i\omega t) \, dt = -(\pi\omega)^{-1} \text{im} \, \chi_{AB}(\omega). \qquad (7.69)$$

All the information of interest is thus obtainable from

$$S_{AB}(\omega) = -\omega^{-1} \text{im} \, \chi_{AB}(\omega). \qquad (7.70)$$

One often refers to $S_{AB}(\omega)$ as the 'line profile', since it plays this role in considering the response to the electromagnetic field. For an Hermitian operator A, it is clear that im $\chi_{AA}(\omega)$ is non-positive. It is easy to trace this property (via eqn (7.66)) to the requirement of causality (eqn (7.46)), i.e. to the absence of response prior to the application of the perturbation. We have seen similar manifestations of causality before, i.e. in the sign of the imaginary parts of the equivalent potential or the rate operator, all of which follow from the identity (2.1.12) and hence from the requirement of 'forward' evolution in time.

3.7.3. *The Liouville operator*

We consider the evaluation of the correlation function for an ensemble of interacting molecules. One of the earliest ideas (Bloombergen, Purcell, and Pound, 1948) has been to regard the mutual interaction as a random perturbation. As a simple example (Kubo 1963) one can consider a dipole d whose natural frequency ω_0 is modulated by a random disturbance of frequency $\omega_1(t)$. The equation of motion for the dipole can be written as

$$\dot{d} = i\omega(t)d, \qquad (7.71)$$

where $\omega(t) = \omega_0 + \omega_1(t)$ and

$$\langle \omega(t) \rangle = \omega_0, \qquad (7.72)$$

the bracket denoting an ensemble average. Equation (7.71) has the solution

$$d(t) = \exp\left[i\omega_0 t + \int_0^t \omega_1(t') \, dt'\right] d(0). \qquad (7.73)$$

Since $\omega(t)$ is random $d(t)$ can also be considered a random variable. Then

$$\langle d(0)d(t) \rangle = |d(0)|^2 \exp(i\omega_0 t)\Phi(t), \qquad (7.74)$$

where

$$\Phi(t) = \left\langle \exp i \int_0^t \omega_1(t') \, dt' \right\rangle \qquad (7.75)$$

and the spectral density function can be obtained as the Fourier transform. Kubo (1959, 1963) has discussed the short- and long-time behaviour of $\Phi(t)$.

This idea can be generalized to the case when the dipole can be in a finite number of states (Anderson 1954, Kubo 1954; see also Sack 1958, Alexander 1962, and Gordon 1965, 1966, 1967). To derive and extend their results it is convenient to introduce the Liouville operator L by

$$LA = \frac{1}{\hbar}[H, A], \qquad (7.76)$$

so that, if A does not depend explicitly on t,

$$\frac{dA}{dt} = iLA. \qquad (7.77)$$

(See Ben-Reuven 1966, Zwanzig 1964, Kubo 1959, and Fano 1957.) The Liouville operator is sometimes referred to as a super-operator (see, for example, Primas 1961, Murray 1962) since it operates on operators rather than on state vectors. A convenient operator basis for it, is suggested by the identity

$$L|m\rangle\langle n| = [(E_m - E_n)/\hbar]|m\rangle\langle n| = \omega_{m,n}|m\rangle\langle n|, \qquad (7.78)$$

where $|m\rangle$ and $|n\rangle$ are eigenstates of H.

In an arbitrary basis

$$\langle i|LA|j\rangle = \frac{1}{\hbar}\sum_k (H_{ik}A_{kj} - A_{ik}H_{kj}), \qquad (7.79\,a)$$

which one can write as

$$= \sum_{kl} L_{ijkl} A_{kl}, \qquad (7.79\,b)$$

where

$$L_{ijkl} = \frac{1}{\hbar}(H_{ik}\delta_{jl} - H_{lj}\delta_{ik}). \qquad (7.80)$$

The advantage in introducing the Liouville operator is that during formal manipulations eqn (7.77) can be handled by techniques made familiar in the study of the time-dependent Schrödinger equation. In particular, the time-correlation function can be written

$$\langle AA(t)\rangle = \mathrm{Tr}\{\rho_0 A \exp(iLt) A\}, \qquad (7.81)$$

where

$$A(t) = \exp(iHt/\hbar) A \exp(-iHt/\hbar) = \exp(iLt) A = A \exp(-iLt) \qquad (7.82)$$

is the Heisenberg solution. In writing the second form we have assumed that A is self-adjoint. An alternative form of eqn (7.81) is

$$\langle AA(t)\rangle = \mathrm{Tr}\{A(t)\rho_0 A\} = \mathrm{Tr}\{A \exp(-iLt)\rho_0 A\}$$
$$= \mathrm{Tr}\{A\rho_0(t)A\}. \qquad (7.83)$$

As was pointed out by Zwanzig (1960, 1961; see also Mori 1965), not all the information contained in $\rho_0(t)$ (or $A(t)$) is required for the evaluation of the correlation function, and the partitioning technique can be used to extract the relevant part. Speaking loosely, and regarding operators as vectors in the linear space on which L operates, we are looking for a projection P, such that $PA(t)$ is 'in the direction of A'. An obvious choice is to define P by

$$PB(t) = A \operatorname{Tr}\{AB(t)\rho_0\}/\operatorname{Tr}\{AA\rho_0\} = A\langle AB(t)\rangle/\langle AA\rangle, \quad (7.84)$$

where $B(t)$ is an arbitrary Heisenberg solution. It is clear that $P^2 = P$ and that

$$\langle APA(t)\rangle = \langle AA(t)\rangle, \qquad PA(0) = A(0). \quad (7.85)$$

The complementary projection to P is Q,

$$Q = I - P, \quad (7.86)$$

where I is the (super) operator identity

$$(I)_{ijkl} = \delta_{ik}\delta_{jl} \quad (7.87\text{a})$$

so that

$$(IA)_{ij} = \sum_{kl} I_{ijkl} A_{kl} = A_{ij}. \quad (7.87\text{b})$$

Using the boundary condition $QA(0) = 0$ we obtain from eqn (7.82)

$$PA(t) = P\exp(iLt)(P+Q)A(0) = P\exp(iLt)PA(0) \quad (7.88)$$

and

$$QA(t) = Q\exp(iLt)PA(0). \quad (7.89)$$

Formally, this problem is rather similar to the problem of partitioning the Schrödinger time-dependent equation, and similar techniques apply. (See also Zwanzig 1964.)

We define the Green (super) operator by

$$R(\lambda) = (\lambda - L)^{-1} \quad (7.90)$$

so that

$$\exp(iLt) = (2\pi i)^{-1} \int_{\text{sp}} \exp(i\lambda t)(\lambda - L)^{-1} \, d\lambda \quad (7.91)$$

and

$$PA(t) = (2\pi i)^{-1} \int_{\text{sp}} \exp(i\lambda t) P(\lambda - L)^{-1} PA(0) \, d\lambda. \quad (7.92)$$

By analogy to eqn (3.14) we can write

$$PR(\lambda)P = [\lambda - PL_0 P - PM(\lambda)P]^{-1}P, \quad (7.93)$$

where we have written $L = L_0 + L_1$, $[L_0, P] = 0$, and

$$PM(\lambda)P = PL_1 P + PL_1 Q(\lambda - QLQ)^{-1} QL_1 P. \quad (7.94)$$

$M(\lambda)$ is the 'relaxation' operator and plays a role similar to that of the equivalent potential (or the rate operator) in the partitioning of the Schrödinger equation (cf. Chapter 3.1)).

It is convenient to introduce the notation
$$PA(\lambda) = PR(\lambda)PA(0), \qquad (7.95)$$
so that
$$PA(t) = (2\pi i)^{-1} \int_{\mathrm{sp}} \exp(i\lambda t) PA(\lambda) \, d\lambda \qquad (7.96)$$
and
$$\frac{d}{dt} PA(t) = (2\pi)^{-1} \int_{\mathrm{sp}} \exp(i\lambda t) \lambda PA(\lambda) \, d\lambda. \qquad (7.97)$$

From eqn (7.93),
$$[\lambda - PL_0 P - PM(\lambda)P] PA(\lambda) = PA(0) \qquad (7.98)$$
and, substituting for $\lambda PA(\lambda)$ in eqn (7.97),
$$\frac{d}{dt} PA(t) = iPL_0 PA(t) - \int_0^t d\tau \, PM(\tau) PA(t-\tau) \, d\tau, \qquad (7.99)$$
where the second term was obtained by using the convolution theorem. The first term vanishes since, by construction, L_0 was selected to commute with P, so that
$$\frac{d}{dt} PA(t) = - \int_0^t d\tau \, PM(\tau) PA(t-\tau). \qquad (7.100)$$
Using the definition of P,
$$PA(t-\tau) = \frac{A}{\langle AA \rangle} \langle AA(t-\tau) \rangle = \frac{A}{\langle AA \rangle} F(t-\tau), \qquad (7.101)$$
where $F(t)$ is the correlation function. From eqns (7.101) and (7.100),
$$\frac{d}{dt} F(t) = \frac{d}{dt} \langle APA(t) \rangle$$
$$= - \int_0^t d\tau \, \langle APM(t) PA(t-\tau) \rangle$$
$$= - \int_0^t d\tau \, \frac{\langle APM(\tau) PA \rangle}{\langle AA \rangle} F(t-\tau), \qquad (7.102)$$
which we can write as
$$\frac{d}{dt} F(t) = - \int_0^t d\tau \, K(\tau) F(t-\tau). \qquad (7.103)$$

Equation (7.103) was first obtained (for a classical correlation function) by Zwanzig (1961a), who refers to $K(\tau)$ as the 'memory', since it relates the value of the correlation function at the time t to its value at earlier $(t-\tau, \tau < t)$ time.

In general, if ρ_0 is stationary,

$$\langle ALA \rangle = \frac{1}{\hbar}\langle A\dot{A}\rangle = \frac{1}{2\hbar}\frac{d}{dt}\langle AA\rangle = 0 \tag{7.104}$$

so that $PLP = 0$ and from eqns (7.102) and (7.94), taking $\langle AA\rangle = 1$,

$$K(\tau) = \left\langle APL(2\pi i)^{-1}\int_{sp}\exp(i\lambda t)(\lambda - QLQ)^{-1}\,d\lambda\, LPA\right\rangle$$
$$= \langle APL\exp[i\tau QL]LPA\rangle \tag{7.105}$$

or, since using eqns (7.84) and (7.77), $LPA = -i\dot{A}$,

$$K(\tau) = \langle \dot{A}\exp[i\tau QL]\dot{A}\rangle. \tag{7.106}$$

The memory is thus a correlation function, but with a modified evolution (super) operator.

In the Markoffian approximation eqn (7.103) is approximated by

$$\frac{d}{dt}F(t) = -\int_0^\infty K(\tau)\,d\tau\,.\,F(t), \tag{7.107}$$

which can be solved to

$$F(t) = \exp\left[-t\int_0^\infty K(\tau)\,d\tau\right] \tag{7.108}$$

and

$$\int_0^\infty F(t)\,dt = \left[\int_0^\infty K(\tau)\,d\tau\right]^{-1}. \tag{7.109}$$

Recalling the definition of $K(\tau)$ we see that the Markoffian approximation corresponds to approximating $M(\lambda)$ by its zero frequency component

$$M(\lambda = 0) = \int_0^\infty K(\tau)\,d\tau. \tag{7.110}$$

To obtain the spectral density function it is convenient to start from the expression

$$F(t) = \mathrm{Tr}\{A\rho_0(t)A\}$$
$$= \mathrm{Tr}\{A\exp(-iLt)\rho_0 A\}$$
$$= (2\pi i)^{-1}\int_{sp}\exp(-i\lambda t)\mathrm{Tr}\{AR(\lambda)\rho_0 A\}\,d\lambda. \tag{7.111}$$

As usual (cf. eqns (3.34) and (6.39)) the contribution to the integral is from the discontinuity of the Green's function as $\mathrm{im}\,\lambda$ changes sign and, comparing eqns (7.111) and (7.17),

$$J(\omega) = -\pi^{-1}\mathrm{im}\,\mathrm{Tr}\{AR(\omega^+)\rho_0 A\}, \tag{7.112}$$

where $\omega^+ = \omega + i\epsilon$ and $\epsilon \to +0$ after the evaluation of the trace. (This procedure was used by Fano 1963 and Ben-Reuven 1966.)

In the linear space of operators in which we are working $J(\omega)$ is a close analogue of the line shape, eqn (2.86),

$$A(E) = -\pi^{-1} \text{im} \, \text{Tr}\{|n\rangle\langle n|G(E^+)|n\rangle\langle n|\}, \qquad (7.113)$$

and can be evaluated using similar techniques.

To see the structure of eqn (7.112) we recall eqn (7.78) which implies

$$R(\omega^+)|n\rangle\langle m| = (\omega^+ - L)^{-1}|n\rangle\langle m| = (\omega^+ - \omega_{n,m})^{-1}|n\rangle\langle m|. \qquad (7.114)$$

Thus

$$R(\omega^+)\rho_0 A = \sum_{n,m} R(\omega^+)|n\rangle p_n A_{n,m}\langle m| = \sum_{n,m} |n\rangle\langle m|(\omega^+ - \omega_{m,n})^{-1} p_n A_{n,m} \qquad (7.115)$$

and

$$J(\omega) = -\pi^{-1} \text{im} \sum_m \langle m|A R(\omega^+)\rho_0 A|m\rangle$$

$$= \pi^{-1} \sum_{m,n} p_n A_{n,m} \delta(\omega - \omega_{m,n}) A_{m,n} \qquad (7.116)$$

in agreement with eqn (7.18). In particular, when the system is initially in a particular state n, we can write

$$\text{Tr}\{A R(\omega^+)|n\rangle\langle n|A\} = \langle n|A G(\omega^+ + E_n)A|n\rangle \qquad (7.117)$$

and

$$J(\omega) = \langle n|A \delta(\omega + E_n - H) A|n\rangle. \qquad (7.118)$$

In sections 3.1.0 and 3.2.2 we have introduced projection operators that centred attention on particular degrees of freedom and have obtained equivalent 'Hamiltonians' which, while operating explicitly on a particular set of degrees of freedom, incorporated exactly the influence of the remaining degrees of freedom. Following Fano (1963, also Baranger 1958) we can define similar projection operators in the present case. The aim is to centre attention on a particular subsystem of the ensemble.

As an example we consider the operator A as the dipole moment of a particular molecule (which we also call A) in an (otherwise optically inactive) host gas, as in pressure broadening. Then

$$\text{Tr}\{A\rho_0(t)A\} = \text{Tr}_A\{A \, \text{Tr}_M\{\rho_0(t)\}A\}, \qquad (7.119)$$

where Tr_A is the trace over the coordinates of the molecule A and Tr_M is the trace over the host gas coordinates.

We assume that at the time $t = 0$ the molecule A is not interacting with the host gas molecules and put

$$\rho_0(0) = \rho_A^0 \rho_M^0, \qquad (7.120)$$

where ρ_A^0 is the density operator for an isolated A molecule. A projection P can now be defined by

$$PB = \rho_M^0 \operatorname{Tr}_M B \qquad (7.121)$$

and $P^2 = P$ if we arrange that $\operatorname{Tr}_M \rho_M^0 = 1$.

The evaluation of the spectral density can now be written as

$$\operatorname{Tr}\{A R(\omega^+)\rho_0(0)A\}$$
$$= \operatorname{Tr}_A\{A \operatorname{Tr}_M\{R(\omega^+)\rho_0(0)\}A\}$$
$$= \operatorname{Tr}\{A\rho_A^0 \rho_M^0 \operatorname{Tr}_M\{R(\omega^+)\rho_M^0\}A\}$$
$$= \operatorname{Tr}\{APR(\omega^+)\rho_0(0)A\}$$
$$= \operatorname{Tr}\{APR(\omega^+)P\rho_0(0)A\}$$
$$= \operatorname{Tr}_A\{A[\omega^+ - L_0 - \langle M(\omega^+)\rangle]^{-1}\rho_A^0 A\},$$

where we have used the normalization $\operatorname{Tr}_M \rho_M^0 = 1$ and $P\rho_0(0) = \rho_0(0)$. Taking H_A as the Hamiltonian of the isolated molecule A,

$$L_0 \equiv \frac{1}{\hbar}[H_A, \ ] \qquad (7.122)$$

and

$$[\omega^+ - L_0 - \langle M(\omega^+)\rangle]^{-1} = \operatorname{Tr}_M\{PR(\omega^+)P\} = \operatorname{Tr}_M\{R(\omega^+)\rho_M^0\}. \qquad (7.123)$$

In the Markoffian approximation $\langle M(\omega)\rangle$ can be replaced by $\langle M(\omega = 0)\rangle$ and the correlation function can be evaluated as

$$\langle AA(t)\rangle = \operatorname{Tr}_A\{A \exp[-\mathrm{i}L_0 t - \mathrm{i}\langle M(\omega = 0)\rangle t]\rho_A^0 A\}$$
$$= \mathbf{A}\exp[-\mathrm{i}\boldsymbol{\omega}_0 t - \mathrm{i}\mathbf{M}t]\mathbf{p}_A^0 \mathbf{A}. \qquad (7.124)$$

For a system with one state, eqn (7.124) reduces to the form of (7.74).

An alternative approach is to note that

$$\rho_A^0(t) = \operatorname{Tr}_M\{\rho_0(t)\} = \operatorname{Tr}_M\{\exp(-\mathrm{i}Lt)\rho_0(0)\}$$
$$= \operatorname{Tr}_M\{\exp(-\mathrm{i}Lt)P\rho_0(0)\}$$
$$= \operatorname{Tr}_M\{P\exp(-\mathrm{i}Lt)P\}\rho_A^0(0). \qquad (7.125)$$

Thus in the Markoffian approximation (cf. eqns (7.103), (7.107), and (7.110))

$$\frac{\mathrm{d}}{\mathrm{d}t}\rho_A^0(t) = -\mathrm{i}L_0\rho_A^0 - \mathrm{i}\langle M(\omega = 0)\rangle \rho_A^0(t), \qquad (7.126)$$

thus identifying $\mathrm{i}\langle M(\omega = 0)\rangle$ as a yield function of the type discussed in section 3.6.2. In $\langle M(\omega = 0)\rangle$ the averaging is over the degrees of freedom of the host gas only. It is still an operator with respect to the molecule A.

A simple illustration of these techniques is provided by the problem of 'motional narrowing'. Consider the isomerization process
$$AB \leftrightharpoons AB',$$
when we observe the nuclear magnetic resonance spectrum of atom A. Letting ω_1 and ω_2 be the resonance frequencies of A in AB and AB' respectively, we obtain in the Markoffian approximation

$$\boldsymbol{\omega}_0 + \mathbf{M} = \begin{pmatrix} \omega_1 + im & -im \\ -im & \omega_2 + im \end{pmatrix}, \tag{7.127}$$

where $m = iM_{12} = iM_{21}$ is the 'rate constant' for the isomerization and we have assumed that AB and AB' are equally probable states of the system. To determine the spectrum we have to invert the matrix $\omega \mathbf{I} - \boldsymbol{\omega}_0 - \mathbf{M}$. This can be done in the present case by evaluating the eigenvalues. Taking for simplicity $\omega_1 = -\omega_2 = \omega_0$ we have the secular equation
$$(\lambda + m)^2 + \omega_0^2 - m^2 = 0, \tag{7.128}$$
where $\operatorname{re}\lambda$ determines the line width and $\operatorname{im}\lambda$ determines the position. For $(\omega_0/m) > 1$,
$$\lambda = \pm i(\omega_0^2 - m^2)^{\frac{1}{2}} - m, \tag{7.129a}$$
so that the two lines are drawn together, but retain their original width m. For $(m/\omega_0) > 1$, λ is purely real and both lines are centred at the mean position (zero),
$$\lambda = m \pm (m^2 - \omega_0^2)^{\frac{1}{2}}. \tag{7.129b}$$
For $(m/\omega_0) \gg 1$ one line has the width of about $2m$ and one line has narrowed to about $\omega_0^2/2m$. It turns out that the narrowed line has most of the amplitude, and is the one observed in practice.

References

ABRAGAM, A. (1961). *Principles of nuclear magnetism.* Clarendon Press, Oxford.
ADACHI, T., and KOTANI, T. (1966). *Prog. theor. Phys., Osaka,* **35,** 485.
AIREY, J. R., GREENE, E. F., KODERA, K., PECK, G. P., and ROSS, J. (1967). *J. chem. Phys.* **46,** 3287.
AKHIEZER, N. I., and GLAZMAN, I. M. (1961). *Theory of linear operators in Hilbert space.* Ungar Publishing Company, New York.
ALEXANDER, S. (1962). *J. chem. Phys.* **37,** 967.
ALLISON, A. C., and DALGARNO, A. (1967). *Proc. Phys. Soc.* **90,** 609.
ANDERSON, P. W. (1949). *Phys. Rev.* **76,** 647.
—— (1954). *J. phys. Soc. Japan,* **9,** 316.
ARGYRES, P. N. (1966). *Lectures in theoretical physics,* viii–A (edited by W. E. Brittin), p. 183. University of Colorado Press, Boulder, Colorado.
ARTHURS, A. M., and DALGARNO, A. (1960). *Proc. R. Soc.* **A256,** 540.
BAKER, M. (1958). *Ann. Phys.* **4,** 271.
BANERJEE, M. K. (1960). *Nuclear spectroscopy* (edited by F. Ajzenbert-Selove), p. 695. Academic Press, New York.
BARANGER, M. (1958). *Phys. Rev.* **111,** 494. Also ibid. **111,** 481; **112,** 855.
BARDSLEY, J. N., and MANDL, F. (1968). *Rep. Progr. Phys.* **21,** part 2, 471.
BASSEL, R. H., and GERJUOY, E. (1960). *Phys. Rev.* **117,** 749.
BATES, D. R. (1960). *Proc. R. Soc.* **A257,** 22.
—— (1962). *Atomic and molecular processes* (edited by D. R. Bates), chap. 14. Academic Press, New York.
—— and HOLT, A. R. (1966). *Proc. R. Soc.* **A292,** 168.
—— MASSEY, H. S. W., and STEWART, A. L. (1953). *Proc. R. Soc.* **A216,** 437.
BECK, D., GREENE, E. F., and ROSS, J. (1962). *J. chem. Phys.* **37,** 2895.
BELLMAN, R., KALABA, R., and WING, G. M. (1960). *J. Math. Phys.* **1,** 280.
BEN-REUVEN, A. (1966). *Phys. Rev.* **141,** 34; ibid. **145,** 7.
BERGER, R. O., O'MALLEY, T. F., and SPRUCH, L. (1965). *Phys. Rev.* **137,** A1068.
—— and SPRUCH, L. (1965). *Phys. Rev.* **138,** B1106.
BERNARD, W., and CALLEN, H. B. (1959). *Rev. mod. Phys.* **31,** 1017.
BERNSTEIN, R. B. (1960). *J. chem. Phys.* **33,** 795.
—— (1961). *J. chem. Phys.* **34,** 361.
—— (1963). *J. chem. Phys.* **38,** 2599.
—— (1965). *Molecular beams* (edited by J. Ross), p. 75. Interscience, New York.

BERNSTEIN, R. B., CURTISS, C. F., IMAM-RAHAJOE, S., and WOOD, H. T. (1966). J. chem. Phys. **44**, 4072.

—— DALGARNO, A., MASSEY, H. S. W., and PERCIVAL, I. C. (1963), Proc. R. Soc. **A274**, 427.

—— and KRAMER, K. H. (1966). J. chem. Phys. **44**, 4473.

—— and LEVINE, R. D. (1968). University of Wisconsin, Theoretical Chemistry Institute Report, 300. J. chem. Phys. **49**, 3872.

BERRY, R. S. (1957). J. chem. Phys. **27**, 1288.

BETHE, H. A. (1937). Rev. mod. Phys. **9**, 69.

—— (1949). Phys. Rev. **76**, 38.

BIRNBAUM, G. (1967). Adv. chem. Phys. **12**, 487.

BLATT, J. M., and BIEDENHARN, L. C. (1952). Rev. mod. Phys. **24**, 258. See also HUBY, R. (1954). Proc. phys. Soc. **67**, 1103.

—— and JACKSON, J. D. (1949). Phys. Rev. **76**, 18.

—— and WEISSKOPF, V. F. (1952). Theoretical nuclear physics. Wiley, New York.

BLOOM, M., and OPPENHEIM, I. (1967). Adv. chem. Phys. **12**, 549.

BLOOMBERGEN, N., PURCELL, E. M., and POUND, R. V. (1948). Phys. Rev. **73**, 679.

BLUM, L. (1966). Molec. Phys. **11**, 63, 105. See also J. phys. Chem., Ithaca, **70**, 2758.

BOHR, N. (1936). Nature, Lond. **137**, 344.

—— PEIERLS, R., and PLACZECK, G. (1939). Nature, Lond. **144**, 200.

—— and WHEELER, J. A. (1939). Phys. Rev. **56**, 426.

BROUT, R. (1954). J. chem. Phys. **22**, 934.

BROWN, G. E. (1959). Rev. mod. Phys. **31**, 893.

BUNKER, D. L. (1966). Theory of elementary gas reaction rates. Pergamon Press, Oxford.

BUTLER, S. T. (1957). Nuclear stripping reactions. Wiley, New York.

CALLAWAY, J., and BAUER, E. (1965). Phys. Rev. **140**, A1072.

CALLEN, H. B., and WELTON, T. A. (1951). Phys. Rev. **83**, 34.

CALOGERO, F. (1967). Variable phase approach to potential scattering. Academic Press, New York.

CASTILLEJO, L., PERCIVAL, I. C., and SEATON, M. J. (1960). Proc. R. Soc. **A254**, 259.

CHEN, J. C. Y. (1964). J. chem. Phys. **40**, 3507, 3513.

—— (1966a). Phys. Rev. **148**, 66.

—— (1966b). Phys. Rev. **152**, 1454.

—— (1967). Phys. Rev. **156**, 12.

—— and MITTLEMAN, M. H. (1966). Ann. Phys. **37**, 264.

CHEW, G. F. (1950). Phys. Rev. **80**, 196.

—— and MANDELSTAM, S. (1960). Phys. Rev. **119**, 467.

CHILD, M. S. (1966). Proc. R. Soc. **A292**, 272.

COESTER, F., and KÜMMEL, H. (1958). Nucl. Phys. **9**, 225.

COOPER, J. (1967). *Rev. mod. Phys.* **39**, 167.

COTTRELL, T. L., and MCCOUBREY, J. C. (1961). *Molecular energy transfer in gases.* Butterworths, London.

COULSON, C. A., and LEVINE, R. D. (1967). *J. chem. Phys.* **47**, 1235.

—— and ZALEWSKI, K. (1962). *Proc. R. Soc.* **A268**, 437.

COZ, M. (1965). *Ann. Phys.* **35**, 53.

CROSS, J. C. JR. (1967). *J. chem. Phys.* **47**, 3724.

DALGARNO, A., and DAVISON, W. D. (1966). *Adv. atom. molec. Phys.* **2**, 1. Academic Press, New York.

DAVISON, W. D. (1962). *Discuss. Faraday Soc.* **33**, 71.

DEMKOV, YU. N. (1963). *Variational principles in the theory of collisions.* Pergamon Press, Oxford.

DEVONSHIRE, A. F. (1937). *Proc. R. Soc.* **A158**, 269.

DEWITT, B. S. (1956). *Phys. Rev.* **103**, 1565.

DIRAC, P. A. M. (1958). *The principles of quantum mechanics*, 4th edn. Clarendon Press, Oxford.

DONATH, W. E. (1965). *J. chem. Phys.* **42**, 118.

DWORETSKY, S., NOVICK, R., SMITH, W. W., and TOLK, N. (1967). *Phys. Rev. Lett.* **18**, 939.

DYSON, F. J. (1949). *Phys. Rev.* **75**, 486.

ECKSTEIN, H. (1956). *Phys. Rev.* **101**, 880.

EDMISTON, C., and KRAUSS, M. (1965). *J. chem. Phys.* **42**, 1119.

EDMONDS, A. R. (1960). *Angular momentum in quantum mechanics.* Princeton University Press.

ELIASON, M. A., and HIRSCHFELDER, J. O. (1959). *J. chem. Phys.* **30**, 1426.

EPSTEIN, S. T. (1957). *Phys. Rev.* **106**, 598.

EU, B. C., and ROSS, J. (1966). *J. chem. Phys.* **44**, 2467.

EVANS, M. G., and POLANYI, M. (1935). *Trans. Faraday Soc.* **31**, 875.

EYRING, H. (1935). *J. chem. Phys.* **3**, 107.

—— GERSHIONOWITZ, H., and SUN, C. E. (1935). *J. chem. Phys.* **3**, 786.

—— HIRSCHFELDER, J. O., and TAYLOR, H. S. (1936). *J. chem. Phys.* **4**, 479.

FADDEEV, L. D. (1961). *Soviet Phys. JETP*, **12**, 1014.

—— (1963). *Soviet Phys. Dokl.* **7**, 600.

FANO, U. (1957). *Rev. mod. Phys.* **29**, 74.

—— (1961). *Phys. Rev.* **124**, 1866.

—— (1963). *Phys. Rev.* **131**, 259.

FESHBACH, H. (1960). *Nuclear spectroscopy* (edited by F. Ajzenberg-Selove), p. 1033. Academic Press, New York.

—— (1962). *Ann. Phys.* **19**, 287.

FETTER, A. L., and WATSON, K. M. (1965). *Adv. theor. Phys.* **1**, 115.

FEYNMAN, R. P., and HIBBS, A. R. (1965). *Quantum mechanics and path integrals.* McGraw-Hill, New York.

REFERENCES

FOLDY, L. L., and TOBOCMAN, W. (1957). *Phys. Rev.* **105**, 1099.

FORD, K. W., and WHEELER, J. A. (1959). *Ann. Phys.* **7**, 259, 287.

FOWLER, R., and GUGGENHEIM, E. A. (1952). *Statistical thermodynamics.* Cambridge University Press.

FRANCIS, N. C., and WATSON, K. M. (1953). *Phys. Rev.* **92**, 291.

FRIEDMAN, B. (1956). *Principles and techniques of applied mathematics.* Wiley, New York.

FRIEDMAN, F. L., and WEISSKOPF, V. F. (1955). *Niels Bohr and the development of physics*, p. 134. Pergamon Press, Oxford.

FRIEDRICHS, K. C. (1948). *Communs. pure appl. Math.* **1**, 361.

FUJITA, S. (1966). *Introduction to non-equilibrium quantum statistical mechanics.* Saunders, Philadelphia.

FURRY, W. H. (1966). *Lectures in theoretical physics*, viii-A (edited by W. E. Brittin), p. 1. University of Colorado Press, Boulder, Colorado.

GAILITIS, M. (1964). *Soviet Phys. JETP*, **17**, 1328.

GELL-MANN, M., and GOLDBERGER, M. L. (1953). *Phys. Rev.* **91**, 398.

GERJUOY, E. (1958a). *Phys. Rev.* **109**, 1806.

—— (1958b). *Ann. Phys.* **5**, 58.

—— (1965). *J. Math. Phys.* **6**, 993. See also ibid. **6**, 1396.

GHIRADI, G. C., and RIMINI, A. (1964). *J. Math. Phys.* **5**, 722.

GIOUMOUSIS, G., and CURTISS, C. F. (1958). *J. chem. Phys.* **29**, 996.

—— —— (1961). *J. Math. Phys.* **2**, 96.

GLARUM, S. H. (1960). *J. chem. Phys.* **33**, 1371.

GLASSGOLD, A. E., HECKROTTE, W., and WATSON, K. M. (1959). *Phys. Rev.* **115**, 1374.

GLASSTONE, S., LAIDLER, K. J., and EYRING, H. (1941). *The theory of rate processes.* McGraw-Hill, New York.

GLAUBER, R. J. (1959). *Lectures in theoretical physics*, i (edited by W. E. Brittin and L. G. Dunham), p. 315. University of Colorado Press, Boulder, Colorado.

—— and SCHOMAKER, V. (1953). *Phys. Rev.* **89**, 667.

GOLDBERGER, M. L., and WATSON, K. M. (1964). *Collision theory.* Wiley, New York.

GOLDEN, S. (1949). *J. chem. Phys.* **17**, 620.

—— (1957). *Nuovo Cim.* Suppl. **5**, 540.

—— (1969). *Quantum statistical foundations of chemical kinetics.* Clarendon Press, Oxford.

GOLDSTEIN, H. (1950). *Classical mechanics.* Addison-Wesley, Reading, Massachusetts.

GORDON, R. G. (1965). *J. chem. Phys.* **43**, 1307.

—— (1966). *J. chem. Phys.* **45**, 1649.

—— (1967). *J. chem. Phys.* **46**, 4399.

GORIN, E. (1938). *Acta phys.-chim. URSS*, **9**, 681.

GOTTFRIED, K. (1966). *Quantum mechanics.* Benjamin, New York.

GOUTERMAN, M. (1965). *J. chem. Phys.* **42**, 351.

GREEN, M. S. (1952). *J. chem. Phys.* **20**, 1281; ibid. **22**, 398.

GREEN, T. A., and JOHNSON, R. E. (1966). *Phys. Rev.* **152**, 9.

GREENE, E. F., MOURSUND, A. L., and ROSS, J. (1965). *Molecular beams* (edited by J. Ross), p. 135. Interscience, New York.

GREIDER, K. R. (1959). *Phys. Rev.* **114**, 786.

—— and GLASSGOLD, A. E. (1960). *Ann. Phys.* **10**, 100.

GROENEWOLD, H. J. (1964). *Nucl. Phys.* **57**, 112.

GURNEE, E. F., and MAGEE, J. L. (1957). *J. chem. Phys.* **26**, 1237.

HAHN, Y. (1966). *Phys. Rev.* **142**, 603; ibid. **148**, 1088.

—— O'MALLEY, T. F., and SPRUCH, L. (1964). *Phys. Rev.* **134**, B394.

HALL, G. G., and LEVINE, R. D. (1966). *J. chem. Phys.* **44**, 1567.

HALMOS, P. B. (1958). *Finite dimensional vector spaces*. Van Nostrand, Princeton, New Jersey.

HAMMER, C. L., and WEBER, T. A. (1967). *J. Math. Phys.* **8**, 494.

HAMMOND, G. S. (1955). *J. Am. chem. Soc.* **77**, 334.

HARRIS, F. E., MICHA, D. A., and POHL, H. A. (1965). *Ark. Fys.* **30**, 259.

HARRIS, R. A. (1963). *J. chem. Phys.* **39**, 978.

HARRISON, H., and BERNSTEIN, R. B. (1963). *J. chem. Phys.* **38**, 2135; ibid. **47**, 1884.

HAUSER, W., and FESHBACH, H. (1952). *Phys. Rev.* **87**, 366.

HEITLER, W. (1954). *The quantum theory of radiation*, 3rd edn. Clarendon Press, Oxford.

—— and MA, S. T. (1949). *Proc. R. Ir. Acad.* **52**, 109.

HENGLEIN, A., LACMANN, K., and KNOLL, B. (1965). *J. chem. Phys.* **43**, 1048.

HERSCHBACH, D. R. (1962). *Discuss Faraday Soc.* **33**, 149.

—— (1965a). *Appl. Optics*, Suppl. No. 2, 128.

—— (1965b). *Molecular beams* (edited by J. Ross), p. 319. Interscience, New York.

HERZBERG, G. (1966). *Molecular spectra and molecular structure*, vol. iii. Van Nostrand, Princeton, New Jersey.

HERZENBERG, A., and MANDL, F. (1963). *Proc. R. Soc.* **A274**, 253.

—— —— (1966). *Proc. Phys. Soc.* **89**, 305, 321.

HERZFELD, K. F. (1963). *Dispersion and absorption of sound by molecular processes* (edited by D. Sette), p. 272. Academic Press, New York.

—— and LITOVITZ, T. A. (1959). *Absorption and dispersion of ultrasonic waves*. Academic Press, New York.

HIRSCHFELDER, J. O., and WIGNER, E. (1939). *J. chem. Phys.* **7**, 616.

—— CURTISS, C. F., and BIRD, R. B. (1954). *Molecular theory of gases and liquids*. Wiley, New York.

HOARE, M. (1961). *Molec. Phys.* **4**, 465.

HOFACKER, G. L. (1963). *Z. Naturf.* **18a**, 607.

—— (1965). *J. chem. Phys.* **43**, S208.

REFERENCES

JACOB, M., and WICK, G. C. (1959). *Ann. Phys.* **7**, 404.

JAUCH, J. M. (1958). *Helv. phys. Acta* **31**, 661.

—— and MARCHAND, J. P. (1966). *Helv. phys. Acta* **39**, 325.

JOACHAIN, C. (1965). *Nucl. Phys.* **64**, 529.

JOHNSON, B. R. (1967). A new method for computing inelastic molecular scattering. Ph.D. Thesis, University of Illinois.

—— and SECREST, D. (1966). *J. Math. Phys.* **7**, 2187; *J. chem. Phys.* **48**, 4682.

JOHNSTON, H. S. (1966). *Gas phase reaction rate theory.* Ronald Press, New York.

JORDAN, T. F. (1962). *J. Math. Phys.* **3**, 429.

KAEMPFFER, F. A. (1965). *Concepts in quantum mechanics.* Academic Press, London.

KAPLAN, J. I. (1958). *J. chem. Phys.* **28**, 278; ibid. **29**, 462.

KAPUR, P. L., and PEIERLS, R. (1938). *Proc. R. Soc.* **A166**, 277.

KARPLUS, M., PORTER, R. N., and SHARMA, R. D. (1965). *J. chem. Phys.* **43**, 3259.

—— and TANG, K. T. (1967). *Discuss. Faraday Soc.* **44**, 56.

KARPLUS, R., and SCHWINGER, J. (1948). *Phys. Rev.* **73**, 1020.

KASSEL, L. S. (1932). *Kinetics of homogeneous reactions.* Chemical Catalog, New York.

KATO, T. (1951). *Prog. theor. Phys., Osaka*, **6**, 394.

—— (1966). *Perturbation theory for linear operators.* Springer-Verlag, Berlin.

KATZ, A. (1964). *Nuovo Cim.* **33**, 1544.

—— (1966). *J. Math. Phys.* **7**, 1802.

KECK, J. C. (1958). *J. chem. Phys.* **29**, 410.

—— and CARRIER, G. (1965). *J. chem. Phys.* **43**, 2284.

KERNER, E. H. (1953). *Phys. Rev.* **91**, 1174.

KHURI, N. N. (1957). *Phys. Rev.* **107**, 1148.

KIRKWOOD, J. G. (1946). *J. chem. Phys.* **14**, 180.

KLOTS, C. E. (1967). *J. chem. Phys.* **46**, 1197.

KOHN, W. (1948). *Phys. Rev.* **74**, 1763.

KONDRATIEV, V. N. (1935). *Zh. éksp. teor. Fiz.* **5**, 250.

KRAMER, K. H., and BERNSTEIN, R. B. (1964). *J. chem. Phys.* **40**, 200.

KRIEGER, T. J. (1967). *Ann. Phys.* **42**, 375.

KRONIG, R. L. (1930). *Band spectra and molecular structure.* Cambridge University Press.

KUBO, R. (1954). *J. Phys. Soc. Japan*, **9**, 935.

—— (1957). *J. Phys. Soc. Japan*, **12**, 570.

—— (1959). *Lectures in theoretical physics*, i (edited by W. E. Brittin and L. G. Dunham), p. 120. University of Colorado Press, Boulder, Colorado.

—— (1963). *J. Math. Phys.* **4**, 174.

—— (1966). *Rep. Prog. Phys.* **19**, 1255.

—— and TOMITA, K. (1954). *J. Phys. Soc. Japan*, **9**, 888.

KUNTZ, P. J., NEMETH, E. M., POLANYI, J. C., ROSNER, S. D., and YOUNG, C. E. (1966). *J. chem. Phys.* **44**, 1168.

LAIDLER, K. J., and POLANYI, J. C. (1965). *Prog. reaction kinetics*, **3**, 1.

LANDAU, L. D. (1932). *Phys. Z. SowjUn.* **2**, 46.

—— (1936). *Phys. Z. SowjUn.* **10**, 67.

—— and LIFSHITZ, E. M. (1965). *Quantum mechanics*, 3rd edn. Pergamon Press, Oxford.

—— and TELLER, E. (1936). *Phys. Z. SowjUn.* **10**, 34.

LANE, A. M., and THOMAS, R. G. (1958). *Rev. mod. Phys.* **30**, 257.

—— and ROBSON, D. (1966). *Phys. Rev.* **151**, 744.

LAWLEY, K. P., and ROSS, J. (1965). *J. chem. Phys.* **43**, 2930.

LAX, M. (1950). *Phys. Rev.* **78**, 306.

—— (1964). *Physics Chem. Solids*, **25**, 487.

LESTER, W. A., and BERNSTEIN, R. B. (1967). *Chem. Phys. Lett.* **1**, 207, 347.

LEVIN, F. S. (1966). *Phys. Rev.* **141**, 858.

LEVINE, R. D. (1966a). *Proc. R. Soc.* **A294**, 467.

—— (1966b). *J. chem. Phys.* **44**, 2029.

—— (1966c). *J. chem. Phys.* **44**, 2035.

—— (1966d). *J. chem. Phys.* **44**, 2046.

—— (1966e). *J. chem. Phys.* **44**, 3597.

—— (1966f). *Molecular relaxation processes*, p. 297. Academic Press, London.

—— (1967a). *J. chem. Phys.* **46**, 331.

—— (1967b). *Intl. J. Quant. Chem.* **1s**, 727.

—— (1967c). *University of Wisconsin, Theoretical Chemistry Institute Report*, 261.

—— (1967d). *University of Wisconsin, Theoretical Chemistry Institute Report*, 269.

—— (1968). *J. chem. Phys.* **49**, 51.

—— and AMOS, A. T. (1967). *Phys. Stat. Sol.* **19**, 587.

—— JOHNSON, B. R., MUCKERMAN, J. T., and BERNSTEIN, R. B. (1968). *J. chem. Phys.* **49**, 56.

LEVINSON, N. (1949). *Det. Kgl. Danske Vidensk. Selsk., Mat.-Fys. Medd.* **25**, no. 9.

LEVY, B. R., and KELLER, J. B. (1963). *J. Math. Phys.* **4**, 54.

LICHTEN, W. (1963). *Phys. Rev.* **131**, 229.

LIGHT, J. C. (1964). *J. chem. Phys.* **40**, 3221. (See also Pechukas and Light (1965) for corrections.)

LIGHTHILL, M. J. (1958). *Fourier analysis and generalized functions*. Cambridge University Press.

LIN, J., and LIGHT, J. C. (1966). *J. chem. Phys.* **45**, 2545.

LINDEMANN, F. A. (1922). *Trans. Faraday Soc.* **17**, 598.

LIPELES, M., NOVICK, R., and TOLK, N. (1965). *Phys. Rev. Lett.* **15**, 815.

LIPPMANN, B. A. (1953). *Phys. Rev.* **91**, 1213.

—— (1956). *Phys. Rev.* **102**, 264.

REFERENCES

LIPPMANN, B. A. (1957). *Ann. Phys.* **1**, 113.
—— (1965a). *Phys. Rev. Lett.* **15**, 11.
—— (1965b). *Physics of electronic and atomic collisions*, p. 210. Scientific Bookcrafters.
—— (1966). *Phys. Rev. Lett.* **16**, 135.
—— and SCHWINGER, J. (1950). *Phys. Rev.* **79**, 469.
LLOYD, P. (1965). *Proc. phys. Soc.* **86**, 825.
LOCKWOOD, G. J., and EVERHART, E. (1962). *Phys. Rev.* **125**, 567.
LONDON, F. (1929). *Z. Elektrochem.* **35**, 552.
—— (1932). *Z. Phys.* **74**, 143.
—— (1937). *Trans. Faraday Soc.* **33**, 8.
LOVELACE, C. (1964). *Phys. Rev.* **135**, B1225.
LOW, F. E. (1959). Brandeis University Summer Institute in Theoretical Physics, Lecture notes.
LÖWDIN, P. O. (1965). *J. chem. Phys.* **43**, S175.
MACDONALD, W. M. (1964). *Nucl. Phys.* **54**, 393.
MAGNUS, W. (1954). *Communs pure appl. Math.* **7**, 649.
MAHAN, B. H. (1960). *J. chem. Phys.* **32**, 362.
—— (1964). *J. chem. Phys.* **40**, 392.
MALFLIET, R., and RUIJGROK, Th. (1967). *Physica*, **33**, 607.
MARCHI, R. P., and SMITH, F. T. (1965). *Phys. Rev.* **139**, A1025.
MARCUS, R. A. (1952). *J. chem. Phys.* **20**, 359.
—— (1964). *J. chem. Phys.* **41**, 610.
—— (1965). *J. chem. Phys.* **43**, 1598.
—— (1966a). *J. chem. Phys.* **45**, 2138, 2630.
—— (1966b). *J. chem. Phys.* **45**, 4493, 4500.
—— (1968). *J. chem. Phys.* **49**, 2610.
—— and RICE, O. K. (1951). *J. Phys. Chem.* **20**, 359.
MASSEY, H. S. W. (1949). *Rep. Progr. Phys.* **12**, 249.
—— and BURHOP, E. H. S. (1952). *Electronic and ionic impact phenomena*. Clarendon Press, Oxford.
—— and MOHR, C. B. O. (1934). *Proc. R. Soc.* **A144**, 188.
MCELROY, M. B., and HIRSCHFELDER, J. O. (1963). *Phys. Rev.* **131**, 1589.
MESSIAH, A. (1961). *Quantum mechanics*. North Holland, Amsterdam.
MICHA, D. A. (1965). *Ark. Fys.* **30**, 411, 425, 437.
MIES, F. H. (1964). *J. chem. Phys.* **40**, 523.
—— and KRAUSS, M. (1966). *J. chem. Phys.* **45**, 4455.
—— and SHULER, K. E. (1962). *J. chem. Phys.* **37**, 177.
MINTURN, R. E., DATZ, S., and BECKER, R. L. (1966). *J. chem. Phys.* **44**, 1149.
MITTLEMAN, M. H. (1961). *Phys. Rev.* **122**, 1930.
—— (1962). *Phys. Rev.* **126**, 373.

MITTLEMAN, M. H. (1965). *Adv. theor. Phys.* **1**, 283.
—— and PU, R. (1962). *Phys. Rev.* **126**, 370.
—— and WATSON, K. M. (1959). *Phys. Rev.* **113**, 198.
MOISEIWITSCH, B. L. (1966). *Variational principles*. Interscience, London.
MØLLER, C. (1945). *Det. K. Danske Vidensk. Selsk. Mat.-Fys. Medd.* **23**, No. 1.
MORI, H. (1965). *Prog. theor. Phys., Osaka*, **33**, 423.
MORI, M., WATANABE, T., and KATSUURA, K. (1964). *J. phys. Soc. Japan* **19**, 380.
MORSE, P. M., and FESHBACH, H. (1953). *Methods of theoretical physics*. McGraw-Hill, New York.
MOSES, H. E. (1955). *Nuovo Cim.* **1**, 103.
MOTT, N. F. (1931). *Proc. Camb. Phil. Soc. math. phys. Sci.* **27**, 553.
—— and MASSEY, H. S. W. (1965). *The theory of atomic collisions*, 3rd edn. Clarendon Press, Oxford.
MULLIKEN, R. S. (1937). *Phys. Rev.* **50**, 1017, 1028.
MURRAY, F. J. (1962). *J. Math. Phys.* **3**, 451.
NEWTON, R. G. (1958). *Ann. Phys.* **4**, 29.
—— (1960). *J. Math. Phys.* **1**, 319.
—— and FONDA, L. (1960). *Ann. Phys.* **9**, 416.
NIELSEN, S. E., and DAHLER, J. S. (1966). *J. chem. Phys.* **45**, 4060.
NIKITIN, E. E. (1962). *Discuss. Faraday Soc.* **33**, 14.
—— (1965a). *Optics Spectrosc. Wash.* **18**, 431.
—— (1965b). *Theor. and Exptl. Chem.* **1**, 83, 90.
—— (1966). *Theory of thermally induced gas phase reactions*. Indiana University Press, Bloomington, Indiana.
NYELAND, C., and BAK, T. A. (1965). *Trans. Faraday Soc.* **61**, 1293.
OHMURA, T. (1964). *Suppl. Prog. theor. Phys., Kyoto* **29**, 108.
O'MALLEY, T. F. (1964). *Phys. Rev.* **134**, A1188.
—— (1966). *Phys. Rev.* **150**, 14.
OSIPOV, A. I., and STUPOCHENKO, E. V. (1963). *Soviet Phys. Dokl.* **6**, 47.
PAULI, W. (1928). *Festschrift zum 60. Geburtstag. A. Sommerfelds*, p. 30. S. Hirzel Verlag, Leipzig.
PECHUKAS, P., and LIGHT, J. C. (1965). *J. chem. Phys.* **42**, 3281.
—— —— (1966a). *J. chem. Phys.* **44**, 3897.
—— —— and RANKIN, C. (1966b). *J. chem. Phys.* **44**, 794.
PELZER, M., and WIGNER, E. (1932). *Z. phys. Chem.* **B15**, 445.
PERCIVAL, I. C. (1960). *Phys. Rev.* **119**, 159.
PETERSON, J. M. (1962). *Phys. Rev.* **125**, 955.
PETERSON, R. L. (1967). *Rev. Mod. Phys.* **39**, 69.
POLANYI, J. C. (1959). *J. chem. Phys.* **31**, 1338.
—— (1967). *Discuss. Faraday Soc.* **44**, 293.
POLANYI, M., and WIGNER, E. (1928). *Z. physik. Chem.* **A139**, 439.

REFERENCES

PORTER, C. E. (1965). *Statistical theories of spectra.* Academic Press, New York.
POWLES, J. G., and STRANGE, J. H. (1964). *Molec. Phys.* **8,** 169.
PRESENT, R. D. (1955). *Proc. natn. Acad. Sci. U.S.A.* **41,** 515.
PRIGOGINE, I., and RESIBOIS, P. (1961). *Physica,* **27,** 629.
PRIMAS, H. (1961). *Helv. phys. Acta* **34,** 331.
RABINOVITCH, B. S., and SETSER, D. W. (1964). *Adv. Photochem.* **3,** 1.
RAMSAUER, C. (1921). *Ann. der Phys.* **64,** 513; ibid. **66,** 545.
RESIBOIS, P. (1961). *Physica,* **27,** 541.
RICE, O. K. (1929). *Phys. Rev.* **33,** 748; ibid. **34,** 1459.
—— (1930). *Phys. Rev.* **35,** 1551.
—— (1961). *J. Phys. Chem.* **65,** 1588.
—— (1967). *Statistical mechanics, thermodynamics and kinetics.* Freeman, San Francisco.
—— and RAMSPERGER, H. C. (1927). *J. Am. chem. Soc.* **49,** 1617.
RICE, S. O. (1945). *Bell Syst. tech. J.* **23,** 1. Reprinted in *Selected papers on noise and stochastic processes.* Dover, New York (1954).
RIESENFELD, W. B., and WATSON, K. M. (1956). *Phys. Rev.* **104,** 492.
ROBERTS, C. S. (1963). *Phys. Rev.* **131,** 209.
ROMAN, P. (1965) *Advanced quantum theory.* Addison-Wesley, Reading, Massachusetts.
ROSEN, N. (1933). *J. chem. Phys.* **1,** 319.
—— and ZENER, C. (1932). *Phys. Rev.* **40,** 502.
ROSENBERG, L. (1963). *Phys. Rev.* **131,** 874.
—— (1964a). *Phys. Rev.* **134,** B937.
—— (1964b). *Phys. Rev.* **135,** B715.
—— (1965). *Phys. Rev.* **140,** B217.
ROSENFELD, J. L. J., and ROSS, J. (1966). *J. chem. Phys.* **44,** 188.
ROSENSTOCK, H. M., WALLENSTEIN, M. B., WAHRHAFTIG, A. L., and EYRING, H. (1952). *Proc. natn. Acad. Sci. U.S.A.* **38,** 667.
ROSS, J., and MAZUR, P. (1961). *J. chem. Phys.* **35,** 19.
ROSS, M. H., and SHAW, G. L. (1961). *Ann. Phys.* **13,** 147.
SACK, R. A. (1958). *Molec. Phys.* **1,** 163.
SCHÖNBERG, M. (1951). *Nuovo Cim.* **8,** 651, 817.
SCHWINGER, J. (1959). *Proc. natn. Acad. Sci. U.S.A.* **45,** 1542.
SECREST, D., and JOHNSON, B. R. (1966). *J. chem. Phys.* **45,** 4556.
SERAUSKAS, R. V., and SCHLAG, E. W. (1965). *J. chem. Phys.* **42,** 3006.
—— —— (1966). *J. chem. Phys.* **45,** 3706.
SHARP, T. E., and RAPP, D. (1965). *J. chem. Phys.* **43,** 1233.
SHIMIZU, H. (1965). *J. chem. Phys.* **43,** 2453.
SIEGERT, A. J. F., and TERAMOTO, E. (1958). *Phys. Rev.* **110,** 1232.
SKINNER, B. G. (1961). *Proc. phys. Soc.* **77,** 551.

SLATER, N. B. (1939). *Proc. Camb. Phil. Soc. math. phys. Sci.* **35**, 56.

—— (1959). *Theory of unimolecular reactions.* Methuen, London.

SMITH, F. J. (1967). *Molec. Phys.* **13**, 121.

—— MASON, E. A., and VANDERSLICE, J. T. (1965). *J. chem. Phys.* **42**, 3257.

SMITH, F. T. (1960). *Phys. Rev.* **118**, 349.

—— (1962a). *J. chem. Phys.* **36**, 248.

—— (1962b). *J. Math. Phys.* **3**, 735.

—— (1963). *J. chem. Phys.* **38**, 1304.

SMITHIES, F. (1958). *Integral equations.* Cambridge University Press.

SNIDER, R. F. (1960). *J. chem. Phys.* **32**, 1051.

SNYDER, H. S. (1951). *Phys. Rev.* **83**, 1154.

SPRUCH, L. (1963). *Lectures in theoretical physics*, iv (edited by W. E. Brittin and B. W. Downs), p. 182. University of Colorado Press, Boulder, Colorado.

STONE, M. H. (1932). *Linear transformations in Hilbert space.* Am. Math. Soc. Publication XV.

STUECKELBERG, E. C. G. (1932). *Helv. phys. Acta*, **5**, 370.

SUGAR, R., and BLANKENBECLER, R. (1964). *Phys. Rev.* **136**, B472.

SUPLINSKAS, R. J., and ROSS, J. (1967). *J. chem. Phys.* **47**, 321.

SWAN, P. (1955). *Proc. R. Soc.* **A228**, 10.

TAKAYANAGI, K. (1963). *Prog. theor. Phys. Kyoto Suppl.* **25**, 1.

—— (1965). *Adv. atom. molec. Phys.* **1**, 149.

TERENIN, A. N., and PRILEZHAEVA, N. A. (1932). *Phys. Z. SowjUn.* **2**, 337.

TER HAAR, D. (1961). *Rep. Prog. Phys.* **24**, 304.

THIELE, E. (1966). *J. chem. Phys.* **45**, 491.

THORSON, W. R. (1961). *J. chem. Phys.* **34**, 1744.

—— (1962). *J. chem. Phys.* **37**, 433.

—— (1963). *J. chem. Phys.* **39**, 1431.

—— (1965). *J. chem. Phys.* **42**, 3878.

TITCHMARSH, E. C. (1948). *Introduction to the theory of Fourier integrals.* Clarendon Press, Oxford.

TOBOCMAN, W. (1961). *Theory of direct nuclear reactions.* Clarendon Press, Oxford.

UNSÖLD, A. Z. (1927). *Z. Phys.* **43**, 563.

VALANCE, W. G., and SCHLAG, E. W. (1966). *J. chem. Phys.* **45**, 216, 4280.

VAN HOVE, L. (1954). *Phys. Rev.* **95**, 249.

—— (1955). *Physica*, **21**, 901.

—— (1956). *Physica*, **22**, 343.

—— (1957). *Physica*, **23**, 441.

VAN NIEUWENHUIZEN, P., and RUIJGROK, TH. (1967). *Physica*, **33**, 595.

VAN WINTER, C. (1965). *Det. K. Danske Vidensk. Selskab., Mat.-Fys.* **2**, No. 10.

REFERENCES

VON NEUMANN, J. (1955). *Mathematical foundations of quantum mechanics.* Princeton University Press.

—— and WIGNER, E. P. (1929). *Physik. Z.* **30,** 467.

WALDMANN, L. (1964). *Physica,* **30,** 17.

WALL, F. T., HILLER, JR., L. A., and MAZUR, J. (1958). *J. chem. Phys.* **29,** 255.

WATSON, K. M. (1952). *Phys. Rev.* **88,** 1163.

—— (1953). *Phys. Rev.* **89,** 575.

—— (1956). *Phys. Rev.* **103,** 489.

—— (1957). *Phys. Rev.* **105,** 1388.

WEINBERG, S. (1964). *Phys. Rev.* **133,** B232.

WEISSKOPF, V. F. (1937). *Phys. Rev.* **52,** 295.

WIDOM, B. (1963). *Adv. chem. Phys.* **5,** 353.

WIGNER, E. P. (1959). *Group theory.* Academic Press, New York.

—— and EISENBUD, L. (1947). *Phys. Rev.* **72,** 29.

WILLIAMS, E. J. (1945). *Rev. mod. Phys.* **17,** 217.

WOLFENSTEIN, L. (1951). *Phys. Rev.* **82,** 690.

YAMAMOTO, T. (1960). *J. chem. Phys.* **33,** 281.

ZEMACH, A. C. (1964). *Nuovo Cim.* **33,** 939.

—— and GLAUBER, R. J. (1956). *Phys. Rev.* **101,** 118.

ZENER, C. (1931). *Phys. Rev.* **38,** 277.

—— (1932). *Proc. R. Soc.* A**137,** 696.

ZUBAREV, D. N. (1960). *Soviet Phys. Usp.* **3,** 320.

ZUMINO, B. (1956). *New York University, Institute of Mathematical Sciences, Report,* CX-23.

ZWANZIG, M. (1960). *J. chem. Phys.* **33,** 1338.

—— (1961a). *Lectures in theoretical physics,* iii (edited by W. E. Brittin, B. W. Downs, and J. Downs), p. 106. University of Colorado Press, Boulder, Colorado.

—— (1961b). *J. chem. Phys.* **34,** 1931.

—— (1964). *Physica,* **30,** 1109.

—— (1965). *A. Rev. phys. Chem.* **16,** 67.

Author Index

Abragam, A., 294.
Adachi, T., 222.
Airey, J. R., 258.
Akhiezer, N. I., 1, 5, 15.
Alexander, S., 294, 307.
Allison, A. C., 63.
Anderson, P. W., 238, 293, 298, 307.
Argyres, P. N., 288.
Arthurs, A. M., 60.

Bak, T. A., 239.
Baker, M., 43, 49, 102.
Banerjee, M. K., 193.
Baranger, M., 311.
Bardsley, J. N., 192.
Bassel, R. H., 194.
Bates, D. R., 175, 176, 231.
Bauer, E., 232.
Beck, D., 230, 258.
Becker, R. L., 192, 193.
Bellman, R., 159.
Ben-Reuven, A., 287, 307, 311.
Berger, R. O., 49.
Bernard, W., 294.
Bernstein, R. B., 23, 38, 39, 44, 49, 50, 63, 99, 105, 106, 206, 226, 227, 230, 232, 233, 237, 252, 253, 257, 282.
Berry, R. S., 175.
Bethe, H. A., 49, 252.
Biedenharn, L. C., 60.
Bird, R. B., 37.
Birnbaum, G., 294.
Blankenbecler, R., 51.
Blatt, J. M., 49, 60, 248, 252, 264.
Bloom, M., 294.
Bloombergen, N., 306.
Blum, L., 166.
Bohr, N., 166, 251, 252.
Brout, R., 64.
Brown, G. E., 261.
Bunker, D. L., 242.
Burhop, E. H. S., 48.
Butler, S. T., 193.

Callaway, J., 232.
Callen, H. B., 293, 294, 299.
Calogero, F., 60, 159.

Carrier, G., 148.
Castillejo, L., 112.
Chen, J. C. Y., 188, 192.
Chew, G. F., 51, 197.
Child, M. S., 242.
Coester, F., 166.
Cooper, J., 287.
Cottrell, T. L., 86.
Coulson, C. A., 121, 176, 180, 218, 237.
Coz, M., 192.
Cross, J. C. Jr., 234.
Curtiss, C. F., 37, 49, 50, 58, 60.

Dahler, J. S., 174.
Dalgarno, A., 60, 63, 171, 252, 253, 257.
Datz, S., 192, 193.
Davison, W. D., 58, 60, 64, 171.
Demkov, Yu. N., 206.
Devonshire, A. F., 100.
DeWitt, B. S., 51, 102.
Dirac, P. A. M., 1, 3.
Donath, W. E., 175.
Dworetsky, S., 172, 228.
Dyson, F. J., 14.

Eckstein, H., 129.
Edmiston, C., 241.
Edmonds, A. R., 24.
Eisenbud, L., 166.
Eliason, M. A., 122.
Epstein, S. T., 114.
Eu, B. C., 166, 267.
Evans, M. G., 252.
Everhart, E., 239.
Eyring, H., 122, 218, 219, 242, 251, 252.

Faddeev, L. D., 108.
Fano, U., 11, 12, 165, 184, 274, 287, 294, 298, 307, 311.
Feshbach, H., 15, 24, 155, 162, 187, 189, 236, 252, 261, 263, 267.
Fetter, A. L., 190.
Feynman, R. P., 1.
Foldy, L. L., 114.
Fonda, L., 165.
Ford, K. W., 44, 228.
Fowler, R., 218.

Francis, N. C., 190.
Friedman, B., 15, 154,
Friedman, F. L., 260.
Friedrichs, K. C., 162, 165.
Fujita, S., 294.
Furry, W. H., 12.

Gailitis, M., 120.
Gell-Mann, M., 159.
Gerjuoy, E., 79, 123, 126, 194.
Gershionowitz, H., 219.
Ghirardi, G. C., 40.
Gioumousis, G., 58, 60.
Glarum, S. H., 294.
Glassgold, A. E., 108, 221.
Glasstone, S., 122, 218, 242.
Glauber, R. J., 123, 215, 294.
Glazman, I. M., 1, 5, 15.
Goldberger, M. L., 33, 159, 190.
Golden, S., 1, 12, 113, 240.
Goldstein, H., 37.
Gordon, R. G., 294, 298, 299, 307.
Gorin, E., 219.
Gottfried, K., 1, 12.
Gouterman, M., 175.
Green, M. S., 293.
Green, T. A., 165.
Greene, E. F., 230, 258.
Greider, K. R., 194, 221.
Groenewold, H. J., 12.
Guggenheim, E. A., 218.
Gurnee, E. F., 238.

Hahn, Y., 192, 206.
Hall, G. G., 269, 273.
Halmos, P. B., 1, 3.
Hammer, C. L., 33.
Hammond, G. S., 242.
Harris, F. E., 241.
Harris, R. A., 165.
Harrison, H., 23, 105.
Hauser, W., 252, 263.
Heckrotte, W., 108.
Heitler, W., 92, 162, 165.
Henglein, A., 193.
Herschbach, D. R., 192, 193, 225, 259.
Herzberg, G., 165.
Herzenberg, A., 166.
Herzfeld, K. F., 86, 226.
Hibbs, A. R., 1.
Hiller, L. A. Jr., 244.
Hirschfelder, J. O., 37, 122, 123, 219, 243.
Hoare, M., 264.
Hofacker, G. L., 242, 244, 269.
Holt, A. R., 231.

Imam-Rahajoe, S., 49, 50.

Jackson, J. D., 49.
Jacob, M., 63.
Jauch, J. M., 113, 129, 139, 159.
Joachain, C., 211.
Johnson, B. R., 62, 160, 229, 282.
Johnson, R. E., 229.
Johnston, H. S., 241, 268.
Jordan, T. F., 159.

Kaempffer, F. A., 1.
Kalaba, R., 159.
Kaplan, J. I., 294.
Kapur, P. L., 166.
Karplus, M., 193, 240, 242, 244.
Karplus, R., 278, 289.
Kassel, L. S., 166, 249, 252.
Kato, T., 159, 212.
Katsuura, K., 239.
Katz, A., 12, 33.
Keck, J. C., 148, 252.
Keller, J. B., 49.
Kerner, E. H., 58, 64, 67, 68.
Khuri, N. N., 43.
Kirkwood, J. G., 293.
Klots, C. E., 150.
Knoll, B., 193.
Kodera, K., 258.
Kohn, W., 212.
Kondratiev, V. N., 148.
Kotani, T., 222.
Kramer, K. H., 232, 233, 237.
Krauss, M., 241, 269, 274.
Krieger, T. J., 253.
Kronig, R. L., 171.
Kubo, R., 293, 294, 299, 301, 306, 307.
Kümmel, H., 166.
Kuntz, P. J., 242.

Lacmann, K., 193.
Laidler, K. J., 122, 218, 241, 242.
Landau, L. D., 95, 176, 180, 237, 252.
Lane, A. M., 60, 166, 267.
Lawley, K. P., 63.
Lax, M., 190, 288.
Lester, W. A., 63, 99.
Levin, F. S., 192.
Levine, R. D., 38, 39, 101, 121, 175, 176, 185, 205, 209, 211, 218, 230, 234, 242, 248, 269, 272, 273, 276, 279, 282, 283, 286, 289, 290.
Levinson, N., 103.
Levy, B. R., 49.
Lichten, W., 177, 239.
Lifshitz, E. M., 180, 237.
Light, J. C., 148, 234, 255, 257, 258.
Lighthill, M. J., 272, 305.
Lin, J., 257, 258.
Lindemann, F. A., 268, 288.

AUTHOR INDEX

Lipeles, M., 172.
Lippmann, B. A., 15, 29, 72, 81, 113, 136, 159, 210, 211.
Litovitz, T. A., 86, 226.
Lloyd, P., 46.
Lockwood, G. J., 239
London, F., 69, 171, 175, 240.
Lovelace, C., 108.
Low, F. E., 33.
Löwdin, P. O., 209.

Ma, S. R., 162, 165.
McCoubrey, J. C., 86.
MacDonald, W. M., 165.
McElroy, M. B., 123.
Magee, J. L., 238.
Magnus, W., 14, 234.
Mahan, B. H., 216.
Malfliet, R., 108.
Mandelstam, S., 51.
Mandl, F., 166, 192.
Marchand, J. P., 113, 139.
Marchi, R. P., 226, 229.
Marcus, R. A., 121, 218, 240, 242, 244, 251, 252.
Mason, E. A., 226.
Massey, H. S. W., 15, 48, 68, 94, 175, 221, 252, 253, 257.
Mazur, J., 244.
Mazur, P., 121.
Messiah, A., 1, 24, 52, 142.
Micha, D. A., 241, 242.
Mies, F. H., 226, 269, 274.
Minturn, R. E., 192, 193.
Mittleman, M. H., 163, 190, 191, 192, 194, 214.
Mohr, C. B. O., 94, 221, 257.
Moiseiwitsch, B. L., 206.
Møller, C., 87.
Mori, H., 308.
Mori, M., 239,
Morse, P. M., 15, 24, 100, 155, 187, 236.
Moses, H. E., 159.
Mott, N. F., 15, 231.
Moursund, A. L., 230.
Muckerman, 282.
Mulliken, R. S., 175.
Murray, F. V., 307.

Newton, R. G., 60, 165.
Nielsen, S. E., 174.
Nikitin, E. E., 176, 237, 251, 255, 269.
Novick, R., 172, 228.
Nyeland, C., 239.

Ohmura, T., 33.
O'Malley, T. F., 49, 192, 206.

Oppenheim, I., 294.
Osipov, A. I., 86.

Pauli, W., 86.
Pechukas, P., 148, 234, 255, 257.
Peck, G., 258.
Peierls, R., 166, 252.
Pelzer, M., 240.
Percival, I C., 112, 206, 252, 253, 257.
Peterson, J. M., 104.
Peterson, R. L., 294.
Placzek, G., 252.
Pohl, H. A., 241.
Polanyi, J. C., 192, 241, 242.
Polanyi, M., 166, 252.
Porter, C. E., 253.
Porter, R. N., 240, 242, 244.
Pound, R. V., 306.
Powles, J. G., 294.
Present, R. D., 216.
Prigogine, I., 288.
Prilezhaeva, N. A., 148.
Primas, H., 307.
Pu, R., 214.
Purcell, E. M., 306.

Rabinovitch, B. S., 251.
Ramsauer, C., 104.
Ramsperger, H. C., 166, 249, 252.
Rapp, D., 99.
Resibois, P., 288.
Rice, O. K., 166, 249, 251, 252, 266, 269.
Rice, S. O., 293.
Riesenfeld, W. B., 165.
Rimini, A., 40.
Robson, D., 166.
Roman, P., 1.
Ronkin, 255.
Rosen, N., 165, 238, 269.
Rosenberg, L., 108, 196, 197, 212.
Rosenfeld, J. L. J., 222, 229.
Rosenstock, H. M., 251, 252, 277.
Rosner, S. D., 242.
Ross, J., 63, 121, 166, 193, 222, 229, 230, 258, 267.
Ross, M. H., 119.
Ruijgrok, Th., 108.

Sack, R. A., 307.
Schlag, E. W., 264, 292.
Schomaker, V., 123.
Schönberg, M., 15, 162.
Schwinger, J., 12, 29, 159, 210, 278, 289.
Seaton, M. J., 112.
Secrest, D., 62, 160.
Serauskas, R. V., 264, 292.
Setser, D. W., 251.
Sharma, R. D., 240, 242, 244.

Sharp, T. E., 99.
Shaw, G. L., 119.
Shimizu, H., 294.
Shuler, K. E., 226.
Siegert, A. J. F., 108.
Skinner, B. G., 238.
Slater, N. B., 166, 218, 249, 252, 268, 269, 279.
Smith, F. J., 81, 226.
Smith, F. T., 102, 121, 141, 205, 226, 229, 242.
Smith, W. W., 172, 228.
Smithies, F., 43.
Snider, R. F., 108.
Snyder, H. S., 159.
Spruch, L., 49, 206.
Stewart, A. L., 175.
Stone, M. H., 1.
Strange, J. H., 294.
Stueckelberg, E. C. G., 69, 175, 237.
Stupochenko, E. V., 86.
Sugar, R., 51.
Sun, C. E., 219.
Suplinskas, R. J., 193.
Swan, P., 103.

Takayanagi, K., 60, 64, 226, 232.
Tang, K. T., 193.
Taylor, H. S., 219.
Teller, E., 95.
Teramoto, E., 108.
Terenin, A. N., 148.
ter Haar, D., 11.
Thiele, E., 269, 279.
Thomas, R. G., 60, 166, 267.
Thorson, W. R., 126, 171, 177.
Titchmarsh, E. C., 157.
Tobocman, W., 64, 114, 193.
Tolk, N., 172, 228.
Tomita, K., 293.

Unsöld, A. Z., 168.

Valance, W. G., 292.
Vanderslice, J. T., 226.
Van Hove, L., 15, 162, 276, 293.
Van Nieuwenhuizen, P., 108.
Van Winter, C., 159.
von Neumann, J., 1, 11, 113, 173.

Wahrhaftig, A. L., 251, 252.
Waldmann, L., 58.
Wall, F. T., 244.
Wallenstein, M. B., 251, 252.
Watanabe, T., 239.
Watson, K. M., 33, 96, 107, 108, 163, 165, 190, 201, 204.
Weber, T. A., 33.
Weinberg, S., 108.
Weisskopf, V. F., 248, 252, 260, 264.
Welton, T. A., 293, 299.
Wheeler, J. A., 44, 228, 252.
Wick, G. C., 63.
Widom, B., 86.
Wigner, E. P., 24, 142, 153, 166, 173, 240, 243.
Williams, E. J., 215.
Wing, G. M., 159.
Wolfenstein, L., 258.
Wood, H. T., 49, 50.

Yamamoto, T., 294.
Young, C. E., 242.

Zalewski, K., 176, 180, 237.
Zemach, A. C., 293.
Zener, C., 69, 176, 180, 237, 238.
Zubarev, D. N., 294, 302.
Zumino, B. 15, 162, 165.
Zwanzig, M., 167, 287, 294, 300, 397, 308, 309.

Subject Index

absolute rate theory, 120.
addition of angular momentum, 21.
adiabatic:
  approximation, 169–78, 211; coupling terms, 174; in the theory of reactive collisions, 239; validity criterion, 171; variational principle for equivalent potential, 207.
  switching, 158.
admittance, 304; and dissipative processes, 305; and spectral density, 305.
angular momentum, 21; addition, 21; internal, 24; orbital, 21, 24, 45, time reversed, 144.

binary collision approximation, 109, 283.
bi-orthogonal set, 187, 277.
Born approximation, 50–3; estimates on validity of, 52; for central potentials, 52; for direct reactions, 197; for dumbbell model, 64; for internal excitation, 64; for local potentials, 51; for phase shifts, 52; for reactive collisions, 117; for rotational excitation, 66; higher orders, 50; in the impact parameter method, 231; in time-dependent collision theory, 132; reciprocity theorem for, 147; violation of the optical theorem, 50.
boundary conditions, in time-dependent collision theory, 134; outgoing, 35; standing waves, 93.
Brillouin–Wigner wave operator, 201.

causal functions, 156.
centrifugal potential, 46; barrier, 47; in the reactive box approximation, 217; in the transition state theory, 219.
channel, 110; orthogonality of states, 138; resolution of wave function, 138.
Clebsch–Gordan (vector coupling) coefficient, 24.
close-coupling approximation, 164, 214.
collision:
  complex, 246, 254; and the optical potential, 261; cross-section for formation, 254, 256; differential cross-section for, 259; in the statistical approximation, 261.
  of identical molecules, 58.
  rate, in ensembles, 80–3.
  rates, 73–80.
commutator, 8.
completeness relation, 6.
conductivity, 304.
constant of motion, 148.
convolution theorem, 155.
cross-section, 33, 37; as a line profile, 185; classical, 38; degeneracy-averaged, 80, 81; differential, 36, 44, 74; for ion-molecule reactions, 217; for separable interactions, 40; glory extrema in elastic, 106; internal excitation, 77; in the impact parameter representation, 223; in the optical model, 221; in the quasi-equilibrium approximation, 251; in the random phase approximation, 221; in the statistical approximation, 256; momentum transfer, 79; reactive collisions, 115; rotational excitation, 67; total, 36, 45, 67.
curve crossing, 175; transition probability for, 236.

decoupling approximation, 214; correction to, 214.
deflexion function, 37; in the semi-classical limit, 226.
degeneracy-averaged cross-section, 80, 81.
density:
  matrix, 12; equation of motion in the Markoffian approximation, 312.
  of states, 31, 101–6; change due to interaction, 102; near a resonance, 103.
  operator, 11.
detailed balance, 85.
diagrammatic methods, 133.
differential cross-section, 36, 77; for potential scattering, 44; for reactive collisions, 115; for rotational excitation, 63, 67; microscopic reversibility of, 149; semi-classical limit, 227; statistical approximation for, 258.
dipole correlation function, 298.

SUBJECT INDEX

direct reactions, 192–200; transition amplitude, 196.
discrete states in the continuum, 165, 178.
distorted wave:
  approximation, for a general distorting potential, 196; for direct collisions, 196; partial wave expansion, 225; semi-classical limit, 226.
  Born approximation, 99; Magnus approximation, 234.
distorting potential, 194.
dumb-bell model, 64.

effective range, 120; theory, 119.
Ehrenfest's theorem, 72, 75; for ensembles, 287; for momentum transfer, 78; reactive collisions, 114.
eigenphase shift, 93, 102; non-crossing rule, 252.
eigenstate, 6.
eigenvalue, 6.
equivalent potential, 95, 169, 201; adiabatic approximation for, 170; for coupled channels, 191; imaginary part of, 190; in the theory of direct reactions, 194; variational principle for, 207.
evolution operator, 13, 19; backward, 134; for macroscopic times, 285; forward, 129, integral equation, 130.
exit state, 282; approximation, 248.
expectation value, 4.
exponential decay, 272.

fluctuation–dissipation theorem, 299.
form factor, 65; for direct reactions, 199.
Fourier transform, 155.
Fredholm:
  determinant, 102, 204.
  method, 43.
free energy, 122.

glory phase, 104.
golden rule, 72.
Green's:
  function, 18, 23; asymptotic form, 34; for interacting molecules, 97; for multi-channel collisions, 127; partial wave expansion, 24.
  operator, 15; coordinate representation, 18; discontinuity across the continuous spectrum, 18; for relative motion, 18; for the time-dependent Schrödinger equation, 20; L-conjugation invariance, 15; spectral resolution, 16.
  (response) function, 302.
  theorem, 123.

Hadamard's inequality, 188.
Hamiltonian, 13; in the centre of mass system, 21.
Heisenberg representation, 14.
Heisenberg's equation of motion, 15; time reversed, 144.
Heitler's equation, 92.
Hellman–Feynman theorem, 101.
Hermitian adjoint, 10.
Hilbert transform, 157.

identical particles, 150.
identity operator, 5; energy representation, 30; momentum representation, 30.
imbedding, 159.
impact parameter, 37, 45; amplitude, 223; method, 231–9; representation, 222.
impulse approximation, 197.
initial state, as a microcanonical ensemble, 81; as a mixture, 80; for internal excitation, 54; in the centre of mass system, 27; in time-dependent collision theory, 129; rotational excitation, 61; scattering by two potentials, 97; symmetrized, 59.
internal excitation, 54–9; coupled equations, 56; scattering amplitude, 56.
intertwining, 157.
ion–molecule reactions, 217
isolated resonance, 182; line shape, 184; scattering amplitude, 182.

Jauch resolution, channel projection operators, 139; reactive collisions, 137.

K matrix, 92.
Kato identity, 213.
Kohn variational principle, 212; for scattering amplitude, 212.
Kramers–Kronig relations, 305.

L-conjugation, 15, 16.
Legendre polynomial, 22.
Levinson's theorem, 48, 103, 105.
limit in the mean, 158.
line:
  profile, 306; Fano parametrization, 184; general definition, 185.
  shape, 184.
linear:
  operator, 4.
  system, 156.
Liouville operator, 307.
Lippmann–Schwinger (L.S.) equation, 29; as eigenfunction of constants of motion, 58; asymptotic form, 34, for a partial wave expansion, 46, for internal excitation, 57; for reactive collisions, 127; boundary conditions, 126; collision of

identical molecules, 59; coordinate representation, 32; for two potentials, 97; in time-dependent collisions, 130; internal excitation, 55; momentum representation, 31; normalization of solutions, 29, 154; reactive collisions, 113; resolvent kernel of, 42–3; rotational excitation, 61; structureless particles, 30; time reversed, 145.
local operator, 9.

M matrix, 119.
Magnus approximation, 234; distorted wave form, 234; for symmetical (resonance) transfer, 238; for two-state problem, 238.
Markoffian approximation, 284, 310.
Massey criterion, 68, 239; adiabatic and sudden regimes, 69; rotational excitation, 69; vibrational excitation, 69.
master equation, 86.
memory function, 309.
microscopic reversibility, 85; for collision rates, 146; for cross-sections, 147; for differential cross-sections, 150; in the statistical approximation, 253.
mixture, 11.
modified wave-number approximation, 226.
momentum transfer cross-section, 79.
motional narrowing, 313.
multichannel collisions, 111.
multiple scattering theory, 106–9.

non-adiabatic transitions, 176.
non-crossing rule, 173, 175.
norm, 3, 158.
normalization:
 on energy scale, 30.
 solutions of L.S. equation, 29, 154.
 to a finite volume, 73.
 to unit density, 23, 30.
number of bound states, 103.

observable, 11; change in a collision, 136.
opacity function, 219, 221; in the distorted wave approximation, 225; in the impact parameter representation, 224, 231; in the statistical approximation, 257, 258, 263; inversion procedure for, 229.
operator, average value, 4; complex conjugation, 142; constant of motion, 148; current, 35, 123; eigenstate of, 6; eigenvalue of, 6; equation of motion, 71; expectation value, 4, 5; functions of, 6, 16; Hermitian, 10, 11; identity, 5; invariance under time reversal, 143;
inverse, 15; kinetic energy, 9; linear, 4, 5; local, 9; matrix representation, 9; measurable, 71; momentum, 21; parity, 149; projection, 11; radial momentum, 21; rate of change in a stationary collision, 71; representation of, 8, 9; spectral resolution, 5; statistical, 11; super-operator, 307; time reversal, 142, time-reversed, 143; trace, 6; translation, 149; transpose, 11.
optical:
 model, 219–31.
 potential, 169, 261, 262; and compound processes, 261; transition rates, 262.
 theorem, 44, 72; for rearrangement collisions, 118, 127; in abstract form, 89; in the coordinate representation, 125.
orthonormal set, 3.
overlapping resonances, 186; in the statistical approximation, 265; scattering amplitude, 188; spectral expansion of Green's function, 187; use of a biorthogonal basis, 187.

parity operator, P, 149.
Parseval's formula, 155.
partial:
 isometry, 88.
 wave analysis, 43–50; asymptotic form, 46; for central potential, 44; for distorted wave amplitudes, 225; for Green's function, 23; for isolated resonance, 183; for rotational excitation, 60.
partition function, 204; for relative motion, 83; internal, 116, 205.
partitioning technique, 162–7; for Heisenberg's equation, 308; for Schrödinger's time-dependent equation, 166; for the Liouville operator, 308; for the Lippmann–Schwinger equation, 163; for the spectral density, 311.
Pauli principle, 48.
permutation symmetry in collisions, 150–3.
perturbed stationary states method, 175.
phase shift, 45, 46; dependence on strength of potential, 101, 105; glory phase, 104; low energy limit, 48; bounds for, 206, 208.
potential energy surface, 240.
pressure broadening, 298, 311.
probability:
 amplitude, 2.
 density amplitude, 3.
product space, 7.
projection operator, 11.
pure state, 2.

quasi-equilibrium approximation, 218, 247.

RRK model, 249, 250.
RRK-M model, 251.
radial Schrödinger equation, 22.
rainbow angle, 38.
Ramsauer effect, 104.
random phase approximation, 221, 255, 257.
rate:
  constant, 83, 116, 122; for ion-molecule reactions, 217; for the reactive box approximation, 217; in the transition state model, 218.
  operator, 201.
Rayleigh–Ritz variational bound principle, 207.
reaction:
  coordinate, 240, 242.
  criterion, 216, 257.
reactive:
  box approximation, 216.
  collisions, 110–28; Born approximation for, 117; channel Hamiltonians, 111; rate of, 114; rate per unit volume, 115; rates in ensembles, 116; resolution of asymptotic states, 113; resolution of Hamiltonian, 111; transition operator for, 117.
rebound reactions, 193.
reciprocity theorem, 16, 143, 145; for reactive collisions, 147; for scattering amplitudes, 146, 149.
recombination of ions, 216.
recurrence time, 248.
relaxation:
  equation, 83–6, 292.
  function, 303.
  operator, 308; in the Markoffian approximation, 310.
representation, angular momentum, 22; coordinate, 4; momentum, 31; total angular momentum, 24, 60; transformation, 7.
resolution of the identity, 5, 16, 18; coordinate representation, 6; product space, 7.
resolvent, 15.
resonance:
  as an increase in the density of states, 103; in inelastic collisions, 189; in subexcitation collisions, 178; in the density of states, 205; orbiting, 49; phase shift, 183; potential, 42.
response function, 301.
Riemann–Lebesgue lemma, 272, 305.
rigid:
  rotor, 24, 66.
  spheres, phase shift, 48, 53.

rotational excitation, 60–9, 261; Born approximation for a dumb-bell model, 66; collision rate, 141; coupled equations, 62; cross-section, 141; scattering amplitude, 63; statistical approximation for, 252, 261.

S-matrix, 90–5; eigenvalues, 93; ensemble for, 253; in time-dependent collision theory, 131, 136; on energy shell, 91; rotational excitation, 62–3; statistical approximation for, 252; unitarity, 90, 92.
S-operator, for reactive collisions, 140, 157; in time-dependent collision theory, 135.
scalar product, 2, 3.
scattering:
  amplitude, 34–5; dumb-bell model, 64; for direct collisions, 182; for internal subexcitation, 182; for isolated resonance, 182; for separable interactions, 40; for overlapping resonances, 188; for two potentials, 98; identical molecules, 59; in the impact parameter representation, 223; in the optical model, 220; partial wave expansion, 44; reactive collisions, 115; rotational excitation, 63; scattering by two centres, 66; semi-classical limit, 226; symmetrized, 152.
  by two potentials, 95–101, 194, 259.
  length, 105, 120.
Schrödinger representation, 14.
Schrödinger's time-dependent equation, 13.
Schwartz inequality, 158.
semi-classical limit, 215, 222, 227.
separable interactions, 39–43.
shadow:
  approximation, 222.
  scattering rate, 254.
shape elastic scattering, 254.
Slater's model, 249.
spectral:
  density, 295; and transition rates, 296; in terms of the Green's operator, 311; in the canonical ensemble, 295; in the Markoffian approximation, 312; using the partitioning technique, 312.
  resolution, 5.
spectrum, 2; continuous, 3.
spherical:
  Bessel function, 23, 44.
  harmonics, 22; addition theorem, 22.
  Neumann function, 23.

state vector, 2; norm, 3; orthogonal, 5; representation, 6.
static approximation, 164, 210.
stationary phase method, 226.
statistical:
  approximation, 252–9; for differential cross-section, 258; for overlapping resonances, 265; for reactive collisions, 255; using energy average, 260.
  operator, 11; after a measurement, 12; equation of motion, 14; for interacting molecules, 80; time evolution, 14.
step-function, 19.
sticking probability, 263.
strength function, 247.
stripping reactions, 193.
sudden:
  approximation, 233, 237.
  limit, 234.
super-operator, 307.
symmetric (resonance) transition, 238.

threshold behaviour, 120.
time-correlation function, 294; and dissipative processes, 298; equation of motion for, 309; equation of motion in the Markoffian approximation, 312; evolution in the Markoffian approximation, 312; in the canonical ensemble, 294; spectral density for, 295.
time-dependent collision theory, 129–41; asymptotic form of wave function, 131; iteration of the Lippmann–Schwinger equation, 132; the Lippmann–Schwinger equation, 130.
time:
  ordering, 130.
  reversal operator, 142.
Titchmarsh's theorem, 157, 305.
trajectory analysis, 242.
transition:
  amplitude density method, 159–60.
  operator, 89; DWB approximation, 99; dependence on a parameter, 100; eigenvectors, 94; for two potentials, 96; Kohn variational principle for, 213; L-conjugation invariance, 89, 117; Lippmann–Schwinger variational principle for, 210; multichannel collisions, 117; partitioned forms for, 202; spectral analysis, 148; threshold behaviour, 120.
  rate, in canonical ensemble, 83, 116, 122; in centre of mass system, 75; in microcanonical ensemble, 82; in terms of Green's theorem, 125; in the statistical approximation, 263; internal excitation, 77; per unit volume, 75–6; reactive collisions, 113; volume dependence, 78.
  state model, 218, 246, 251.
transmission coefficient, 247, 263; for overlapping resonances, 266.
transpose, 11.
turning-point, 47.

unimolecular:
  breakdown, 268, 269–82; decay width, 275; decoupling approximation, 273; evolution in time, 270; evolution of products, 274; exit state, 282; exponential decay, 272; in an ensemble, 276; line shape, 270; resolution into components, 273.
  reactions, averaged density matrix approach, 278; excitation rate, 290; high pressure limit, 279; low pressure limit, 279; relaxation equation, 292; the Lindemann mechanism, 289.

variable phase method, 159.
variational principle, 205; for equivalent Hamiltonian, 208; for equivalent potential, 211; for L-conjugation invariant operators, 209; for scattering amplitude, 212; for transition operator, 210, 213; Kohn type, 212.
vector space, dual, 3; linear, 3; product space, 7.
vibrational excitation, 95.

wave:
  function, 4.
  operator, 87; as a partial isometry, 88; DWB approximation, 100; Hermitian adjoint, 87; in time-dependent collision theory, 134; multichannel collisions, 117.
  packet, 33.
Wronskian, 123.

yield function, 120, 121, 137; and reaction rates, 122; for reactive collisions, 121; for the reactive box approximation, 217; for unimolecular reactions, 292; in the quasi-equilibrium approximation, 251; in the transition state model, 218.

PRINTED IN GREAT BRITAIN
AT THE UNIVERSITY PRESS, OXFORD
BY VIVIAN RIDLER
PRINTER TO THE UNIVERSITY